COMPUTABLE STRUCTURES AND THE
HYPERARITHMETICAL HIERARCHY

STUDIES IN LOGIC

AND

THE FOUNDATIONS OF MATHEMATICS

VOLUME 144

Honorary Editor:

P. SUPPES

Editors:

S. ABRAMSKY, *London*
S. ARTEMOV, *Moscow*
D.M. GABBAY, *London*
R.A. SHORE, *Ithaca*
A.S. TROELSTRA, *Amsterdam*

ELSEVIER
AMSTERDAM • LAUSANNE • NEW YORK • OXFORD • SHANNON • SINGAPORE • TOKYO

COMPUTABLE STRUCTURES AND THE HYPERARITHMETICAL HIERARCHY

C.J. ASH [†]

J. KNIGHT
University of Notre dame
Department of Mathematics
Notre Dame, IN 46556-0398, U.S.A.

2000

ELSEVIER
AMSTERDAM • LAUSANNE • NEW YORK • OXFORD • SHANNON • SINGAPORE • TOKYO

ELSEVIER SCIENCE B.V.
Sara Burgerhartstraat 25
P.O. Box 211, 1000 AE Amsterdam, The Netherlands

© 2000 Elsevier Science B.V. All rights reserved.

This work is protected under copyright by Elsevier Science, and the following terms and conditions apply to its use:

Photocopying
Single photocopies of single chapters may be made for personal use as allowed by national copyright laws. Permission of the Publisher and payment of a fee is required for all other photocopying, including multiple or systematic copying, copying for advertising or promotional purposes, resale, and all forms of document delivery. Special rates are available for educational institutions that wish to make photocopies for non-profit educational classroom use.

Permissions may be sought directly from Elsevier Science Global Rights Department, PO Box 800, Oxford OX5 1DX, UK; phone: (+44) 1865 843830, fax: (+44) 1865 853333, e-mail: permissions@elsevier.co.uk. You may also contact Global Rights directly through Elsevier's home page (http://www.elsevier.nl), by selecting 'Obtaining Permissions'.

In the USA, users may clear permissions and make payments through the Copyright Clearance Center, Inc., 222 Rosewood Drive, Danvers, MA 01923, USA; phone: (978) 7508400, fax: (978) 7504744, and in the UK through the Copyright Licensing Agency Rapid Clearance Service (CLARCS), 90 Tottenham Court Road, London W1P 0LP, UK; phone: (+44) 171 631 5555; fax: (+44) 171 631 5500. Other countries may have a local reprographic rights agency for payments.

Derivative Works
Tables of contents may be reproduced for internal circulation, but permission of Elsevier Science is required for external resale or distribution of such material.
Permission of the Publisher is required for all other derivative works, including compilations and translations.

Electronic Storage or Usage
Permission of the Publisher is required to store or use electronically any material contained in this work, including any chapter or part of a chapter.

Except as outlined above, no part of this work may be reproduced, stored in a retrieval system or transmitted in any form or by any means, electronic, mechanical, photocopying, recording or otherwise, without prior written permission of the Publisher.
Address permissions requests to: Elsevier Science Rights & Permissions Department, at the mail, fax and e-mail addresses noted above.

Notice
No responsibility is assumed by the Publisher for any injury and/or damage to persons or property as a matter of products liability, negligence or otherwise, or from any use or operation of any methods, products, instructions or ideas contained in the material herein. Because of rapid advances in the medical sciences, in particular, independent verification of diagnoses and drug dosages should be made.

First edition 2000

Library of Congress Cataloging in Publication Data
A catalog record from the Library of Congress has been applied for.

ISBN: 0 444 50072 3

∞ The paper used in this publication meets the requirements of ANSI/NISO Z39.48-1992 (Permanence of Paper).

Transferred to digital printing 2006

Preface

0.1 Introduction

This book explores the relation between definability and computability in mathematical structures. Much of the material that we wish to present grew out of work on the three problems below.

Problem 1 *For a computable structure $\mathcal{A}$, give syntactical conditions guaranteeing that for all computable copies $\mathcal{B}$, there is a computable isomorphism from $\mathcal{A}$ onto $\mathcal{B}$.*

Problem 2 *For a computable structure $\mathcal{A}$, give syntactical conditions guaranteeing that for all computable copies $\mathcal{B}$, all isomorphisms from $\mathcal{A}$ onto $\mathcal{B}$ are computable.*

Problem 3 *For a relation R on a computable structure $\mathcal{A}$, give syntactical conditions guaranteeing that for all computable copies $\mathcal{B}$ and all isomorphisms F from $\mathcal{A}$ onto $\mathcal{B}$, $F(R)$ is computably enumerable.*

We may vary the problems by passing to higher levels in the hyperarithmetical hierarchy. Varying Problem 3 in this way, we ask for conditions guaranteeing that in computable copies of $\mathcal{A}$, the image of R is Σ^0_α. We may also vary the problems by considering *arbitrary* (instead of computable) copies of $\mathcal{A}$, and relativizing. Varying Problem 3 in this way, we ask for conditions guaranteeing that in an arbitrary copy $\mathcal{B}$ of $\mathcal{A}$, the image of R is computably enumerable, or more generally Σ^0_α, *relative to $\mathcal{B}$*. We may ask when the image of R in computable copies is d-c.e. (a difference of computably enumerable sets), or, relativizing, we may ask when the image of R in an arbitrary copy $\mathcal{B}$ is d-c.e. relative to $\mathcal{B}$ (a difference of relations that are c.e. relative to $D(\mathcal{B})$). We shall mention some further variants later.

The reader may or may not find such problems immediately appealing. The results gain interest when applied to various familiar kinds of mathematical structures. In addition, the problems form nicely graded families, some immediately approachable, some not; and they have proved to be valuable in suggesting technical advances. We believe that the ideas and methods developed for these problems should be useful in a wider context.

The background material needed for a thorough understanding of the results includes certain topics from computability (ordinal notation, the hyperarithmetical hierarchy) and model theory (infinitary formulas, consistency criteria) that are often omitted in basic texts and have not previously been gathered together. We devote several chapters to this background material.

We give a detailed account of the results on the three basic problems and the specific variants described above. We include other results that can be obtained using the same techniques, and we mention some further variants of the basic problems that have required significant modifications of the techniques. We do not attempt to give an exhaustive list of results. Our aim is to present the main ideas.

In all of the results on the basic problems and their variants, the syntactical conditions involve computable infinitary formulas. We shall discuss infinitary formulas in general, proving some basic results, such as the Scott Isomorphism Theorem. Then we describe the computable infinitary formulas. Taken all together, the computable infinitary formulas have the same expressive power as the hyperarithmetical infinitary formulas; i.e., those in the least admissible fragment of $L_{\omega_1\omega}$. (We mention this for those who happen to be familiar with admissible sets.)

The computable infinitary formulas are classified as computable Σ_α or Π_α, for various computable ordinals α, and they are assigned indices within these classes. The importance of the formulas is tied to the classification. In particular, in computable structures, satisfaction of computable Σ_α (or computable Π_α) formulas is Σ_α^0 (or Π_α^0). In arbitrary structures $\mathcal{B}$, satisfaction of computable Σ_α formulas is Σ_α^0 relative to $\mathcal{B}$. Moreover, any formula with this feature is logically equivalent to a computable Σ_α formula. This fact is one of several *expressive completeness* results, indicating that for the kinds of problems we are interested in, the computable infinitary formulas say everything that we need to say.

We prove a version of the Barwise-Kreisel Compactness Theorem, saying that a Π_1^1 set of computable infinitary sentences has a model if every Δ_1^1 subset does. We regard this as a theorem of computability theory, and we give a proof using only basic facts taken from the old standard reference on computability by Rogers [134]. Our treatment makes no mention of admissible sets or Kripke-Platek set theory.

Results on the three basic problems are obtained using finite-injury priority constructions. For the variants involving arbitrary copies of a given structure, we use forcing. The higher level variants for computable structures call for priority constructions in which the information about the requirements is Δ_α^0, for an arbitrary computable ordinal α.

We prove a *metatheorem*, due to Ash, giving abstract conditions (on a tree and some related objects) guaranteeing the success of a nested priority construction. The proof of the metatheorem involves a rather complicated construction. A proof *using* the metatheorem has the advantage of being modular. It calls for defining the tree and other objects, and then verifying the list of conditions. Thus, the metatheorem certainly simplifies the process of *checking* a proof. The process of *discovering* what might be true—before trying to write out a proof—

involves informal, often non-verbal, thinking, together with testing of low-level cases. We do not claim that the metatheorem makes proofs simple. However, it has brought within reach several results that seemed impossible without it, and it has been used successfully by students with little prior experience in writing up priority constructions.

The general results on the three basic problems, and related results on computable structures, have hypotheses that may at first appear impossibly strong. However, the results do apply to various familiar kinds of structures. The hypotheses involve a single nice copy of the structure. What is required is a thorough understanding of certain "back-and-forth" relations. We give results calculating these relations for certain classes of structures—in particular, for well orderings and superatomic Boolean algebras. We can show that the computable structures in these classes have nice copies.

The thesis of this book is that syntactical conditions account for bounds on complexity that are "intrinsic" (bounds that persist under isomorphism). We emphasize "positive" results—results supporting this thesis. We illustrate the results with examples involving linear orderings, Boolean algebras, groups, vector spaces, models of arithmetic, etc. We mention only a few "negative" results. There are by now quite a number of these, involving structures invented to show that certain hypotheses cannot be dropped, certain syntactical conditions are not sufficient, etc. The negative results are certainly interesting, and they involve a great deal of ingenuity in the constructions.

This book is not a survey of computable structure theory. We focus mainly on results obtained using a particular collection of ideas and techniques. Our hope is that the reader who has mastered the background material and understood the results that we have chosen to present will be in a position to apply these techniques, and to vary them as needed, to solve quite different-looking problems. For a broader view of computable structure theory, we recommend the books by Ershov and Goncharov [43] and Harizanov [64].

0.2 Constructivizations

Structures here are countable unless we specifically say otherwise. Moreover, it should be assumed that the language of a structure $\mathcal{A}$ is computable, and the universe is a subset of ω (the set of natural numbers), usually computable. This means that we can assign Gödel numbers to elementary first order sentences in the language of $\mathcal{A}$ augmented by the elements. Then the *atomic diagram* of $\mathcal{A}$, denoted by $D(\mathcal{A})$, and the *complete diagram*, denoted by $D^c(\mathcal{A})$, can be identified with sets of natural numbers. We say that $\mathcal{A}$ is *computable* if $D(\mathcal{A})$ is computable. This is equivalent to saying that the universe of $\mathcal{A}$ is computable, and the relations and operations are uniformly computable. We say that $\mathcal{A}$ is *decidable* if $D^c(\mathcal{A})$ is computable.

Some of the basic questions that interest us were first considered by Ershov [42], Goncharov [53], [54], Nurtazin [129], and others at Novosibirsk, who spoke of *constructivizations* instead of computable structures, and *strong construc-*

tivizations instead of decidable structures. The following definitions are due to Ershov.

A *constructivization* is a pair $(\mathcal{A}, \nu)$, where $\mathcal{A}$ is an abstract structure, countable, but with arbitrary objects as elements, and ν is a function from ω onto the universe of $\mathcal{A}$ such that the relation

$$\mathcal{A} \models \varphi(\nu(\overline{m})),$$

between *atomic* formulas φ and tuples $\overline{m}$ from ω, is computable. A *strong constructivization* is a pair $(\mathcal{A}, \nu)$, as above, in which the relation

$$\mathcal{A} \models \varphi(\nu(\overline{m})),$$

between *arbitrary* formulas φ and tuples $\overline{m}$ from ω, is computable. The abstract structure $\mathcal{A}$ is said to be *constructivizable*, or *strongly constructivizable* if there exists ν as above.

Proposition 0.1 *(a) A (non-empty) structure $\mathcal{A}$ has a constructivization if and only if it has a computable copy.*

(b) A (non-empty) structure $\mathcal{A}$ has a strong constructivization if and only if it has a decidable copy.

We give the proof for (a)—the proof for (b) is the same. First, suppose that $\mathcal{A}$ has a computable copy $\mathcal{B}$. We obtain a constructivization $\nu = f \circ g$, where f is an isomorphism from $\mathcal{B}$ onto $\mathcal{A}$, and g is a computable function from ω onto $\mathcal{B}$. Now, suppose that ν is a constructivization of $\mathcal{A}$. The relation $\nu(m) = \nu(m')$ is an equivalence relation on ω, and it is computable, since the formula $x = y$ is atomic. Taking the first representative of each equivalence class, we obtain a computable subset B of ω such that ν maps B $1-1$ onto $\mathcal{A}$. The induced structure $\mathcal{B}$ with universe B is a computable copy of $\mathcal{A}$.

0.3 Authorship and acknowledgements

This book was not meant to be joint. Chris Ash was supposed to write it. I expected to participate only by nagging him occasionally until it was done. Sadly, Chris died, having written parts of five chapters, plus a tentative list of titles for further chapters. I wanted the book finished. I hope much of the material that Chris would have included is here.

I am grateful to John Crossley, Sara Miranda, and other colleagues and friends of Ash at Monash University for their hospitality and support. I am grateful to Anne-Marie Vandenberg for locating and typing the part of the book that Chris had written, and for typing my initial drafts of chapters. I am grateful to Manny Lerman and Richard Shore for their encouragement and suggestions.

I am grateful to my students for their many contributions. Carlos Alegria, Sarah Oates, Kelleen Hurlburt, John Thurber, Alan Vlach, Grzegorz Michalski, Alex McAllister, Charles McCoy, and Andrew Arana have all proved interesting results in computable structure theory. Several of these results are discussed in the book. Some students, Alex McAllister, in particular, suggested numerous corrections and improvements on an early draft.

Finally, I am grateful to Tim McNicholl for detailed comments on a near-final draft. The errors that remain are all mine.

Julia F. Knight

Contents

Preface   **v**
     0.1   Introduction   v
     0.2   Constructivizations   vii
     0.3   Authorship and acknowledgements   viii

1   Computability   **1**
     1.1   Informal concepts   1
     1.2   Church's Thesis   4
     1.3   Proof by Church's Thesis   5
     1.4   Some basic results   5
     1.5   Coding functions   7
     1.6   Kleene's Normal Form Theorem   8
     1.7   Facts about c.e. sets and relations   8
     1.8   The standard list of c.e. sets   12
     1.9   The $s-m-n$ Theorem   12
     1.10   The Recursion Theorem   13
     1.11   Relative computability   14
     1.12   The relativized $s-m-n$ Theorem   16
     1.13   Turing reducibility   16
     1.14   Other reducibilities   17

2   The arithmetical hierarchy   **21**
     2.1   Jumps   21
     2.2   Basic definitions   22
     2.3   Basic theorems   24
     2.4   Combining arithmetical relations   26
     2.5   Alternative definitions   27
     2.6   Approximations   28
     2.7   Trees   30

3   Languages and structures   **33**
     3.1   Propositional languages and structures   33
     3.2   Predicate languages   34
     3.3   Structures for a predicate language   36

3.4	Satisfaction	37
3.5	Enlarging a structure	38
3.6	Theories and models	39
3.7	Prenex normal form	40
3.8	Isomorphism	40
3.9	Quotient structures	41
3.10	Model existence	42
3.11	Compactness	44
3.12	Some special kinds of structures	45
3.13	Computable sets of sentences	49
3.14	Complexity of structures	50
3.15	Complexity of definable relations	51
3.16	Copies of a given structure	53
3.17	Complexity of quotient structures	55

4 Ordinals   57
4.1	Set theoretic facts	57
4.2	Inductive proofs and definitions	58
4.3	Operations on ordinals	59
4.4	Cantor normal form	60
4.5	Constructive and computable ordinals	60
4.6	Kleene's O	61
4.7	Constructive and computable ordinals	62
4.8	Transfinite induction on ordinal notation	66

5 The hyperarithmetical hierarchy   71
5.1	The hyperarithmetical hierarchy	71
5.2	The analytical hierarchy	75
5.3	The Kleene-Brouwer ordering	82
5.4	Relativizing	84
5.5	Ershov's hierarchy	86

6 Infinitary formulas   89
6.1	Predicate formulas	89
6.2	Sample formulas	90
6.3	Subformulas and free variables	92
6.4	Normal form	93
6.5	Model existence	94
6.6	Scott's Isomorphism Theorem	96
6.7	Ranks and special Scott families	98
6.8	Rigid structures and defining families	100
6.9	Definability of relations	101
6.10	Propositional formulas	102

CONTENTS

7 Computable infinitary formulas — 105
- 7.1 Informal definitions — 105
- 7.2 Formal definition — 107
- 7.3 Sample formulas — 108
- 7.4 Satisfaction and the hyperarithmetical hierarchy — 109
- 7.5 Further hierarchies of computable formulas — 110
- 7.6 Computable propositional formulas — 112
- 7.7 The simplest language — 115
- 7.8 Hyperarithmetical formulas — 116
- 7.9 X-computable formulas — 118

8 The Barwise-Kreisel Compactness Theorem — 121
- 8.1 Model existence and paths through trees — 121
- 8.2 The Compactness Theorem — 123
- 8.3 Hyperarithmetical saturation — 127
- 8.4 Orderings and trees — 129
- 8.5 Boolean algebras — 131
- 8.6 Groups — 132
- 8.7 Priority constructions — 133
- 8.8 Ranks — 136

9 Existence of computable structures — 137
- 9.1 Equivalence structures — 137
- 9.2 Abelian p-groups — 139
- 9.3 Linear orderings — 142
- 9.4 Boolean algebras — 150
- 9.5 Results of Wehner — 152
- 9.6 Decidable homogeneous structures — 155

10 Completeness and forcing — 159
- 10.1 Images of a relation — 159
- 10.2 Ershov's hierarchy — 167
- 10.3 Images of a pair of relations — 169
- 10.4 Isomorphisms — 173
- 10.5 Expansions — 177
- 10.6 Sets computable in all copies — 177
- 10.7 Copies of an arbitrary structure — 179

11 The Ash-Nerode Theorem — 181
- 11.1 Simple examples — 181
- 11.2 Results of Ash and Nerode — 182
- 11.3 Applications — 185
  - 11.3.1 Vector spaces — 185
  - 11.3.2 Algebraically closed fields — 186
- 11.4 Expansions — 187
- 11.5 Results of Harizanov — 188

11.6 Ershov's hierarchy	189
11.7 Pairs of relations	192

12 Computable categoricity and stability 197
    12.1 Simple examples 197
    12.2 Relations between notions 198
    12.3 Computable categoricity 199
    12.4 Decidable structures 201
    12.5 Computable stability 203
    12.6 Computable dimension 205
    12.7 One or infinitely many 206
    12.8 Quotient structures 209

13 n-systems 213
    13.1 Introduction 213
    13.2 Statement of the metatheorem 214
    13.3 Some examples 216
    13.4 Proof of the metatheorem 220
    13.5 Looking ahead 223
    13.6 Michalski's Theorem 224

14 α-systems 227
    14.1 Statement of the metatheorem 227
    14.2 Organizing the construction 228
    14.3 Derived systems 230
    14.4 Simultaneous runs 234
    14.5 Special (α_n)-systems 235
    14.6 Further metatheorems 237

15 Back-and-forth relations 239
    15.1 Standard back-and-forth relations 239
    15.2 α-friendly families 241
    15.3 Examples 242
        15.3.1 Arithmetic 242
        15.3.2 Vector spaces 242
        15.3.3 Linear orderings 244
        15.3.4 Boolean algebras 246
    15.4 Ranks 247
    15.5 Stronger back-and-forth relations 250
    15.6 Abstract back-and-forth relations 251
    15.7 Open problems 252

16 Theorems of Barker and Davey 253
    16.1 Barker's Theorems 253
    16.2 Davey's Theorems 259

17 Δ_α^0 stability and categoricity — 263
17.1 Relations between notions 263
17.2 Δ_α^0 stability 264
17.3 Well orderings 268
17.4 Δ_α^0 categoricity 269
17.5 Superatomic Boolean algebras 272

18 Pairs of computable structures — 275
18.1 Simple examples 275
18.2 General results 278
18.3 Results of Feiner and Thurber 281
18.4 Limit structures 284
18.5 Quotient orderings 287

19 Models of arithmetic — 291
19.1 Scott sets 291
19.2 Enumerations 294
19.3 Structures representing a Scott set 296
19.4 Harrington's Theorem 305
19.5 Solovay's Theorems 307
19.6 Sets computable in all models 308
19.7 Open problems 309

A Special classes of structures — 311
A.1 Vector spaces 311
A.2 Fields 312
A.3 Orderings 314
    A.3.1 Operations on orderings 314
A.4 Boolean algebras 315
A.5 Equivalence structures 317
A.6 Abelian p-groups 317
A.7 Models of arithmetic 318
    A.7.1 Matijasevic's Theorem 319
    A.7.2 Definability of satisfaction 320
    A.7.3 Independence 321

Bibliography — 323

Index — 335

Chapter 1

Computability

1.1 Informal concepts

The intuitive notion of an effective procedure for answering a class of questions pervades all of mathematics. With the common use of computers, it scarcely needs explaining, but there are some points which need to be established.

First, the objects with which one can compute all turn out to be, essentially, finite strings, or sequences, of symbols from a finite alphabet. A finite sequence from a finite alphabet can always be *coded* as a single natural number. For example, letting 1, 2, 3, and 4 be codes for the symbols (, !,), and ∗, respectively, and making use of the uniqueness of prime factorizations, we can code the string ((!)∗ by the single number $2^1 3^1 5^2 7^3 11^4$. As a result of coding, we may assume that all computations are performed on natural numbers. We shall deal so exclusively with natural numbers that unless otherwise specified, "set" will mean "set of natural numbers", "relation" will mean "relation on natural numbers", and "(partial) function" will mean "(partial) function on natural numbers". The set of natural numbers is denoted by ω.

Second, it is useful to consider computations with a k-tuple of natural numbers as input. For example, on input (n_1, n_2, n_3), we may compute $(n_1 + n_2) \cdot n_3$. In principle, the input can be coded as a single number $2^{n_1} 3^{n_2} 5^{n_3}$. Sometimes we wish to do this, but sometimes we do not.

Third, in actually programming computers, one is concerned with efficiency. Throughout what follows, we are *not*. In particular, the method of coding we used above is grossly inefficient. We chose it because of tradition and because it is easy to understand. A related point is that a real computer has only a finite capacity. We shall consider *idealized* computers, with no limit on the size of the input, or the amount of memory space available for use, or the size of the output.

Fourth, it is essential to recognize that there are effective procedures that on some inputs yield an output and on others do not. Typical of these procedures are those which involve a search, and yield an output if the search halts, but not otherwise. To describe such procedures, we take as our basic objects of study *partial* functions, of one or more variables (where the variables range over natural numbers). The term "partial function" includes, by the way, "total function" (one defined for all values of the variables) as a special case. We are interested in partial *computable* functions.

We illustrate these notions in a mathematical setting. Hilbert, in the 10th problem on his famous list of 23, asked for an effective procedure for determining, given a polynomial $p(x_1, \ldots, x_k)$, with integer coefficients, and any number of variables, whether the equation

$$p(x_1, \ldots, x_k) = 0$$

has a solution in the integers. There is a natural procedure which gives the answer "Yes" just in case there is a solution. The procedure involves systematically testing assignments of integers to the variables until a solution appears. If there is no solution, then the testing continues forever.

Having settled on a method of coding, we may define a function f such that if x is the code of a polynomial, and y is the code of a sequence of integers appropriate to substitute for the variables, then $f(x,y)$ is the absolute value of the integer which results from making the substitution. If it is not the case that x and y are codes of the correct sorts, let $f(x,y) = 1$. Now, f is a total function which is clearly computable.

In terms of f, we may define a partial function g, where $g(x)$ is the least y for which $f(x,y) = 0$, if any, and undefined otherwise. Then g is a partial computable function. Given x, we need only calculate

$$f(x,0), f(x,1), f(x,2), \ldots ,$$

in turn, until we come to the least y for which $f(x,y) = 0$, if there is one, while if there is no such y, the search goes on forever and no output results. Let

$$h(x) = \begin{cases} 1 & \text{if there exists } y \text{ such that } f(x,y) = 0 , \\ 0 & \text{otherwise} \end{cases}.$$

Then h is a total function. From what we have said so far, it is not clear whether h is computable. Given input x, we can compute $f(x,0), f(x,1), f(x,2), \ldots$, and so produce the answer 1 if there is an appropriate y. However, to compute h, we need some other kind of search, to be performed simultaneously, to produce the answer 0 in case the first search does not end.

A fifth and last point arises on considering a single mathematical statement whose truth or falsity is presently unknown. One example is Goldbach's Conjecture, the statement that every even number greater than 4 is the sum of two prime numbers. Let us define the function k, where $k(x) = 1$ if Goldbach's Conjecture is true, and 0 otherwise. Any constant function is computable, and k is either the constant function 1 or the constant function 0. Therefore, k is computable. Admittedly, we do not know *which* procedure works, but that does not come into our reckoning.

1.1. INFORMAL CONCEPTS

In general, there is always an effective procedure for answering a single question, or any finite number of questions. The procedure consists of consulting a table of answers. We may not know which is the correct table, but that is a separate issue. The non-trivial questions of computability ask whether there is a procedure which applies to all inputs in some *infinite* set.

In addition to *k-ary functions* (that is, functions of k variables), we consider *k-ary relations*. For example, we may define a 2-ary, or *binary*, relation R which holds for a pair of numbers (x, y) just in case $x + y$ is even. We write $R(\overline{x})$ if the relation R holds for the tuple $\overline{x}$, and $\neg R(\overline{x})$, otherwise. We are interested in computability of relations. We say that a k-ary relation R is *computable* if there is an effective procedure which on input $\overline{x} = (x_1, \ldots, x_k)$, yields output "Yes" or "No" according to whether the relation holds or not. We may convert the output to numbers and deal with the *characteristic function* of R, where this is the total k-ary function χ_R such that

$$\chi_R(\overline{x}) = \begin{cases} 1 & \text{if } R(\overline{x}), \\ 0 & \text{if } \neg R(\overline{x}). \end{cases}$$

To say that R is computable is the same as to say that χ_R is computable.

A 1-ary, or *unary*, relation R is identified with the set for which it holds, so we may write $R(x)$ or $x \in R$ interchangeably. Similarly, a k-ary relation is identified with the set of k-tuples $\overline{x} = (x_1, \ldots, x_k)$ for which it holds, so we may write $\overline{x} \in R$ instead of $R(\overline{x})$.

Varying the notion of computability, we say that R is *semi-computable* if there is an effective procedure which on input $\overline{x}$, yields output "Yes" if $R(\overline{x})$, and otherwise continues forever without giving an output.[1] Converting the output to numbers, we define the *semi-characteristic function* of R to be the partial function δ_R such that

$$\delta_R(\overline{x}) = \begin{cases} 1 & \text{if } R(\overline{x}), \\ \text{undefined} & \text{if } \neg R(\overline{x}). \end{cases}$$

Then R is semi-computable if and only if its semi-characteristic function is computable. Note that if R is computable, then it is semi-computable, because an effective procedure for computing χ_R can always be modified so that instead of giving output 0, it performs some endless calculation.

We illustrate, again in the setting of Hilbert's 10th Problem. Let $R(x)$ hold if x is the code for a polynomial such that the corresponding equation has a solution in the integers, and recall the functions f, g, and h defined above. Clearly, R is semi-computable; we have

$$\delta_R(x) = \begin{cases} 1 & \text{if } g(x) \text{ is defined}, \\ \text{undefined} & \text{otherwise}. \end{cases}$$

We do not claim that R is computable; $\chi_R = h$, where the computability of h has been left in doubt.

---
[1] There is another, unrelated meaning of the term "semi-computable". The term, with the definition that we gave, is descriptive and has some history, but we shall switch to another term, with a different definition, once we have shown that the two definitions are equivalent.

1.2 Church's Thesis

So far, we have not shown that there is any function or relation which is not in fact computable. To do so, we need a precise definition, or at least some basic properties, of the class of computable functions. There are a number of different definitions. The one due to Turing, in terms of *Turing machines*, is perhaps the most convincing. (See the classic text by Kleene [80] for a discussion of several definitions, and proofs that they are all equivalent.)

We refrain here from describing the ingredients of a Turing machine, since we need so little of this information to proceed. Turing machines, and all known computers, operate by combining very simple basic steps according to some finite list of instructions. Turing was responsible for the idea, commonplace today, that it is not necessary to build a new machine whenever we wish to compute a new function. He described a *universal* machine, taking a program as part of the input, and capable of computing any function which could be computed by another machine.

The various definitions of the class of computable functions are all equivalent. For a discussion of several different-looking definitions, together with proofs of equivalence, see [80]. Since Turing, no new computation methods have been found which yield functions outside the class captured by Turing's definition. The inescapable conclusion is that any function which is computable by any means whatever is computable by a Turing machine. This statement, although not in its original form, is attributed to Church.

> Church's Thesis: Computable = Computable by a Turing machine

Church's Thesis asserts that the intuitive notion of being effectively computable agrees with the formal notion of being computable by a Turing machine. It is hard to see how the statement could ever be proved, but it is easy to see how it might be refuted—by discovering a really new method of computation, one that computes new functions. This has not happened in approximately 70 years, and it is generally accepted that Church's Thesis is true beyond all reasonable doubt.

Many people adopted a different vocabulary to indicate that their results about computability refer to the formal notion.

Informal name	Formal name
(total) computable function	(total) recursive function
partial computable function	partial recursive function
computable relation	recursive relation
semi-computable relation	recursively enumerable relation

The word "recursive" comes from the definition of computability due to Kurt Gödel and Stephen Kleene. This definition, more algebraic than Turing's, involved, among other things, a scheme for obtaining new functions by *recursion*. The phrase "general recursive function" sometimes appears in the literature instead of "recursive function". The reason for the word "general" is to distinguish the notion from the more restrictive class of "primitive recursive functions", which Gödel originally called "recursive".

The name "recursively enumerable", abbreviated by "r.e.", arises from a different characterization of the semi-computable relations as the relations R such that there is a computable function which lists, or *enumerates* (in some order) all k-tuples in R. We shall see shortly that this characterization is equivalent to the one that we chose as a definition—the semi-characteristic function of R is a partial computable function.

Recently, there has been a move to use the term "computable" instead of "recursive", on the grounds that it is more descriptive. Similarly, "recursively enumerable" and "r.e." are being replaced by "computably enumerable" and "c.e."

1.3 Proof by Church's Thesis

Having emphasized that we are dealing with Turing computable functions, we follow the apparently bizarre course of *not* showing rigorously that the functions that we claim are computable can in fact be computed by a Turing machine. We accept any description of a computation which accords with our intuitive notion of computability as evidence that the function is Turing computable. This form of argument is known as a *proof by Church's Thesis*.

In the early years of the subject, painstaking efforts were made to show that functions were Turing computable, and Kleene's characterization was very useful for this purpose. Our justification for omitting all of this effort is that the work has indeed been done, and if there is any failure of Church's Thesis, it must involve a totally new concept in computation, very different from those which we use. Rogers [134] took this approach, with considerable gains in brevity and clarity. The simplest viewpoint is just to assume that Church's Thesis is true. In any event, we claim that our results on computable functions hold for Turing computable functions.

1.4 Some basic results

In the remainder of this chapter we review some very basic results from computability that we shall be using later. These results can all be found in [134]. We need only a few facts about Turing machines. Each Turing machine is determined by a program which is a finite sequence of symbols from a finite language, and so can be coded by a single number, traditionally denoted by e. We adopt some convention so that *every* number codes some Turing machine, and we refer to the eth Turing machine. On input n, the machine may eventually halt, giving a natural number value, or it may not.

The eth Turing machine can be regarded as computing a unary partial function, denoted by φ_e. Thus, we have a list $\varphi_0, \varphi_1, \varphi_2, \ldots$, of all unary partial functions that are Turing computable. There are repetitions, of course, since different Turing machines (corresponding to different programs) may compute the same function. We say that φ_e is the partial Turing computable function with *index e*.

Notation: If the computation of Turing machine e on input n halts, then we write $\varphi_e(n) \downarrow$, and otherwise we write $\varphi_e(n) \uparrow$. If $n < r$ and the computation halts within r steps, then we may write $\varphi_{e,r}(n) \downarrow$, and otherwise, we may write $\varphi_{e,r}(n) \uparrow$.

The partial computable function $\varphi_e(n)$, of the two variables e and n, can be computed by first decoding e as a Turing machine and then running this machine with input n. As an example of a proof by Church's Thesis, having just indicated informally how to compute $\varphi_e(n)$, we conclude that it is Turing computable. A Turing machine that computes this function is said to be *universal*.

Note that there can be no list $\psi_0, \psi_1, \psi_2, \ldots$ of all Turing computable *total* functions such that the function $\psi_e(n)$, of the two variables e and n, is computable. For, suppose there is such a list. Then the function f defined by $f(n) = \psi_n(n) + 1$ is computable, and hence, by Church's Thesis, Turing computable. It follows that f is one of the functions ψ_e. Then we have both $f(e) = \psi_e(e)$ and $f(e) = \psi_e(e) + 1$, a contradiction.

For partial functions, a diagonal argument similar to the one above yields, not a contradiction, but an interesting fact about the function $\varphi_n(n)$, which, by Church's Thesis, is a perfectly good, Turing computable, partial function of one variable.

Theorem 1.1 *The partial computable function $\varphi_n(n)$ cannot be extended to any total computable function.*

Proof: Suppose that $\varphi_n(n)$ can be extended to a total computable function $f(n)$. Then $f(n) + 1$ is a computable total function, which must therefore be equal to $\varphi_e(n)$, for some e. Since φ_e is total, $\varphi_e(e) \downarrow$, so we have

$$\varphi_e(e) = f(e) + 1 = \varphi_e(e) + 1 \ ,$$

a contradiction.

We can argue the existence of non-computable functions purely on the basis of cardinality—there are uncountably many functions and only countably many Turing machines. Applying the previous theorem, we may obtain a specific example of a non-computable function f. Let

$$f(n) = \begin{cases} \varphi_n(n) & \text{if this is defined , and} \\ 0 & \text{otherwise .} \end{cases}$$

Actually, we described another specific non-computable function earlier. Hilbert's 10th Problem was to find an effective procedure for computing a certain total

1.5 Coding functions

As we mentioned, a pair of numbers can be coded as a single number. We may assume that this is done by some standard computable *pairing function* that maps $\omega \times \omega$ $1-1$ onto ω. One such pairing function is

$$2^x(2y+1) - 1 .$$

Another is

$$\frac{1}{2}(x^2 + 2xy + y^2 + 3x + y) .$$

For a satisfactory pairing function, the two functions that reclaim the components of a pair, in addition to the pairing function itself, should be reasonably simple to compute. Fixing one such pairing function (to use from here on), we write $\langle x, y \rangle$ for the value of the pairing function at (x, y). We denote the component functions by $(\ )_1$ and $(\ )_2$, so that $z = \langle (z)_1, (z)_2 \rangle$. A final property of the two pairing functions above, which may occasionally be helpful, is that

$$(z)_1, (z)_2 \leq z .$$

We may use the pairing function to encode triples by defining $\langle x, y, z \rangle$ to be $\langle \langle x, y \rangle, z \rangle$ and, repeating this device, we may encode k-tuples for any k. Where it is clear from the context what k is, we may use $(\ )_1, (\ )_2, \ldots, (\ )_k$ to denote the corresponding decoding functions. The coding functions for k-tuples allow us, in principle, to consider only *unary* computable functions, since the k-ary function

$$f(x_1, x_2, \ldots, x_k)$$

is clearly computable if and only if the unary function

$$g(x) = f((x)_1, (x)_2, \ldots, (x)_k)$$

is computable. Nevertheless, it is often more natural to work with several variables. We use $\bar{x}$ to denote the tuple $(x_1, x_2, \ldots, x_k)$, where k is arbitrary or determined by the context.

In some situations, it is useful to encode finite sequences of *all* lengths simultaneously. We could do this using the uniqueness of prime factorization. Here is another method, making use of the coding functions for k-tuples for various fixed k. The empty sequence is encoded by 0, and a sequence $(x_1, x_2, \ldots, x_k)$, for $k \geq 1$, is encoded by

$$\langle x_1, x_2, \ldots, x_k, k-1 \rangle + 1 = \langle \langle x_1, x_2, \ldots, x_k \rangle, k-1 \rangle + 1 .$$

---

[2] In addition to [116], [35], see the beautiful expositions [34], [113].

Finally, we may wish to encode finite sets. One method is to encode the set $\{x_1, x_2, \ldots, x_k\}$, where $x_1 < x_2 < \ldots < x_k$, by the number

$$x = 2^{x_1} + 2^{x_2} + \ldots + 2^{x_k},$$

and encode the empty set by 0. We refer to the code for a set as the *canonical index* of the set. Noting that each number x is the canonical index for some finite set, we write D_x for the set having canonical index x.

1.6 Kleene's Normal Form Theorem

For any relation $R(\overline{x}, y)$, we let $\mu y\, R(\overline{x}, y)$ denote the least y, if any, for which $R(\overline{x}, y)$. Thus, $\mu y\, R(\overline{x}, y)$ is a partial function of $\overline{x}$. We refer to this operator, transforming relations into partial functions, as the "μ-operator". Clearly, if $R(\overline{x}, y)$ is a computable relation, then $\mu y\, R(\overline{x}, y)$ is a partial computable function, which we compute by testing the values of y, in increasing order. Here is Kleene's Normal Form Theorem.

Theorem 1.2 *There exist a computable unary function U and, for each n, an $(n+2)$-ary computable relation T_n, such that the n-ary partial function computed by the eth Turing machine is $U[\mu z\, T_n(e, \overline{x}, z)]$.*

Proof: Our proof is necessarily only a sketch, since we have refrained from giving the precise details of Turing machines. We can encode a Turing machine "configuration", giving the current state and contents of the location being accessed, plus the contents of all storage locations used so far. Since we can encode a single configuration, we can also encode a sequence of configurations.

We define the relation $T_n(e, \overline{x}, z)$ to hold if z is the code for a sequence of configurations, starting with the *initial* configuration giving the input $\overline{x}$, passing from one configuration to the next by following the instructions for the Turing machine whose code is e, and ending with a *halting* configuration that gives numerical output. We define the function $U(z)$ to decode the number given as output, or 0, if this is inapplicable.

Theorem 1.2 is useful in a variety of contexts. For one thing, it implies that a *single* application of the μ-operator is enough for computing any partial computable function.

1.7 Facts about c.e. sets and relations

According to our definition, a relation is c.e. if and only if its semi-characteristic function is a partial computable function. The next result gives another possible definition, saying that a relation is is c.e. if and only if it is a *projection* of a computable relation.

1.7. FACTS ABOUT C.E. SETS AND RELATIONS

Theorem 1.3 *An n-ary relation $S(\overline{x})$ is c.e. if and only if there is a computable relation $R(\overline{x}, y)$ such that*

$$S(\overline{x}) \Leftrightarrow \exists y\, R(\overline{x}, y) \ .$$

Proof: First, suppose that S is c.e. By our definition, the semi-characteristic function $\delta_S(\overline{x})$ is a partial computable function. By Theorem 1.2, we have

$$\delta_S(\overline{x}) = U(\mu z\, T_n(e, \overline{x}, z))\ ,$$

for some e. Then

$$\delta_S(\overline{x}) = 1 \Leftrightarrow \delta_S(\overline{x}) \downarrow \Leftrightarrow \exists z\, T_n(e, \overline{x}, z)\ .$$

Therefore, we have

$$S(\overline{x}) \Leftrightarrow \exists y\, R(\overline{x}, y)\ ,$$

where $R(\overline{x}, y)$ is the computable relation $T_n(e, \overline{x}, y)$.

Now, suppose there is a computable relation R such that

$$S(\overline{x}) \Leftrightarrow \exists y\, R(\overline{x}, y) \ .$$

To compute $\delta_S(\overline{x})$, we check, in turn, whether $R(\overline{x}, 0)$, $R(\overline{x}, 1)$, $R(\overline{x}, 2)$, etc., and if the answer is ever "Yes", then we give output 1.

The next result characterizes computable relations in terms of c.e. relations.

Theorem 1.4 *A relation $S(\overline{x})$ is computable if and only if both $S(\overline{x})$ and $\neg S(\overline{x})$ are c.e.*

Proof: First, suppose that $S(\overline{x})$ is computable. We have already indicated how a procedure for computing the total function χ_S can be converted into a procedure for computing the partial function δ_S. We obtain a procedure for computing $\delta_{\neg S}$ in a similar way. Now, suppose that both δ_S and $\delta_{\neg S}$ are partial computable functions. We can compute $\chi_S(\overline{x})$ by simultaneously applying procedures for computing $\delta_S(\overline{x})$ and $\delta_{\neg S}(\overline{x})$ until one of them halts, and then giving the appropriate output 1 or 0.

A set or relation is said to be *co-c.e.* if its complement is c.e. Thus, Theorem 1.4 says that a relation is computable if and only if it is both c.e. and co-c.e.

The next result characterizes partial computable functions in terms of c.e. relations.

Theorem 1.5 *An n-ary partial function $f(\overline{x})$ is computable if and only if the $(n+1)$-ary relation $f(\overline{x}) = y$ is c.e.*

Proof: If the relation $f(\overline{x}) = y$ is c.e., then by Theorem 1.3, there is a computable relation R such that

$$f(\overline{x}) = y \Leftrightarrow \exists z\, R(\overline{x}, y, z) .$$

Then we can compute $f(\overline{x})$ by searching for a pair (y, z) such that $R(\overline{x}, y, z)$, and giving output y if the search halts. Conversely, if $f(\overline{x})$ is a partial computable function, then by Theorem 1.2, there is some e such that

$$f(\overline{x}) = U[\mu z\, T_n(e, \overline{x}, z)] .$$

By the definition of T_n,

$$f(\overline{x}) = y \Leftrightarrow \exists z\, [T_n(e, \overline{x}, z)\, \&\, U(z) = y] ,$$

so by Theorem 1.3, the relation $f(\overline{x}) = y$ is c.e.

The *domain* of an n-ary partial function $f(\overline{x})$ is an n-ary relation $S(\overline{x})$, where $S(\overline{x})$ if and only if $f(\overline{x})$ is defined. In fact, the usual definition says that a relation is c.e. if it is the domain of a partial computable function. The next result says that this definition is equivalent to ours.

Theorem 1.6 *A relation S is c.e. if and only if it is the domain of a partial computable function.*

Proof: By definition, if S is c.e., then it is the domain of the partial computable function δ_S. Conversely, if S is the domain of a partial computable function f, then a procedure for computing f can be converted into a procedure for computing δ_S, replacing any output by 1.

As we have mentioned, if R is a k-ary relation and S is the unary relation such that

$$S(x) \Leftrightarrow R((x)_1, \ldots, (x)_k) ,$$

then R is c.e. if and only if S is. So, in principle, we need only deal with c.e. sets. We have just seen that a set is c.e. if and only if it is the domain of a partial computable function. The next result gives further characterizations.

Theorem 1.7 *A set S is c.e. if and only if any one of the following conditions holds:*
  (a) S is the range of a partial computable function,
  (b) S is empty or is the range of a total computable function,
  (c) S is finite or is the range of a total computable function which is $1 - 1$.

Proof: To see that (a), (b) or (c) implies that S is c.e., it is sufficient to observe that every finite set is computable, and therefore c.e., and also to show that the range of any partial computable function is c.e. Suppose that S is the range of a partial computable function f. Then

$$y \in S \Leftrightarrow \exists x\, f(x) = y ,$$

1.7. FACTS ABOUT C.E. SETS AND RELATIONS

and by Theorems 1.5 and 1.3, there is a computable relation R such that

$$\exists x\, f(x) = y \iff \exists x\, \exists z\, R(x, y, z)\,.$$

Then

$$y \in S \iff \exists w\, R((w)_1, y, (w)_2)\,,$$

so by Theorem 1.3, S is c.e.

Conversely, suppose that S is c.e. We must show that (a), (b), and (c) hold. If S is empty, then it is the range of the empty partial computable function. If S is finite and non-empty, say $S = \{a_0, a_1, \ldots, a_k\}$, then S is the range of the computable function defined by

$$f(n) = \begin{cases} a_n & \text{if } n < k \\ a_k & \text{if } n \geq k\,. \end{cases}$$

Finally, suppose that S is infinite. We show that S is the range of a total computable $1-1$ function. By Theorem 1.3, there is a computable relation R such that

$$y \in S \iff \exists z\, R(y, z)\,.$$

We define the required f by recursion, letting $f(n) = y$ for the least pair (y, z) such that

$$R(y, z)\ \&\ y \notin \{f(0), f(1), \ldots, f(n-1)\}\,.$$

Remark: We now have several characterizations of the class of *computably enumerable* sets. The name is justified by Condition (b) (or Condition (a)) of Theorem 1.7. If the set S is the range of a total computable function f, then we have an effective procedure for enumerating S. We compute and print the value of $f(0)$, then $f(1)$, then $f(2)$, and so on.

A c.e. set S can be enumerated without repetition. However, we cannot enumerate the elements of S in *increasing* order unless S is a computable set. To see this, note that if S is infinite and we have a computable procedure for enumerating the elements in increasing order, then we can determine whether $n \in S$ by proceding with the enumeration until we see a number greater than n. We have $n \in S$ if and only if n has appeared by this stage.

The next result is simple but useful.

Theorem 1.8 *For any c.e. sets X and Y, there exist disjoint c.e. sets $X^* \subseteq X$ and $Y^* \subseteq Y$ such that $X^* \cup Y^* = X \cup Y$. Moreover, if $X \cup Y$ is computable, then X^* and Y^* are computable.*

Proof: We enumerate X and Y. At any given stage, we put x into X^* if x appears in X without having appeared in Y at a strictly earlier stage, and we put x into Y^* if it appears in Y without having appeared in X at the same stage or earlier. If $X \cup Y$ is computable, then to determine whether x is in X^*, we first determine whether $x \in X \cup Y$. If, so, then we watch the enumerations of X^* and Y^*, knowing that x is in just one of the two.

We say that S *separates* X and Y if S contains all elements of X and no elements of Y. It is possible to have disjoint c.e. sets X and Y with no computable separator.

Example: Let $X = \{e : \varphi_e(e) = 0\}$ and let $Y = \{e : \varphi_e(e) = 1\}$. Let S be a separator for X and Y, and suppose that $\chi_S = \varphi_e$. Then $\varphi_e(e)$ must have value either 0 or 1. If $\varphi_e(e) = 1$, then $e \in Y$, so $e \notin S$ and $\varphi_e(e) = 0$. Similarly, if $\varphi_e(e) = 0$, then $e \in X$, so $e \in S$ and $\varphi_e(e) = 1$. In either case, we have a contradiction, so there is no computable separator for X and Y.

1.8 The standard list of c.e. sets

We let W_e denote the domain of φ_e. Then, by Theorem 1.6, the list

$$W_0, W_1, W_2, \ldots$$

consists of exactly the c.e. sets (with repetitions). We call e an *index* of the set W_e. Note that the binary relation $x \in W_e$ is the domain of the partial computable function $\varphi_e(x)$ (of the two variables e and x), and so, by Theorem 1.6, it is a c.e. binary relation. Let

$$K = \{x : x \in W_x\} \, .$$

This set, called the *halting set*, will be referred to often in what follows.

Theorem 1.9 *The set K is c.e. but not computable.*

Proof: By definition, K is the domain of the partial computable function $\varphi_x(x)$. Hence, it is c.e. If K were computable, then we could extend $\varphi_x(x)$ to a total computable function f such that

$$f(x) = \begin{cases} \varphi_x(x) & \text{if } x \in K \, , \\ 0 & \text{otherwise} \, . \end{cases}$$

By Theorem 1.1, this is a contradiction.

An alternative proof is to show that $\neg K$ is not c.e. If $W_e \subseteq \neg K$, then $e \notin W_e$. Therefore, $e \in \neg K - W_e$.

1.9 The $s - m - n$ Theorem

We use $\varphi_e^{(n)}$ to denote the n-ary partial function computed by the eth Turing machine. Thus, φ_e is $\varphi_e^{(1)}$. As we indicated earlier, $\varphi_e^{(n)}(x_1, x_2, \ldots, x_n)$

1.10. THE RECURSION THEOREM

may be thought of as a partial computable function of the $n+1$ variables $e, x_1, x_2, \ldots, x_n$. We say that e is an *index* for the n-ary partial computable function $\varphi_e^{(n)}$. When the variables are displayed, we may omit the superscript. For example, $\varphi_e(x, y, z)$ can only mean $\varphi_e^{(3)}(x, y, z)$.

Theorem 1.10 (a) *For each m and n, and each partial computable function f, in variables $\overline{y} = (y_1, \ldots, y_m)$ and $\overline{x} = (x_1, \ldots, x_n)$, there is an m-ary total computable function $k(\overline{y})$ such that*

$$\varphi_{k(\overline{y})}(\overline{x}) = f(\overline{y}, \overline{x}) \; ,$$

for all $\overline{y}$ and $\overline{x}$.

Proof: The function k needs only to yield, for each $\overline{y}$, the code for a Turing machine which, on input $\overline{x}$, simulates the action of some Turing machine computing f on input $\overline{y}, \overline{x}$. Clearly, this operation on Turing machines, and their codes, is computable.

We note that in this proof, the value of $k(\overline{y})$ is effectively determined from $\overline{y}$ and an index for f. Thus, we have an alternative version of the theorem.

Theorem 1.10 (b) *For each m and n, there is an $(m+1)$-ary total computable function k such that*

$$\varphi_{k(e, \overline{y})}^{(n)}(\overline{x}) = \varphi_e^{(m+n)}(\overline{y}, \overline{x}) \; ,$$

for all e, $\overline{y}$, and $\overline{x}$.

Comments: Theorem 1.10 (a) follows immediately from Theorem 1.10 (b)—let e be an index for f and let the new $k(\overline{y})$ be the old $k(e, \overline{y})$. The converse is also immediate—apply Theorem 1.10 (a), letting $f(e, \overline{y}, \overline{x}) = \varphi_e(\overline{y}, \overline{x})$, with $m+1$ playing the role of m, to get an $(m+1)$-ary function $k(e, \overline{y})$. Either of the two versions of Theorem 1.10 may be referred to as the "$s - m - n$ Theorem". The name comes from Kleene's original version of Theorem 1.10 (b), which involved a function denoted by S_n^m.

1.10 The Recursion Theorem

The remarkable result below, which will be used frequently later, is due to Kleene. Both it and the corollary given next may be called the "Recursion Theorem".

Theorem 1.11 *For each m and each total computable function f, there is an n for which*

$$\varphi_n^{(m)} = \varphi_{f(n)}^{(m)} \; .$$

Moreover, n can be computed effectively from an index for f.

Proof: Consider the $(m+1)$-ary partial computable function $\varphi^{(m)}_{f(\varphi_x(x))}(\overline{y})$. By the $s-m-n$ Theorem (Theorem 1.10 (a)), there is a total computable function k such that
$$\varphi^{(m)}_{k(x)}(\overline{y}) = \varphi^{(m)}_{f(\varphi_x(x))}(\overline{y}) .$$
Now, k is φ_e for some e, giving
$$\varphi^{(m)}_{\varphi_e(x)}(\overline{y}) = \varphi^{(m)}_{f(\varphi_x(x))}(\overline{y}) .$$
Taking $x = e$, we obtain
$$\varphi^{(m)}_{\varphi_e(e)}(\overline{y})) = \varphi^{(m)}_{f(\varphi_e(e))}(\overline{y}) ,$$
so for $n = \varphi_e(e)$, we have
$$\varphi^{(m)}_n = \varphi^{(m)}_{f(n)} .$$
(Note that $\varphi_e(e)$ is defined, since $k = \varphi_e$ is total.)

Corollary 1.12 *For any $(m+1)$-ary partial computable function $g(e, \overline{x})$, there is some n such that $g(n, \overline{x}) = \varphi^{(m)}_n(\overline{x})$.*

Proof: By the $s-m-n$ Theorem, there is a total computable function f such that
$$g(e, \overline{x}) = \varphi^{(m)}_{f(e)}(\overline{x}) .$$
The result then follows from Theorem 1.11.

Note that Theorem 1.11 follows quickly from Corollary 1.12, by taking $g(e, \overline{x}) = \varphi^{(m)}_{f(e)}(\overline{x})$.

Remark: Taking $g(e, x) = e$ in Corollary 1.12 yields a number n for which $\varphi_n(x) = n$. Thus, n is the code for a Turing machine that on any input yields its own code. Reflecting on the difficulty of finding such a Turing machine from first principles gives some indication of the power of the Recursion Theorem. It seems that some trickery is necessary, such as that used in the proof of the theorem.

1.11 Relative computability

Everyone is familiar with interactive programs—programs that pause occasionally to ask a question, and continue running when the answer is supplied. We imagine running such a program, where the questions have the form "Is $n \in X$?", and the machine is equipped with an X-*oracle* that can provide answers to these questions. Actually, we think of the program itself as being independent of the set X, and, reflecting this, we might more precisely have oracle questions of the form "Is n in your set?" Obviously, if X is not a computable set, then the

1.11. RELATIVE COMPUTABILITY

X-oracle cannot be replaced by an actual computing device. We do not attempt to say how the oracle determines the answers to the questions, regarding the oracle as an indivisible device which performs these miracles.

This idea is very useful. We may ask which partial functions can be computed by an effective procedure, using the X-oracle, for a fixed set X. A (partial) function that can be computed in this way is said to be *computable relative to X*. A set or relation is *computable relative to X* if the characteristic function is computable relative to X. Similarly, a set or relation is *computably enumerable*, or *c.e., relative to X* if the semi-characteristic function is a partial function computable relative to X. We may say computable, or c.e., *in X*, instead of computable, or c.e., *relative to X*.

Our results so far use only the facts that if certain procedures are effective, then various modifications of these procedures, with additional steps, are also effective. If, in such a result, we replace *computable* everywhere by *computable relative to X*, we obtain another true statement, known as the *relativization* of the original result *to the set X*. For example, by relativizing Theorem 1.3 to X, we obtain the fact that $S(\overline{x})$ is c.e. relative to X if and only if there is a relation $R(\overline{x}, y)$, computable relative to X, such that

$$S(\overline{x}) \Leftrightarrow \exists y\, R(\overline{x}, y)\,.$$

By relativizing Theorem 1.4 to X, we obtain the fact that R is computable relative to X if and only if R and $\neg R$ are both c.e. relative to X.

We have not given a precise definition of a Turing machine. Nevertheless, we now claim that the definition can easily be modified to allow steps asking questions of, and receiving answers from, an X-oracle. We call the modified Turing machine an *oracle machine*. We emphasize that a given oracle machine operates effectively. Numbering the machines and their steps gives us, for any given set X, notions φ_e^X, and W_e^X. Thus, we have indices for partial functions computable relative to X and for sets and relations c.e. relative to X.

Notation: We write $\varphi_e^X(n) \downarrow$, or $\varphi_e^X(n) \uparrow$, to indicate that the computation of oracle machine e halts, or fails to halt, on input n. If the computation of oracle machine e on input n halts within r steps, using oracle information given in τ (where τ is a finite initial segment of χ_X for some set X), then we write $\varphi_{e,r}^\tau(n) \downarrow$, and otherwise we write $\varphi_{e,r}^\tau(n) \uparrow$.

As before, to show that a (partial) function is formally computable relative to X; i.e., computable by an oracle machine using an X-oracle, we simply describe an effective procedure that involves asking questions of the oracle. We claim that we could implement the procedure by an oracle machine. We are using Church's Thesis. This is not a *new* Church's Thesis for oracle machines. The steps taken between appeals to the oracle follow a procedure which is effective in the ordinary sense, and so can be carried out by a Turing machine with no added powers.

1.12 The relativized $s-m-n$ Theorem

The straightforward relativization of Theorem 1.10 (a) is that for each m, n, and X, for any partial function f computable relative to X, there is a total function k, computable relative to X, such that for all $\overline{y}$ and $\overline{x}$,

$$\varphi^X_{k(\overline{y})}(\overline{x}) = f(\overline{y}, \overline{x}) \ .$$

Looking at the proof of the theorem, however, we see that $k(\overline{y})$ is the result of manipulating codes for oracle machines in an effective way, so that we have a stronger relativization in which the function k is actually *computable*.

Essentially the same argument was given for Theorem 1.10 (b) as for Theorem 1.10 (a). In the relativized version, we see that $k(e, \overline{y})$ is again the result of manipulating codes for oracle machines in a computable way, and also that this manipulation is independent of X. Therefore, instead of the most obvious relativization of Theorems 1.10 (a) and 1.10 (b), we have the following stronger result.

Theorem 1.13 *(a) If $f(\overline{y}, \overline{x})$ is partial computable relative to X, then there is a total computable function $k(\overline{y})$ such that*

$$\varphi^X_{k(\overline{y})}(\overline{x}) = f(\overline{y}, \overline{x}) \ .$$

(b) For each m and n, there is a total $(m+1)$-ary computable function $k(e, \overline{y})$ such that

$$\varphi^X_{k(e,\overline{y})}(\overline{x}) = \varphi^X_e(\overline{y}, \overline{x}) \ ,$$

for all sets X, for all e and all m-tuples $\overline{y}$ and n-tuples $\overline{x}$.

1.13 Turing reducibility

If X and Y are sets such that Y is computable relative to X, then we also say that Y is *Turing reducible* to X, and we write $Y \leq_T X$. We note that Turing reducibility is a *partial ordering* on the class of sets; i.e.,

$$X \leq_T X \ ,$$

and

$$X \leq_T Y \ \& \ Y \leq_T Z \ \Rightarrow \ X \leq_T Z \ .$$

We define the relation of *Turing equivalence*, denoted by $\equiv_T$, letting

$$X \equiv_T Y \ \Leftrightarrow \ X \leq_T Y \ \& \ Y \leq_T X \ .$$

From the fact that $\leq_T$ is a partial ordering, it follows that $\equiv_T$ is an equivalence relation. The equivalence classes are the *Turing degrees*.

We can generalize the notion of computability relative to a single set to that of computability relative to a tuple of relations $S_1, \ldots, S_k$. Here the oracle machine may ask questions of the form "Does $S_i(y_1, \ldots, y_m)$ hold?" This generalization can be reduced to that of computability relative to a tuple of sets $A_1, \ldots, A_k$, where for each i,

$$A_i = \{\langle y_1, \ldots, y_m \rangle : S_i(y_1, \ldots, y_m)\} \,,$$

for some m.

This, in turn, can be reduced to the notion of computability relative to a single set

$$A_1 \oplus A_2 \oplus \ldots \oplus A_k \,,$$

where $B \oplus C$ denotes the disjoint union

$$\{2n : n \in B\} \cup \{2n+1 : n \in C\} \,,$$

and $A_1 \oplus A_2 \oplus \ldots \oplus A_k$ is regarded as bracketed so that

$$A_1 \oplus A_2 \oplus \ldots \oplus A_{i+1} = (A_1 \oplus A_2 \oplus \ldots \oplus A_i) \oplus A_{i+1} \,.$$

We may also consider computability relative to a *function* f. If f is a total function, then we identify f with its graph. We write φ_e^f for the partial function computed by oracle machine e with oracle f, and we write W_e^f for the domain of this function.

We may relativize a result to a tuple of relations. For example, the relativization of Theorem 1.4 says that a relation R is computable relative to the tuple of relations $S_1, \ldots, S_k$ if and only if both R and $\neg R$ are computably enumerable relative to $S_1, \ldots, S_k$. This is not a new result, because S is computable (or c.e.) relative to $S_1, \ldots, S_k$ if and only if it is computable (or c.e.) relative to the single set

$$A_1 \oplus \ldots \oplus A_k$$

described above.

1.14 Other reducibilities

There are several stronger notions of reducibility. One is the notion of "many-one" reducibility. For sets X and Y, we say that Y is *many-one reducible to* X *via* f provided that f is a total computable function, and for all x,

$$x \in Y \Leftrightarrow f(x) \in X \,.$$

We write $Y \leq_m X$. The relation $\leq_m$ is again a partial ordering on the class of sets, and we can define *many-one equivalence* and *many-one degrees* in the same way that we defined Turing equivalence and Turing degrees. It is clear that

$$Y \leq_m X \Rightarrow Y \leq_T X \,,$$

so that many-one equivalence is a refinement of Turing equivalence. We shall be interested in many-one reducibility primarily because of this fact.

We have seen, in Theorem 1.9, that the halting set

$$K = \{x : x \in W_x\}$$

is computably enumerable but not computable. A set X is said to be *complete for a class* $\mathcal{C}$ if $X \in \mathcal{C}$ and all sets in $\mathcal{C}$ are Turing reducible to X. The result below says that K is a complete c.e. set.

Theorem 1.14 *For every c.e. set* Y, $Y \leq_T K$. *Moreover, from an index for* Y *as a c.e. set, we can effectively determine an index for* Y *as a set computable relative to* K. *(In fact, we can find an index for a computable function* f *such that* $Y \leq_m K$ *via* f.)

Proof: Let g be the partial function such that

$$g(x,y) = \begin{cases} 1 & x \in Y , \\ \text{undefined} & \text{otherwise} . \end{cases}$$

To compute $g(x,y)$, we need only enumerate Y, so, by Church's Thesis, g is partial computable. By the $s-m-n$ Theorem (Theorem 1.10 (a)), there is a total computable function k such that $\varphi_{k(x)}(y) = g(x,y)$. Then

$$W_{k(x)} = \begin{cases} \omega & \text{if } x \in Y , \\ \emptyset & \text{otherwise} . \end{cases}$$

Thus,

$$\begin{aligned} x \in Y &\Leftrightarrow k(x) \in W_{k(x)} \\ &\Leftrightarrow k(x) \in K . \end{aligned}$$

Therefore, $Y \leq_m K$ via k.

We need to show that that it is possible to compute an index for such a function k, given a c.e. index for Y. If $Y = W_e$, we need only carry along e as an extra variable in the argument above. Let

$$g(e,x,y) = \begin{cases} 1 & \text{if } x \in W_e , \\ \text{undefined} & \text{otherwise} . \end{cases}$$

Since the binary relation $x \in W_e$ (on e and x) is c.e., g is partial computable. By the $s-m-n$ Theorem again, there is a total computable function k such that

$$\varphi_{k(e,x)}(y) = g(e,x,y) .$$

Then, as before,

$$x \in Y \Leftrightarrow k(e,x) \in K ,$$

and $W_e \leq_m K$ via the computable function $k(e,x)$ (of the variable x).

Another notion that appears naturally in certain contexts is "enumeration" reducibility. We define the notion here. For more information, see [32]. The

1.14. OTHER REDUCIBILITIES

definition says that Y is *enumeration reducible* to X if there is a c.e. relation R (between codes for finite sets and elements of ω) such that

$$y \in Y \Leftrightarrow \exists x \, [D_x \subseteq X \ \& \ R(x,y)] \,.$$

We write $Y \leq_e X$. The next result gives another possible definition of enumeration reducibility, more closely tied to the name.

Theorem 1.15 *For any sets X and Y, the following are equivalent:*
(a) $Y \leq_e X$,
(b) *for any total function f with range X, Y is c.e. relative to f.*

The theorem says that the sets enumeration reducible to X are the ones whose semi-characteristic function is computable in every enumeration f of X. We write $f|s$ for the restriction of the function f to $s = \{n : n < s\}$.

Proof: (a) $\Rightarrow$ (b) If Y is enumeration reducible to X via R, and f is a total function with range X, then

$$y \in Y \Leftrightarrow \exists x, s \, [D_x \subseteq ran(f|s) \ \& \ R(x,y)] \,.$$

(b) $\Rightarrow$ (a) Supposing (b), we attempt (unsuccessfully) to construct an enumeration f of X satisfying the following requirements:

$$R_e : W_e^f \neq Y$$

In the rest of the proof, we let p, q, etc. range over finite partial enumerations of X. When we write $\varphi_e^p(y) \downarrow$, we mean that p gives sufficient information for a halting computation; moreover, if the fact $(n, k) \notin p$ is used in the computation, then $p(n)$ is defined, with value different from k. If, for some y, either

$$\varphi_e^p(y) \downarrow \ \& \ y \notin Y \ \text{ or else}$$

$$\neg \, (\exists q \supseteq p) \, \varphi_e^q(y) \downarrow \ \& \ y \in Y \,,$$

then for any enumeration $f \supseteq p$, f satisfies R_e. There must exist e and p with no $q \supseteq p$ taking care of R_e in this way—otherwise, we could determine an enumeration f satisfying all R_e. Then we have

$$y \in Y \Leftrightarrow (\exists q \supseteq p) \, \varphi_e^q(y) \downarrow \,.$$

Let R be the relation

$$\{(x,y) : (\exists q \supseteq p)[q \text{ maps some finite subset of } \omega \text{ onto } D_x \ \& \ \varphi_e^q(y) \downarrow]\} \,.$$

Then R is c.e., and Y is enumeration reducible to X via R.

Chapter 2

The arithmetical hierarchy

In this chapter, we continue our review of basic computability theory. We define the "jump" of an arbitrary set X. We show that the jump operation on sets induces an operation on Turing degrees. We define the hierarchy of *arithmetical* relations, built up from computable relations by the operations of projection and complementation. We show how the levels in the arithmetical hierarchy are related to the sets obtained by iterating the jump function on $\emptyset$. We end the chapter with some useful results on trees.

Most of the results may be found in standard texts such as [134] or [144].

2.1 Jumps

In Chapter 1, we defined the halting set

$$K = \{x : x \in W_x\} .$$

Theorem 1.14 says that K is a complete c.e. set. Relativizing the definition to an arbitrary set X, we define

$$X' = \{x : x \in W_x^X\},$$

called the *jump* of X. Iterating, we define $X^{(n)}$ for all $n \in \omega$, where

$$X^0 = X , \text{ and } X^{(n+1)} = (X^{(n)})' .$$

Clearly, X' is c.e. relative to X. This and the result below yield the fact that X' is complete for the class of sets c.e. relative to X.

Theorem 2.1 *If Y is c.e. relative to X, then $Y \leq_T X'$. In fact, from an index for Y as a set c.e. relative to X (independent of X), we can effectively determine an index for Y as a set computable relative to X'.*

Proof: We relativize the proof of Theorem 1.14 (the result for K), substituting Theorem 1.13 (a) for the first application of Theorem 1.10.

One consequence of Theorem 2.1 is that $X \leq_T X'$, and, in fact, there is some e such that for all X, $\varphi_e^{X'} = \chi_X$. More generally, we have the following.

Corollary 2.2 *If $Y \leq_T X$, then $Y \leq_T X'$. Furthermore, given an index for Y as a set computable in X, we can effectively determine, independent of X, an index for Y as a set computable in X'.*

Proof: We note that Y is c.e. relative to X and apply Theorem 2.1.

Theorem 2.3 *If $Y \leq_T X$, then $Y' \leq_T X'$. Furthermore, from an index for Y as a set computable in X, we can effectively determine (independent of X), an index for Y' as a set computable in X'.*

Proof: We have a procedure (independent of Y) for enumerating Y' using a Y-oracle. Given an index for Y as a set computable in X, we can convert the procedure for enumerating Y' using a Y-oracle into a procedure for enumerating Y' using an X-oracle. Answers to the questions about membership in Y are obtained by running the procedure for computing χ_Y using the X-oracle. Then the conclusion follows from Theorem 2.1.

Theorem 2.3 implies that there is a well-defined jump operation on Turing degrees, induced by the jump operation on sets. The same reasoning used to show that K is not computable also shows that X' is not computable relative to X. Therefore, whenever we apply the jump operation to a Turing degree, we get a strictly greater degree.

The jump operation on Turing degrees is not $1-1$—there exist sets X and Y such that
$$X' \equiv_T Y' \ \& \ X \not\equiv_T Y \ .$$
For example, we shall see that there is a completion T of PA with $T' \equiv_T \emptyset'$, and a completion of PA cannot be computable.

It is useful sometimes to classify sets according to their jumps, or n^{th} jumps. A set X is said to be *low* if $X' \equiv_T \emptyset'$. More generally, X is low_n if $X^{(n)} \equiv_T \emptyset^{(n)}$.

2.2 Basic definitions

The *arithmetical* relations are those obtained from computable relations by finitely many steps of projection and complementation. These relations form a hierarchy, based on the number of steps of the two different kinds. We define classes of Σ_n^0, Π_n^0, and Δ_n^0 relations, for $n \geq 1$. To start off, a relation is Σ_1^0 if it can be expressed in the form

$$\exists y\, R(\overline{x}, y) \ , \text{ where } R(\overline{x}, y) \text{ is computable } .$$

2.2. BASIC DEFINITIONS

A relation $S(\overline{x})$ is Π_1^0 if the complementary relation, $\neg S(\overline{x})$, is Σ_1^0. Proceeding by induction, for $n > 1$, a relation is Σ_n^0 if it can be expressed in the form

$$\exists y\, R(\overline{x}, y)\,, \text{ where } R(\overline{x}, y) \text{ is } \Pi_{n-1}^0\,.$$

A relation $S(\overline{x})$ is Π_n^0 if $\neg S(\overline{x})$ is Σ_n^0. Finally, for all $n \geq 1$, a relation is Δ_n^0 if it is both Σ_n^0 and Π_n^0. A relation is said to be *arithmetical* if it is Δ_n^0, for some n. The connection between *arithmetical* relations and *arithmetic* will be made precise in the next chapter.

The classes at the first level in the arithmetical hierarchy—Σ_1^0, Π_1^0, and Δ_1^0—are already familiar. By Theorem 1.3, a relation $S(\overline{x})$ is computably enumerable if and only if there is a computable relation R such that

$$S(\overline{x}) \Leftrightarrow \exists y\, R(\overline{x}, y)\,.$$

From this, it follows that the Σ_1^0 relations are just the c.e. relations. We said that the co-c.e. relations are the complements of c.e. relations. Hence, the Π_1^0 relations are just the co-c.e. relations. Note that the Π_1^0 relations can be expressed in the form $\forall y\, R(\overline{x}, y)$, where $R(\overline{x}, y)$ is computable.

By unwinding the inductive definition above, we see that a relation is Σ_n^0 just in case it can be expressed in the form

$$Q_1 y_1\, Q_2 y_2 \cdots Q_n y_n\, R(\overline{x}; y_1, y_2, \ldots, y_n)$$

where R is a computable relation and $Q_1, Q_2, \cdots, Q_n$ is a sequence of quantifiers, alternating between $\exists$ and $\forall$, such that Q_1 is $\exists$—so, Q_n will be $\forall$ if n is even and $\exists$ if n is odd. For example, if $R(x, y_1, y_2, y_3)$ is computable, then the relation is Σ_3^0. Dually, a relation is Π_n^0 just in case it can be expressed in the form

$$Q_1 y_1\, Q_2 y_2 \cdots Q_n y_n\, R(\overline{x}; y_1, y_2, \ldots, y_n)$$

where $R(\overline{x}; y_1, \ldots, y_n)$ is computable and $Q_1, Q_2, \ldots, Q_n$ is an alternating sequence of quantifiers in which Q_1 is $\forall$.

Remarks on notation and terminology: Above, we have used the quantifiers $\forall$ and $\exists$ not as symbols of a formal language, but as everyday abbreviations for "for all" and "there exists". The superscript zero in the notation Σ_n^0 and Π_n^0 indicates that the variables associated with the quantifiers are intended to range over natural numbers, as opposed to sets or functions. (The natural numbers are objects of *type* 0, while sets of numbers and functions on numbers are objects of *type* 1.)

Clearly, if a relation is either Σ_n^0 or Π_n^0, then it is both Σ_{n+1}^0 and Π_{n+1}^0—we obtain an expression of the desired form by inserting redundant quantifiers. We can show that the result of applying *Boolean* operations (i.e., connectives) and quantifiers to arithmetical relations is always arithmetical. We use familiar tricks, moving the quantifiers to the outside, changing variables as necessary, and then contracting adjacent pairs of like quantifiers.

For example,
$$\forall y\, \exists z\, R(x,y,z) \ \& \ \exists y\, \forall z\, S(x,y,z)$$
can be re-expressed as
$$\forall y\, \exists z\, R(x,y,z) \ \& \ \exists u\, \forall v\, S(x,u,v) ,$$
and then as
$$\forall y\, \exists u\, \exists z\, \forall v\, [R(x,y,z) \ \& \ S(x,u,v)] ,$$
and finally as
$$\forall y\, \exists w\, \exists v\, [R(x,y,(w)_1) \ \& \ S(x,(w)_2,v)] .$$
Thus, the original relation is Π_3^0.

Here again we are using $\&$, and we will use $\vee$, and $\neg$, not as symbols from a formal language but as abbveviations for the everyday operations "and", "or", and "not" on relations.

2.3 Basic theorems

We have seen that the sets at level n in the arithmetical hierarchy are included among those at level $n+1$. The hierarchy is *proper*; i.e., there are new sets at each level. To prove this, we first prove an *Enumeration Theorem* saying that for each n, there is a single Σ_n^0 relation enumerating all Σ_n^0 relations in a given number of variables. There is a similar result for Π_n^0 relations.

Theorem 2.4 (Kleene) *For each n and each tuple of variables $\overline{x}$,*

(a) there is a Σ_n^0 relation $E(e,\overline{x})$ which, for different values of e, yields all of the m-ary Σ_n^0 relations in the variables $\overline{x}$,

(b) there is a Π_n^0 relation $E(e,\overline{x})$ which, for different values of e, yields all Π_n^0 relations in the variables $\overline{x}$.

Proof of (a): The Σ_n^0 relations (in $\overline{x}$) are those of the form
$$Q_1 y_1 \cdots Q_n y_n\, R(\overline{x}; y_1, \ldots, y_n)$$
where $R(\overline{x}; y_1, \ldots, y_n)$ is computable and $Q_1, \ldots, Q_n$ is an appropriate alternating sequence of quantifiers.

Case 1: Suppose Q_n is $\exists$.

Then $Q_n y_n\, R(\overline{x}; y_1, \ldots, y_n)$ is $\exists y_n\, R(\overline{x}; y_1, \ldots, y_n)$, which, by Theorem 1.3, is a c.e. relation in $\overline{x}; y_1, \ldots, y_n$. By Theorem 1.6, this is the domain of the partial computable function
$$\varphi_e^{(m+n-1)}(\overline{x}; y_1, \ldots, y_{n-1}) ,$$
for some e. By Theorem 1.2, this can be expressed as
$$\exists y_n\, T_{m+n-1}(e, \overline{x}; y_1, \ldots, y_{n-1}, y_n) ,$$

2.3. BASIC THEOREMS

(where T_{m+n-1} is a computable relation). Therefore, in this case, we can take $E(e, \overline{x})$ to be

$$Q_1 y_1 \cdots Q_n y_n \, T_{m+n-1}(e, \overline{x}; y_1, \ldots, y_n) \ .$$

Case 2: Suppose Q_n is $\forall$.

Then $Q_n y_n \, R(\overline{x}; y_1, \ldots, y_n)$ is $\forall y_n \, R(\overline{x}; y_1, \ldots, y_n)$, which is logically equivalent to

$$\neg \, \exists y_n \, \neg R(\overline{x}; y_1, \ldots, y_n) \ .$$

By the same reasoning as in Case 1, this can be expressed as

$$\neg \, \exists y_n \, T_{m+n-1}(e, \overline{x}; y_1, \ldots, y_n) \ ,$$

for some e. Therefore, in this case, we can take $E(e, \overline{x})$ to be

$$Q_1 y_1 \cdots Q_n y_n \, \neg T_{m+n-1}(e, \overline{x}; y_1, \ldots, y_n) \ .$$

We omit the proof of (b), as it is similar.

Using the enumeration relations from Theorem 2.4, we can show that the arithmetical hierarchy does not collapse.

Corollary 2.5 *For each $n \geq 1$, there is a Σ_n^0 relation that is not Π_n^0.*

Proof: Consider the Σ_n^0 enumeration relation $E(u; x_1, x_2, \ldots, x_m)$. Substituting x_1 for u, we have a Σ_n^0 relation $E(x_1; x_1, x_2, \ldots, x_m)$. If this is also Π_n^0, then the complement is Σ_n^0. Then for some e, we have

$$E(e; x_1, x_2, \ldots, x_m) \Leftrightarrow \neg E(x_1; x_1, x_2, \ldots, x_m) \ .$$

Substituting this e for x_1, we obtain

$$E(e; e, x_2, \ldots, x_m) \Leftrightarrow \neg E(e; e, x_2, \ldots, x_m) \ ,$$

a contradiction.

Corollary 2.5 yields the fact that the arithmetical hierarchy does not collapse.

Corollary 2.6 *For each n, there is a relation which is Δ_{n+1}^0 but not Δ_n^0.*

2.4 Combining arithmetical relations

We note that if relations $A(\overline{x})$ and $B(\overline{x})$ are both Σ_n^0, then the quantifiers in the conjunction
$$A(\overline{x}) \ \& \ B(\overline{x})$$
can be brought to the front alternately, so that the resulting expression still has only n alternations of quantifiers, beginning with $\exists$. Therefore,
$$A(\overline{x}) \ \& \ B(\overline{x})$$
is Σ_n^0. The same remark applies to the disjunction
$$A(\overline{x}) \vee B(\overline{x}) \ .$$
Likewise, if $A(\overline{x})$ and $B(\overline{x})$ are both Π_n^0, then so are
$$A(\overline{x}) \ \& \ B(\overline{x}) \quad \text{and} \quad A(\overline{x}) \vee B(\overline{x}) \ .$$
Similarly, if $A(\overline{x})$ is Σ_n^0 and $B(\overline{x})$ is Π_n^0, then
$$A(\overline{x}) \ \& \ B(\overline{x}) \quad \text{and} \quad A(\overline{x}) \vee B(\overline{x})$$
are both Δ_{n+1}^0.

Remark: A relation that is arithmetical can be replaced by a set at the same level in the arithmetical hierarchy. If $S(y_1, \ldots, y_m)$ is expressed in the form $Q_1 z_1 \cdots Q_n z_n \, R(y_1, \ldots, y_m; z_1, \ldots, z_n)$, and
$$A = \{\langle y_1, \ldots, y_m \rangle : S(y_1, \ldots, y_m)\} \ ,$$
then we have
$$u \in A \Leftrightarrow Q_1 z_1 \cdots Q_n z_n \, R((u)_1, \ldots, (u)_m; z_1, \ldots, z_n) \ .$$
Thus, if S is Σ_n^0 (or Π_n^0, or Δ_n^0), then so is A.

Theorem 2.7 *For a relation S,*
  (a) S is Σ_1^0 if and only if it is c.e. relative to some computable set (or some finite tuple of computable relations).
  (b) for $n > 1$, the following are equivalent:
    (i) S is Σ_n^0,
    (ii) S is c.e. relative to some Π_{n-1}^0 set (or some finite tuple of relations, each of which is Π_{n-1}^0 or Σ_{n-1}^0),
    (iii) S is c.e. relative to some Δ_n^0 set.

Proof: Statement (a) clear. We prove (b).
  (i) $\Rightarrow$ (ii) If S is Σ_n^0, then S is the projection of some Π_{n-1}^0 relation R. By the relativization of Theorem 1.3 to R, S is c.e. relative to R.

2.5. ALTERNATIVE DEFINITIONS

(ii) $\Rightarrow$ (iii) This is clear, since a Π_{n-1}^0 set is Δ_n^0, and any tuple of relations, each Π_{n-1}^0 or Σ_{n-1}^0, can be replaced by a single Δ_n^0 set.

(iii) $\Rightarrow$ (i) Suppose S is c.e. relative to A, where A is Δ_n^0. Say that $S = W_e^A$. We have a c.e. relation $R(\overline{x}, u, v)$ saying that there exists a halting computation of the eth oracle machine such that the questions put to the oracle all involve elements of D_u and D_v, with positive answers for elements of D_u and negative answers for elements of D_v. Then S can be expressed in the form

$$\exists v\, \exists u\, [R(\overline{x}; u, v)\ \&\ D_u \subseteq A\ \&\ D_v \subseteq \neg A]\ .$$

Therefore, it is enough to show that if a set B is Σ_n^0, then so is the set $\{u : D_u \subseteq B\}$. To see this, we note that in the statement

$$(\forall z \in D_u)\, B(z)\ ,$$

the bounded quantifier $(\forall z \in D_u)$ can be brought inside without altering the quantifier prefix of B. Certainly, $(\forall z \in D_u)$ can be interchanged with any quantifier $\forall$. In addition,

$$(\forall z \in D_u)\, \exists y\, C(u, z, y, \ldots)$$

is equivalent to

$$\exists y\, (\forall z \in D_u)\, C(u, z, (y)_z, \ldots)\ .$$

Corollary 2.8 *For $n \geq 1$, if a relation R is computable relative to some Δ_n^0 set, then it is Δ_n^0.*

Proof: Both of the relations R and $\neg R$ are c.e. relative to the Δ_n^0 set. Thus, by Theorem 2.7, R and $\neg R$ are both Σ_n^0.

2.5 Alternative definitions

It is possible to define the arithmetical hierarchy in terms of the sets $\emptyset^{(n)}$, which are obtained by iterating the jump function on the empty set. We shall show that a relation is Σ_n^0 (or Π_n^0, or Δ_n^0) just in case it is c.e. (or co-c.e., or computable) relative to $\emptyset_n^0$.

Lemma 2.9 *(a) For $n \geq 1$, $\emptyset^{(n)}$ is a complete Σ_n^0 set.*
(b) For $n \geq 1$, $\emptyset^{(n-1)}$ is a complete Δ_n^0 set.

Proof: (a) We use induction. For $n = 1$ the statement holds because K is a complete c.e. set, and $\emptyset^1$ is essentially the same as K. Supposing that the statement holds for n, where $n \geq 1$, consider $n + 1$. Since $\emptyset^{(n+1)}$ is c.e. relative to $\emptyset^{(n)}$, and by the Induction Hypothesis (H.I.), $\emptyset^{(n)}$ is Σ_n^0, we can apply Theorem 2.7 to conclude that $\emptyset^{(n+1)}$ is Σ_{n+1}^0. If A is Σ_{n+1}^0, then A is c.e. relative to some Π_n^0 set, and also c.e. relative to the complement, a Σ_n^0 set which we call B. By H.I., $B \leq_T \emptyset^{(n)}$. Then A is c.e. relative to $\emptyset^{(n)}$, and by Theorem 2.1, $A \leq_T \emptyset^{(n+1)}$.

(b) For $n = 1$, the statement is clear. For $n > 1$, since $\emptyset^{(n-1)}$ is Σ^0_{n-1}, it is Δ^0_n. Moreover, if A is Δ^0_n, then A and $\neg A$ are each c.e. relative to some Σ^0_{n-1} set, and using H.I., we may suppose (for both) that the set is $\emptyset^{(n-1)}$. Therefore, $A \leq_T \emptyset^{(n-1)}$.

The result below gives the alternative definition for the classes of Σ^0_n, Π^0_n, and Δ^0_n relations.

Theorem 2.10 *For $n \geq 1$,*
  (a) the Σ^0_n relations are those which are c.e. relative to $\emptyset^{(n-1)}$,
  (b) the Π^0_n relations are those which are co-c.e. relative to $\emptyset^{(n-1)}$,
  (c) the Δ^0_n relations are those which are computable relative to $\emptyset^{(n-1)}$.

Proof: This is immediate from Theorem 2.7 and Lemma 2.9.

We can locate functions as well as relations at various levels in the arithmetical hierarchy. Theorem 1.5 said that a partial function is computable if and only if its graph is c.e. We extend this idea, saying, for $n > 0$, that a function f is *partial* Δ^0_n if the graph of f is Σ^0_n. By Theorem 2.10, this holds just in case the graph of f is c.e. relative to $\emptyset^{(n-1)}$. By the relativization of Theorem 1.5, the latter holds if and only if f is partial computable relative to $\emptyset^{(n-1)}$.

Notation: By Theorem 2.10, $\emptyset^{(n-1)}$ is a complete Δ^0_n set; i.e., $\emptyset^{(n-1)}$ is Δ^0_n, and every Δ^0_n set is computable from it. We find it convenient to use the notation Δ^0_n not only for the class of Δ^0_n sets and relations, but also for a specific complete Δ^0_n set. For now, we choose $\emptyset^{(n-1)}$.

2.6 Approximations

A c.e. set is computably approximated in a natural way by finite sets. If $A = W_e$, then A is approximated by the computable sequence of finite sets

$$A_s = \{n \leq s : \varphi_{e,s}(n) \downarrow\}.$$

We can approximate a Δ^0_2 function, total or partial, by a total computable function. If $g(\overline{x}, s)$ is a total function, we write $\lim_{s \to \infty} g(\overline{x}, s)$ for the partial function f that is defined, with value y, just in case $g(\overline{x}, s)$ has value y for all sufficiently large s.

Theorem 2.11 (Limit Lemma) *A function $f(\overline{x})$ is partial Δ^0_2 if and only if there exists a total computable function $g(\overline{x}, s)$ such that $f(\overline{x}) = \lim_{s \to \infty} g(\overline{x}, s)$.*

Proof: First, suppose that $f(\overline{x}) = \lim_{s \to \infty} g(\overline{x}, s)$, where g is a total computable function. Then

$$f(\overline{x}) = y \iff \exists t \, (\forall s \geq t) \, g(\overline{x}, s) = y.$$

2.6. APPROXIMATIONS

From this, it is clear that the graph of f is Σ_2^0, so, according to the definition above, f is partial Δ_2^0.

Now, suppose that f is partial Δ_2^0. Then f is computable relative to $\emptyset'$, or K. Say $f = \varphi_e^K$. We must define $g(\overline{x}, s)$ so that if $\varphi_e^K(\overline{x}) \downarrow$, then $g(\overline{x}, s)$ has the same value for all sufficiently large s, and if $\varphi_e^K(\overline{x}) \uparrow$, then $g(\overline{x}, s)$ has no limiting value. Consider $\varphi_{e,s}^{K_s}(\overline{x})$—the result of computing s steps using oracle K_s. Note that for different s, we may have different halting computations of $\varphi_{e,s}^{K_s}(x)$. We vow to distrust any new computation. We let $g(\overline{x}, s)$ be s, if either $\varphi_{e,s}^{K_s}(\overline{x}) \uparrow$ or else $\varphi_{e,s}^{K_s}(\overline{x}) \downarrow$, with a computation that is new, and we let $g(\overline{x}, s) = \varphi_{e,s}^{K_s}(\overline{x})$, otherwise. Then g is the required total computable function.

The next result gives a useful method for approximating a Σ_2^0 or Π_2^0 relation.

Proposition 2.12 *(a) A relation $S(\overline{x})$ is Σ_2^0 if and only if there is a total computable function $g(\overline{x}, s)$, taking values 0 and 1, such that $S(\overline{x})$ if and only if*

$$\text{for all sufficiently large } s, \; g(\overline{x}, s) = 1 \;.$$

(b) A relation $S(\overline{x})$ is Π_2^0 if and only if there is a total computable function $g(\overline{x}, s)$, taking values 0 and 1, such that $S(\overline{x})$ if and only if

$$\text{for infinitely many } s, \; g(\overline{x}, s) = 1 \;.$$

Proof: It is easy to see that (a) implies (b) (looking at complements). We prove (a). If we have g as described, then S is clearly Σ_2^0. If S is Σ_2^0, then by Theorem 2.10, S is c.e. relative to $\emptyset'$, or K. Let $S = dom(\varphi_e^K)$, and take a total computable function $h(\overline{x}, s)$ approximating the partial Δ_2^0 function φ_e^K, as in the Limit Lemma. Let

$$g(\overline{x}, s) = \begin{cases} 1 & \text{if } s > 0 \; \& \; h(\overline{x}, s) = h(\overline{x}, s-1) \;, \\ 0 & \text{otherwise} \;. \end{cases}$$

If $\varphi_e^K(\overline{x}) \downarrow$, then $h(\overline{x}, s)$ is eventually constant, so $g(\overline{x}, s)$ has value 1 for all but finitely many s. If $\varphi_e^K(\overline{x}) \uparrow$, then $h(\overline{x}, s)$ changes value infinitely often, so there are infinitely many s for which $g(\overline{x}, s)$ has value 0.

We can apply the preceding result to approximate certain functions given by Π_2^0 relations.

Proposition 2.13 *Let $S(n, k)$ be a Π_2^0 relation. Suppose that for each n, there is a unique k such that $S(n, k)$. Let f be the function such that $S(n, f(n))$, for all n. Then there is a computable function $h(n, s)$ such that $f(n) = k$ if and only if there are infinitely many s such that $h(n, s) = k$.*

Proof: Let $g(n, k, s)$ be as in Proposition 2.12—having value 1 or 0 for all n, k, and s, and with the feature that $S(n, k)$ if and only if there are infinitely many s such that $g(n, k, s) = 1$. Let

$$h(n, s) = \begin{cases} k & \text{if } s = \langle k, t \rangle \; \& \; g(n, k, t) = 1 \; \& \; \forall i < k \; g(n, i, t) = 0 \\ s & \text{otherwise} \;. \end{cases}$$

2.7 Trees

Trees are often considered as abstract structures. In most of what follows, we take a different point of view. A *tree* is a set P of finite sequences, closed under initial segments (or, at certain times later on, *non-empty* initial segments). Level n of P consists of the sequences in P of length n. A *path* through P is an infinite sequence π whose finite initial segments are all in P (we ignore the empty sequence if it is not in P). A *subtree* of P is a set $Q \subseteq P$ such that Q is also a tree; i.e., Q is closed under (appropriate) initial segments.

Notation: We write $2^{<\omega}$ for the tree consisting of all finite sequences of 0's and 1's, and we write $\omega^{<\omega}$ for the tree consisting of all finite sequences of natural numbers. We write 2^n, ω^n for the set of sequences of length n—at level n—in the trees $2^{<\omega}$, $\omega^{<\omega}$, respectively.

The next result, König's Lemma, gives the existence of paths through certain trees.

Theorem 2.14 *If P is an infinite subtree of $2^{<\omega}$, then P has a path.*

Proof: There is a sequence of tree elements $(\sigma_n)_{n \in \omega}$, such that for each n, σ_n is at level n, σ_n has infinitely many extensions in P, and $\sigma_{n+1} \supseteq \sigma_n$. Then $\pi = \cup_n \sigma_n$ is a path.

A computable infinite subtree of $2^{<\omega}$ need not have a computable path.

Example: Let A and B be disjoint c.e. sets with no computable separator—for an explicit example, see the end of Section 1.7. Let P consist of those $\sigma \in 2^{<\omega}$ such that if $length(\sigma) = s$, then

$$k \in A_s \;\Rightarrow\; \sigma(k) = 1$$

and

$$k \in B_s \;\Rightarrow\; \sigma(k) = 0$$

Clearly, P is an infinite computable subtree of $2^{<\omega}$. Moreover, if π is a path through P, then π is the characteristic function of a set S separating A and B.

For a computable infinite subtree of $2^{<\omega}$, we can easily produce a Δ_2^0 path. The result below, the "Low Basis Theorem" of Jockusch and Soare [70], says that we can do better.

Theorem 2.15 (Jockusch-Soare) *If P is a computable infinite subtree of $2^{<\omega}$, then P has a low path.*

Proof: Using Δ_2^0 (the complete Δ_2^0 oracle), we determine a sequence $(\sigma_n)_{n \in \omega}$ as in the proof of König's Lemma, together with a nested sequence $(P_n)_{n \in \omega}$ of infinite computable subtrees of P. Let $P_0 = P$, and let $\sigma_0 = \emptyset$. Suppose we

2.7. TREES

have determined P_n, a computable subtree of P, with an element σ_n at level n having extensions at all higher levels in P_n. We use Δ_2^0 to check whether there exists $r \geq n$ such that

$$(\forall \tau \in P_n \cap 2^r)(\tau \supseteq \sigma_n \rightarrow \varphi_{n,r}^\tau(n) \downarrow) .$$

If so, then $P_{n+1} = P_n$, and if not, then P_{n+1} consists of all τ such that either $\tau \subseteq \sigma$ or else $\tau \supseteq \sigma$, $\tau \in P_n$, and

$$length(\tau) = r \rightarrow \varphi_{n,r}^\tau(n) \uparrow .$$

Let $\sigma_{n+1} \in P_{n+1}$, where $length(\sigma_{n+1}) = n+1$, and σ_{n+1} has extensions at all levels. Now, having determined P_n and σ_n for all n, we let

$$\pi = \cup_n \sigma_n .$$

Clearly, π is a path through all P_n. Given Δ_2^0, we can determine whether $n \in \pi'$, since we can recover the steps leading to P_{n+1}. Therefore, π is low.

Corollary 2.16 *There exists a low non-computable set.*

Proof: Let P be as in the example preceding Theorem 2.15, an infinite computable tree with no computable path. By Theorem 2.15, there is a low path.

While a computable tree need not have a computable path, there can be no one non-computable set that is computable in all paths. This is a special case of a result of Jockusch and Soare [71].

Theorem 2.17 *Suppose P is a computable infinite subtree of $2^{<\omega}$. If X is a set computable in all paths through P, then X is computable.*

Proof: We attempt to construct a nested sequence of computable infinite subtrees $(P_n)_{n \in \omega}$ such that $P_0 = P$, and for each n, there is some x such that for all paths π through P_{n+1}, $\varphi_n^\pi(x) \neq \chi_X(x)$. We must fail, or there would be a common path π in which X is not computable. Therefore, there exist an infinite computable subtree Q and an n such that for all x, the set

$$\{\sigma \in Q : \neg(\varphi_{n,length(\sigma)}^\sigma(x) \downarrow = \chi_X(x))\}$$

is finite. From this, it follows that X is computable. For each x, we can find r such that for all $\sigma \in Q$ of length r, $\varphi_{e,r}^\sigma(x) \downarrow$, and the value is always the same. That value must be $\chi_X(x)$.

If X is a Δ_2^0 set, then by the Limit Lemma, there is a computable function approximating χ_X. The result below says that if we can effectively bound the stage at which the approximation is first correct on the initial segment of a given length, then X is actually computable. This fact will be used in Chapter 9.

Proposition 2.18 *Let X be a Δ_2^0 set, and suppose g is a computable function such that for all x,*
$$\lim_{s \to \infty} g(x,s) = \chi_X(x)$$
(as in the Limit Lemma). Suppose also that there is a computable function b such that
$$\forall u \; \exists s \leq b(u) \; \forall x < u \; g(x,s) = \chi_X(x) \; .$$
Then X is computable.

Proof: We may suppose that if $x \leq y$, then $b(x) \leq b(y)$ (replacing the given b by another, if necessary). We define a computable subtree P of $2^{<\omega}$. For $\sigma \in 2^u$, let $\sigma \in P$ if and only if
$$\exists s \leq b(u) \; \forall x < u \; g(x,s) = \sigma(x) \; .$$
It is not difficult to see that χ_X is the unique infinite path through P. From this, it follows that X is computable.

Chapter 3

Languages and structures

In this chapter, we review some definitions and state our conventions about languages and structures. The basic reference is [26]. Mathematical structures such as vector spaces and linear orderings are already familiar. (Some basic facts about these and other special classes of structures are given in the Appendix.)

We recall such general algebraic notions as *isomorphism* and *quotient structure*. Finally, we indicate how we plan to measure complexity of structures.

Probably, the reader has experience with formal some languages. There are different kinds of languages. We shall use both "predicate" and "propositional" languages. *Propositional* languages have symbols which, all by themselves, are statements, to be interpreted simply as "true" or "false". *Predicate* languages include symbols standing for relations and operations, and variables. The predicate languages that we consider will all be *first order*; i.e., the variables are thought of as ranging over individuals, as opposed to sets or relations. In this chapter, we consider only *finitary*, or elementary, formulas. Later, however, we shall have more use for *infinitary*, non-elementary, formulas.

3.1 Propositional languages and structures

For propositional languages, the *logical* symbols, those common to all propositional languages, are the *connectives*, *logical constants*, and *parentheses*. The *non-logical* symbols, those distinguishing one propositional language from another, are the *propositional variables*. Thus, we identify a propositional language with its set of propositional variables.

Formally, we use as connectives only & (and), ∨ (or), and ¬ (not). Informally, we may also use → (implies), and ↔ (if and only if), as abbreviations for more complicated constructions. The logical constants are ⊥ (falsity) and ⊤ (truth). For a given propositional language P, the *atomic formulas* are ⊥, ⊤, and the elements of P. Further finitary propositional formulas are built up from the atomic formulas, using connectives (and parentheses), according to the usual rules. For propositional languages, we may use the terms "formula" and "sentence" interchangeably. (For predicate languages, we shall make a distinction.)

A segment of a formula that is itself a formula is called a *subformula*. For a formula φ, the set of subformulas, denoted by $sub(\varphi)$, is defined by induction.

(i) if φ is atomic, then $sub(\varphi) = \{\varphi\}$,

(ii) if φ is either $\psi \mathbin{\&} \chi$ or $\psi \vee \chi$, then $sub(\varphi) = sub(\psi) \cup sub(\chi) \cup \{\varphi\}$.

A *structure* for the propositional language P is a set $A \subseteq P$. *Satisfaction*, or *truth* of P-formulas is defined in the obvious way. In particular, $\top$ is always true, $\bot$ is never true, and for a propositional variable p, p is true in A just in case $p \in A$. As usual, we write $A \models \varphi$ to mean that φ is *true* in A. Two propositional formulas φ and ψ are said to be *logically equivalent* if they are true in exactly the same structures.

3.2 Predicate languages

For predicate languages, the *logical symbols* (common to all) are the connectives, logical constants, and parentheses, together with *quantifiers*, *variables* (individual), and *equality*. The *non-logical* symbols (distinguishing one language from another) are *relation*, or *predicate* symbols, *operation*, or *function* symbols, and (non-logical) *constants*. These stand for the relations, operations, and special elements that we wish to talk about. Again, we identify a language with its set of non-logical symbols.

The connectives and logical constants are the same as for propositional languages. The quantifiers are $\forall$ (for all) and $\exists$ (there exists). Formally, the variables are v_0, v_1, v_2, etc., but informally, we shall also use x, y, z, etc. For equality, we use $=$. Each relation or operation symbol has a definite *arity*, or number of places.

For a given predicate language L, the definition of *formula* is based on that of *term*. The L-terms are built up according to the usual rules. The simplest terms are variables and constants, and more complicated terms are built up inductively by applying an n-ary operation symbol to an n-tuple of known terms. There are two kinds of *atomic formulas*. First, if τ_1 and τ_2 are terms, then $\tau_1 = \tau_2$ is an atomic formula. Second, if R is an n-ary relation symbol and $\tau_1, \ldots, \tau_n$ is an n-tuple of terms, then $R(\tau_1, \ldots, \tau_n)$ is an atomic formula. Further finitary formulas are built up inductively by applying a logical connective to a known formula, or pair of formulas, or by applying a quantifier to a known formula.

Having precise rules for forming terms and formulas is useful in proving facts about formulas. However, when we wish to make clear the meaning of a formula, we violate the formation rules in the usual ways. For example, when we are talking about addition, our language will include the binary operation symbol $+$, and we may write $x + y$, although the rule for building terms says to write $+(x, y)$. Similarly, when we are talking about linear orderings, our language will include the binary relation symbol $<$, and we may write $x < y$, although the rule for building atomic formulas says to write $<(x, y)$. We drop parentheses where no ambiguity results.

3.2. PREDICATE LANGUAGES

We mention some terminology used to describe special kinds of formulas. A formula is *basic* if it is either an atomic formula or the negation of one. A formula is *open* if it has no quantifiers. In order to define "sentence", we must first say something about "free" and "bound" variables.

A variable, or occurrence of a variable in a formula, is *bound* if it is introduced by a quantifier, Otherwise, it is *free*. For example, in the formula

$$\forall y\, (x < y \ \lor \ x = y)\ ,$$

x occurs free, while y occurs only bound. There are some formulas such as

$$P(x)\ \&\ \forall x\, Q(x)\ ,$$

in which the same variable occurs both bound and free.

Formally, we define the set of free variables of φ, denoted by $fv(\varphi)$, by induction.
 (i) if φ is atomic, $fv(\varphi)$ is the set of all variables appearing in φ,
 (ii) if φ is either $(\psi \& \chi)$ or $(\psi \lor \chi)$, then $fv(\varphi) = fv(\psi) \cup fv(\chi)$,
 (iii) if φ is $\neg \psi$, then $fv(\varphi) = fv(\psi)$,
 (iv) if φ is either $\forall x\, \psi$ or $\exists x\, \psi$, then $fv(\varphi) = fv(\varphi) - \{x\}$.

Now, a *sentence* is a formula with no free variables. For example, $\forall y\, (0 < y \ \lor \ 0 = y)$ is a sentence (0 is a constant).

In general, a formula *says* something about its free variables. In particular, in a linear ordering, the formula

$$\forall y\, (x < y \ \lor \ x = y)$$

says that x is the least element. Note that we could change the bound variable from y to z and obtain a formula

$$\forall z\, (x < z \ \lor \ x = z)$$

saying the same thing about x. Similarly, instead of the formula

$$P(x)\ \&\ \forall x\, Q(x)\ ,$$

in which x occurs both bound and free, we could write

$$P(x)\ \&\ \forall y\, Q(y)\ ,$$

which would say the same thing about x.

We may write $\tau(x_1, \ldots, x_n)$ to indicate that the variables in the term τ are among $x_1, \ldots, x_n$. Similarly, we may write $\varphi(x_1, \ldots, x_n)$ to indicate that the free variables in the formula φ are among $x_1, \ldots, x_n$.

The set $sub(\varphi)$ of subformulas of a predicate formula φ is defined inductively, as for propositional formulas, but with the obvious new clause for the quantifiers.

3.3 Structures for a predicate language

Let L be a predicate language. An L-structure is a pair $\mathcal{A} = (A, I)$, where A, called the *universe*, is a set, and I, called the *interpretation*, is a function on L such that

    (a) for each constant symbol c in L, $I(c) \in A$,
    (b) for each n-ary relation symbol R in L, $I(R) \subseteq A^n$,
    (c) for each n-ary operation symbol f in L, $I(f)$ is a function from A^n to A.

A convention usually adopted is that the universe of a structure must be nonempty. It will occasionally be useful for us to violate this convention. For each language L with no constant symbols, there is exactly one *empty L-structure*, with empty universe, and with the relations and operations defined vacuously in the only way possible.

In denoting a structure $\mathcal{A} = (A, I)$, we may write $|\mathcal{A}|$ for the universe A and $s^\mathcal{A}$ for $I(s)$, where s is a symbol in L. More often than not, we denote the structure simply by a tuple listing the universe and the range of the interpretation. To make clear which symbol corresponds to which item on the list, we may use the same letter to denote a symbol and its interpretation. For example, as the language of ordered fields, we take $L = \{+, \cdot, <, 0, 1\}$, where + and $\cdot$ are binary operation symbols, $<$ is a binary relation symbol, and 0 and 1 are constants. We may then denote the ordered field of rationals by $(Q, +, \cdot, <, 0, 1)$.

We sometimes follow common mathematical practice and blur the distinction between a structure and its universe—we speak of the elements of a field F, etc. Having described a structure $\mathcal{A}$ without naming the language in advance, we may denote the language of $\mathcal{A}$ by $L^\mathcal{A}$. When we speak of the *cardinality* of a structure $\mathcal{A}$, we mean the cardinality of $|\mathcal{A}|$.

Below are some sample languages, used for certain classes of structures of particular interest.

1. The usual language for *linear orderings* consists of a single binary relation symbol $<$.

2. The usual language for *Boolean algebras* consists of two binary operation symbols— $\wedge$ (for meet) and $\vee$ (for join), a unary operation symbol $'$ (for complement), and constants 0 and 1.

3. The usual language for *groups* consists of a binary operation symbol $\circ$ (for the group operation), a unary operation symbol $^{-1}$ (for inverse) and a constant e (for the identity).. For Abelian groups, we may use + for the group operation, − for inverse, and 0 for the identity.

4. The usual language for *vector spaces* over a field F consists of a binary operation symbol + (for addition of vectors), a constant 0 (for the zero vector), and a family of unary operation symbols $\cdot_s$ (for multiplication by the scalar s), one for each $s \in F$.

5. The language of *arithmetic* consists of binary operation symbols + and $\cdot$, a unary operation symbol S (for successor), and a constant 0.

3.4. SATISFACTION

If L is a language with no relation symbols, only operation symbols and constants, then the L-structures are called *algebras*. If L is a language with no operation symbols or constants, only relation symbols, then the L-structures are called *relational structures*.

Note that with these languages, linear orderings are relational structures, while Boolean algebras, groups, vector spaces, and models of arithmetic are all algebras. We may vary the languages. At certain points later, in working with linear orderings, it will be convenient for us to replace the binary relation symbol $<$ by a binary operation symbol, making the linear orderings into algebras.

Remarks: It is possible to construe a constant as a 0-ary operation. It is also possible to construe an n-ary operation as an $(n+1)$-ary relation. Thus, we could, if we wish, deal only with relational structures.

3.4 Satisfaction

Satisfaction is the all-important notion which connects formulas with structures. For languages with operation symbols and constants, we cannot define satisfaction of formulas without first defining "denotation" of terms. We fix a language L, and let $\mathcal{A}$ be an L-structure. For a term $\tau(\overline{x})$, with variables among those in an n-tuple $\overline{x}$, the *denotation* is an n-ary operation $\tau^{\mathcal{A}}(\overline{x})$ on $|\mathcal{A}|$ whose value at an n-tuple $\overline{a}$ is computed in the natural way. If τ is one of the variables x_i in $\overline{x}$, then $\tau^{\mathcal{A}}(\overline{a})$ is the corresponding a_i in $\overline{a}$, and if τ is a constant c, then $\tau^{\mathcal{A}}(\overline{a}) = c^{\mathcal{A}}$. For more complicated terms $\tau = f(\tau_1, \ldots, \tau_m)$, $\tau^{\mathcal{A}}(\overline{x})$ is the result of composing $f^{\mathcal{A}}$ with the m-tuple of operations $\tau_1^{\mathcal{A}}(\overline{x}), \ldots, \tau_m^{\mathcal{A}}(\overline{x})$.

For a formula $\varphi(\overline{x})$, we write $\mathcal{A} \models \varphi(\overline{a})$ to indicate that the formula $\varphi(\overline{x})$ is satisfied in $\mathcal{A}$ when $\overline{a}$ is substituted for $\overline{x}$—we may say that the "assignment" that takes $\overline{x}$ to $\overline{a}$, satisfies the formula in the structure. We may occasionally write

$$\mathcal{A} \models \varphi \left(\frac{\overline{b}}{\overline{u}} \right) \left(\frac{\overline{a}}{\overline{x}} \right)$$

to indicate that φ is satisfied in $\mathcal{A}$ by the assignment that first takes $\overline{u}$ to $\overline{b}$ and then takes the elements of of $\overline{x}$ that are *not* in $\overline{u}$ to the corresponding elements of $\overline{a}$.

As in propositional logic, $\top$ is always satisfied, and $\bot$ is never satisfied. If $\varphi(\overline{x})$ says that two terms are equal, then $\mathcal{A} \models \varphi(\overline{a})$ if the denotations of the terms agree at $\overline{a}$. If $\varphi(\overline{x})$ says that some m-tuple of terms $\tau_i(\overline{x})$ is in the relation R, then $\mathcal{A} \models \varphi(\overline{a})$ if $R^{\mathcal{A}}$ contains the tuple of denotations $b_i = \tau_i^{\mathcal{A}}(\overline{a})$. The inductive clauses in the definition of satisfaction, for formulas obtained by taking a negation or conjunction, or adding a quantifier, should be clear. We write $\varphi^{\mathcal{A}}(\overline{x})$ for the set of tuples $\overline{a}$ (corresponding to $\overline{x}$) such that $\mathcal{A} \models \varphi(\overline{a})$. A sentence is said to be *true* in $\mathcal{A}$ if it is satisfied, and *false* otherwise.

3.5 Enlarging a structure

Let $\mathcal{A}$ and $\mathcal{B}$ be L-structures. We say that $\mathcal{B}$ is an *extension* of $\mathcal{A}$, or $\mathcal{B}$ is a *substructure* of $\mathcal{A}$ if the relations and operations of $\mathcal{A}$ are the restrictions of the corresponding operations of $\mathcal{B}$ (this implies, in particular, that for any operation $f^\mathcal{B}$, the value on any tuple from $|\mathcal{A}|$ is in $|\mathcal{A}|$), and for any constant symbol c, $c^\mathcal{A} = c^\mathcal{B}$. We may write $\mathcal{A} \subseteq \mathcal{B}$.

Remark: If $\mathcal{A}$ is a relational structure, then every subset of $\mathcal{A}$ is the universe of a substructure.

There is another way to enlarge a structure, without adding elements. Let $\mathcal{A}$ be a structure, and let $\mathcal{B}$ be a new structure formed by adding new constants, relations, or operations to $\mathcal{A}$, without adding new elements. We call $\mathcal{B}$ an *expansion* of $\mathcal{A}$, and we call $\mathcal{A}$ a *reduct* of $\mathcal{B}$. We may denote the expansion $\mathcal{B}$ by a list consisting of $\mathcal{A}$ first, and then the new constants, relations, and operations. For example, if $\mathcal{B}$ is the expansion of $\mathcal{A}$ obtained by adding a single new relation R, then we may write $\mathcal{B} = (\mathcal{A}, R)$.

Suppose $\mathcal{A}$ is a substructure of $\mathcal{B}$, so $\mathcal{B}$ is an extension of $\mathcal{A}$. We say that $\mathcal{A}$ is an *elementary substructure* of $\mathcal{B}$, or $\mathcal{B}$ is an *elementary extension* of $\mathcal{A}$, if for all formulas $\varphi(\overline{x})$ and all appropriate tuples $\overline{a}$ in $\mathcal{A}$,

$$\mathcal{A} \models \varphi(\overline{a}) \iff \mathcal{B} \models \varphi(\overline{a}) .$$

We write $\mathcal{A} \prec \mathcal{B}$. We say that $\mathcal{A}$ and $\mathcal{B}$ are *elementarily equivalent* if they satisfy the same sentences. We write $\mathcal{A} \equiv \mathcal{B}$.

Theorem 3.1 (Tarski Criterion) *Let $\mathcal{A}$ be a structure and let B be a subset of $|\mathcal{A}|$. Suppose that for all formulas $\varphi(\overline{u}, x)$ and all tuples $\overline{a}$ in B, appropriate to substitute for $\overline{u}$, if there exists b such that $\mathcal{A} \models \varphi(\overline{a}, b)$, then there is such a b in B. Then B is the universe of a structure $\mathcal{B}$ such that $\mathcal{B} \prec \mathcal{A}$.*

Proof sketch: First, note that B is the universe of a substructure $\mathcal{B}$. If the language of $\mathcal{A}$ is relational, this is trivial, but it is true in any case. It is easy to show, by induction on formulas $\varphi(\overline{u})$, that for all tuples $\overline{b}$ in B,

$$\mathcal{A} \models \varphi(\overline{b}) \iff \mathcal{B} \models \varphi(\overline{b}) .$$

The hypothesis takes case of the only clause that is not immediate from the definition of satisfaction.

We are interested in countable languages and countable structures. The following result, an application of Theorem 3.1 gives a way of passing from an uncountable structure to a countable one, preserving satisfaction of various sentences.

Theorem 3.2 (Downward Löwenheim-Skolem Theorem) *Suppose $\mathcal{A}$ is a structure for a countable language. Then there is a countable structure $\mathcal{B} \prec \mathcal{A}$.*

Proof: We form an increasing sequence of countable subsets of $\mathcal{A}$. Let B_0 be a countable subset of $\mathcal{A}$. Given B_n, we form B_{n+1} as follows. For each formula $\varphi(\overline{u}, x)$ and each tuple $\overline{a}$ in B_n such that for some b in $\mathcal{A}$, $\mathcal{A} \models \varphi(\overline{a}, b)$, choose one such b, and let B_{n+1} be the result of adding these chosen elements to B_n. Now, by the Tarski Criterion, $\cup_n B_n$ is the universe of the desired countable substructure $\mathcal{B} \prec \mathcal{A}$.

Exercise: Suppose $(\mathcal{A}_n)_{n \in \omega}$ is a sequence of structures such that for all n, $\mathcal{A}_n \prec \mathcal{A}_{n+1}$, and let $\mathcal{B}$ be the "union"; i.e., $\mathcal{B}$ is the common extension with universe equal to $\cup_n |\mathcal{A}_n|$. Show that $\mathcal{A}_n \prec \mathcal{B}$, for all n. [The Tarski Criterion is no help. Proceed by induction on formulas.]

3.6 Theories and models

A *model* for a sentence φ is a structure in which φ is true. Similarly, if Γ is a set of sentences, a model for Γ is a structure in which all of the sentences of Γ are true. A sentence, or set of sentences, is *consistent* if it has a non-empty model. (We exclude the empty structure here because the usual proof-theoretic notion of consistency corresponds to existence of a non-empty model.)

We fix a predicate language L and consider L-sentences and L-structures. A *theory* is the set of all sentences true in some non-empty structure, or class of non-empty structures. We write $Th(\mathcal{A})$, $Th(K)$, for the theory of the structure $\mathcal{A}$, or the class K. For example, groups are the models of the usual group axioms, and the theory of groups is the set of sentences true in all groups. The theory of finite groups has as models all of the finite groups, and some infinite groups as well. If K is the empty class of structures, then $Th(K)$ consists of *all* sentences in the language L.

A theory T is *complete* if it is consistent, and for each sentence φ in the language, either $\varphi \in T$ or $\neg\varphi \in T$.[1] Equivalently, T is complete if it has the form $Th(\mathcal{A})$, for some non-empty structure $\mathcal{A}$. If Γ is a consistent set of sentences, then a *completion* of Γ is a complete theory $T \supseteq \Gamma$ (in the same language). Every consistent theory has a completion; some theories have many completions.

For a structure $\mathcal{A}$, we think of the universe A as a set of constants. The *complete diagram* of $\mathcal{A}$, denoted by $D^c(\mathcal{A})$, is the set of all $(L \cup A)$-sentences true in the expansion $(\mathcal{A}, a_{a \in A})$. The *atomic diagram* of $\mathcal{A}$, denoted by $D(\mathcal{A})$, is the set of basic sentences in $D^c(\mathcal{A})$. We may define the *open diagram*, etc., in a similar way. Sometimes it is convenient to broaden the definitions slightly to allow multiple names for the elements of a structure. Let B be a set of constants, and let i be a function from B onto A. Now, i interprets the constants, making each $(L \cup B)$-sentence true or false in $\mathcal{A}$. We may write $D^c(\mathcal{A})$, and use the term "complete diagram", for the set of all $(L \cup B)$-sentences that i makes true in $\mathcal{A}$, and we may write $D(\mathcal{A})$, and use the term "atomic diagram", for the set of basic sentences in $D^c(\mathcal{A})$.

---

[1] there is an inconsistent theory—take the theory of the empty class of structures.

3.7 Prenex normal form

Again, we fix a language L and consider only L-formulas and L-structures. Formulas $\varphi(\overline{x})$ and $\psi(\overline{x})$ are said to be *logically equivalent* if they are satisfied by the same tuples in all non-empty structures. (Again we exclude the empty structure in order to retain a match with the corresponding proof-theoretic notion.) We consider also a notion weaker than logical equivalence. If Γ is a set of sentences, then formulas $\varphi(\overline{x})$ and $\psi(\overline{x})$ are *equivalent over* Γ if they are satisfied by the same tuples in all non-empty models of Γ. For example, the formulas $\neg\, \exists y\ y < x$ and $\forall y\ \neg\, y < x$ are logically equivalent. The formulas $\neg\, y < x$ and $y = x \vee x < y$ are not logically equivalent, but they are equivalent over the set of axioms for linear orderings.

A formula φ is in *prenex normal form* provided that the quantifiers (if any) are all at the front. Starting with an arbitrary formula, we may bring the quantifiers across the connectives, maintaining logical equivalence. (In doing this, we may need to change some of the variables, to avoid clashes.)

Proposition 3.3 *Given a formula φ, we can find a formula ψ, in prenex normal form, such that ψ is logically equivalent to φ.*

Let φ be a formula in prenex normal form. We say that φ is Σ_n if there are n blocks of like quantifers, beginning with $\exists$, in front of the open (quantifier-free) part of the formula. Similarly, φ is Π_n if there are n blocks of like quantifiers, beginning with $\forall$, before the open part.

3.8 Isomorphism

Let L be a language, and let $\mathcal{A}$ and $\mathcal{B}$ be L-structures. An *isomorphism* from $\mathcal{A}$ onto $\mathcal{B}$ is a $1-1$ function F, from $|\mathcal{A}|$ onto $|\mathcal{B}|$, such that F preserves the relations, operations, and constants; i.e.,

  (a) for any constant c in L, $F(c^{\mathcal{A}}) = c^{\mathcal{B}}$,
  (b) for any n-ary operation symbol f in L, if $F(a_i) = b_i$ for $i = 1, \ldots, n$, and $F(c) = d$, then

$$f^{\mathcal{A}}(a_1, \ldots, a_n) = c \ \Rightarrow\ f^{\mathcal{B}}(b_1, \ldots, b_n) = d\ ,$$

  (c) for any n-ary relation symbol R in L, if $F(a_i) = b_i$, for $i = 1, \ldots, n$, then

$$R^{\mathcal{A}}(a_1, \ldots, a_n) \ \Leftrightarrow\ R^{\mathcal{B}}(b_1, \ldots, b_n)\ .$$

We write $\mathcal{A} \cong_F \mathcal{B}$ to indicate that F is an isomorphism from $\mathcal{A}$ onto $\mathcal{B}$, and we write $\mathcal{A} \cong \mathcal{B}$ to indicate that there exists an isomorphism. An *automorphism* of $\mathcal{A}$ is an isomorphism from $\mathcal{A}$ onto itself.

In many mathematical contexts, isomorphic structures are identified, on the grounds that they are "the same" except for the names of the elements, and, as a consequence, they share the same abstract mathematical properties. This is appropriate when the *isomorphism type* is all that matters. We think of isomorphism as an equivalence relation on structures—under the conventions that all structures have universe a subset of ω, and all languages are computable, it actually is an equivalence relation. The equivalence classes are the isomorphism types. We are interested in computability properties, and we distinguish among isomorphic copies since they may have different computability properties.

Cantor showed that any two (non-empty) countable dense linear orderings without endpoints are isomorphic. The proof involved making lists of the elements of both orderings and producing a chain of finite partial isomorphisms by finding an appropriate image for the next element first on one side and then the other. There are a number of other results proved in a similar way. The following definition describes the essential feature of the family of finite partial isomorphisms.

Let $\mathcal{A}$ and $\mathcal{B}$ be structures for the same language. A non-empty family $\mathcal{F}$ of finite partial functions from $\mathcal{A}$ to $\mathcal{B}$ has the *back-and-forth property* if for all $p \in \mathcal{F}$, p preserves satisfaction of atomic formulas, and

$$(\forall a \in \mathcal{A})(\exists q \in \mathcal{F})[q \supseteq p \ \& \ a \in dom(q)],$$

$$(\forall b \in \mathcal{B})(\exists q \in \mathcal{F})[q \supseteq p \ \& \ b \in ran(q)].$$

Cantor had observed that if $\mathcal{A}$ and $\mathcal{B}$ are dense linear orderings without endpoints, then the set $\mathcal{F}$ of finite order-preserving partial functions from $\mathcal{A}$ to $\mathcal{B}$ has the back-and-forth property. Cantor's isomorphism construction generalizes to give the following.

Proposition 3.4 *Let $\mathcal{A}$ and $\mathcal{B}$ be countable structures for the same language, and let $\mathcal{F}$ be a family of finite partial functions from $\mathcal{A}$ to $\mathcal{B}$ such that $\mathcal{F}$ has the back-and-forth property. Then for any $p \in \mathcal{F}$, there is an isomorphism f from $\mathcal{A}$ onto $\mathcal{B}$ such that $F \supseteq p$.*

3.9 Quotient structures

An *equivalence relation* on a set A is a binary relation that is transitive, reflexive (on A), and symmetric (see the Appendix). A *congruence relation* on a structure $\mathcal{A}$ is an equivalence relation $\sim$ on $|\mathcal{A}|$ that "respects" the relations and operations of $\mathcal{A}$, as follows:

(a) if R is an n-ary relation symbol $a_i \sim b_i$ for $i = 1, \ldots, n$, then

$$(a_1, \ldots, a_n) \in R^{\mathcal{A}} \Leftrightarrow (b_1, \ldots, b_n) \in R^{\mathcal{A}},$$

(b) if F is an n-ary operation symbol and $a_i \sim b_i$ for $i = 1, \ldots, n$, then

$$F^{\mathcal{A}}(a_1, \ldots, a_n) \sim F^{\mathcal{A}}(b_1, \ldots, b_n).$$

Suppose $\sim$ is a congruence relation on $\mathcal{A}$. Then we can form the *quotient structure* $\mathcal{A}/\sim$, with universe equal to the set of $\sim$-equivalence classes, and with well-defined relations and operations on the equivalence classes, induced, in the obvious way, by the relations and operations from $\mathcal{A}$.

Examples

1. Suppose G is a group and N is a normal subgroup of G. Let
$$a \sim b \Leftrightarrow b^{-1} \circ a \in N.$$
Then $G/\sim$ is the familiar quotient group G/N.

2. Let F be the free group generated by constants $a_1, a_2, \ldots$. Each element of F is named by a term τ in the language of groups with the added constants. There are many terms naming the same element, but by crossing off adjacent occurrences of a_i and a_i^{-1}, and occurrences of e (unless τ itself is e), we obtain a unique shortest term. Let R be a set of atomic sentences, each asserting the equality of two terms. We define a congruence relation $\sim$, where $\tau_1 \sim \tau_2$ just in case there is a proof of $\tau_1 = \tau_2$ from the group axioms and R. Then $F/\sim$ is the group generated by the elements $a_1, a_2, \ldots$ under the relations R. In this context, the set of generators, together with the set of relations, form what is called a *presentation* of the group.

3. Let $\mathcal{A}$ be a linear ordering. It is not difficult to see that equality is the only actual congruence relation on $\mathcal{A}$. However, if $\sim$ is an equivalence relation on $\mathcal{A}$ for which the equivalence classes are intervals, then the ordering on $\mathcal{A}$ induces a natural ordering on the set of equivalence classes, something we would like to call a . We justify this by the (admittedly artificial) device of replacing the usual binary relation $<$ by the binary operation min, where $min(x, y)$ is the minimum of x and y.

3.10 Model existence

Probably, the reader is familiar with some formal system of proof. The definitions of "theory" and "completion" are normally given in terms of proofs. The definition of consistency may also be given in terms of proofs. Our definition did not mention proofs. We said that a set of sentences is consistent just in case it has a non-empty model. We want a useful test for model existence, and we shall give one that does not mention proofs. We are not doing this out of perversity. In Chapter 6, we shall develop a consistency criterion for infinitary logic, and we could not assume that the reader is familiar with a formal system of proof in that context. What we do here will generalize.

We begin by describing the features of a set D which can serve as the atomic diagram for a structure $\mathcal{B}$. For simplicity, we suppose that the underlying language is relational.

Lemma 3.5 *Let L be a relational language, and let B be a set of constants. For a set D of basic $(L \cup B)$-sentences, the following are equivalent:*

3.10. MODEL EXISTENCE

(a) D is the atomic diagram of an L-structure with universe B,

(b) for each atomic sentence φ in the language $(L \cup B)$, exactly one of φ, $\neg\varphi$ is in D, $\top \in D$, $\bot \notin D$, and for $b, b' \in B$, $b = b'$ is in D if and only if b is equal to b'.

We have characterized the atomic diagrams of structures with universe B. Similarly, we can characterize the diagrams of L-structures such that B is the set of names for the elements, allowing multiple names for a given element. Instead of saying that the sentence $b = b'$ is in D if and only if b is equal to b', we say that that $b = b' \in D$ is an equivalence relation which "respects" the relations named in L. The structure with atomic diagram D will in fact be the quotient of a structure $\mathcal{A}$, with universe equal to B, by the relation $\sim$, where

$$b \sim b' \Leftrightarrow b = b' \in D .$$

We require that $\sim$ is a congruence relation.

Lemma 3.6 *Suppose L is a relational language, and let B be a set of new constants. For a set D of basic $(L \cup B)$-sentences, the following are equivalent:*

(a) D is the diagram of an L-structure with B as the set of names for the elements,

(b) for each atomic sentence φ, exactly one of φ, $\neg\varphi$ is in D, $\top \in D$, $\bot \notin D$, for $b \in B$, $b = b$ is in D, and if φ and $b = b'$ are in D and φ' is the result of replacing one or more occurrences of b in φ by b', then φ' is in D.

Fact: If B is a non-empty set of constants, and Γ is a consistent set of basic $(L \cup B)$, where L is a relational language, then Γ has a model for which B is the set of names for the elements. The same is true for a set of open $(L \cup B)$-sentences, since satisfaction of any open formula is determined by the satisfaction of its atomic subformulas.

Theorem 3.7 (Consistency Criterion) *Let Γ be a set of sentences in a relational language L, and let B be a set of new constants. Then Γ has a model with B as the set of names for elements if and only if Γ can be extended to a set Γ^* of $(L \cup B)$-sentences with the following properties:*

(a) for each sentence φ, exactly one of φ, $\neg\varphi$ is in Γ^*,

(b) $(\varphi \ \& \ \psi)$ is in Γ^* if and only if both φ, ψ are in Γ^*,

(c) $(\varphi \lor \psi$ is in Γ^* if and only if at least one of φ, ψ is in Γ^*,

(d) $\forall x \, \varphi$ is in Γ^* if and only if for all $b \in B$, $\varphi(b)$ is in Γ^*,

(e) $\exists x \, \varphi$ is in Γ^* if and only if for some $b \in B$, $\varphi(b)$ is in Γ^*.

(f) the set of basic sentences in Γ^* is the atomic diagram of a structure $\mathcal{B}$.

Proof: If $\mathcal{B}$ is a model of Γ, where $\mathcal{B}$ has universe B, or the elements of $\mathcal{B}$ have names in B, then $\Gamma^* = D^c(\mathcal{B})$ clearly satisfies (a)–(f). Now, suppose that Γ has an extension Γ^* satisfying (a)–(f). Let $\mathcal{B}$ be the structure such that $D(\mathcal{B})$ is the set of basic sentences in Γ^*. It is easy to show, by induction on L-formulas $\varphi(\overline{x})$, that for all tuples $\overline{b}$ from B,

$$\mathcal{B} \models \varphi(\overline{b}) \Leftrightarrow \varphi(\overline{b}) \in \Gamma^* .$$

Therefore, $\mathcal{B}$ is a model of Γ.

If Γ is a consistent set of sentences in a countable language, then, by the Downward Löwenheim-Skolem Theorem, Γ has model that is either finite or countably infinite. It follows that for any countably infinite set B of new constants, Γ has a model for which B is the set of names of the elements.

Since every formula is logically equivalent to one in prenex normal form, we could deal only with such formulas. The Consistency Criterion is somewhat simpler cast in this way. Let L be a countable relational language, and let B be a countably infinite set of new constants. Suppose that Γ is a set sentences in prenex normal form. Let $NS(\Gamma, B)$ be the countable set of sentences obtained by substituting constants from B for the free variables in various subformulas of the sentences of Γ and open L-formulas. The sentences in $NS(\Gamma, B)$ are all in prenex normal form.

Proposition 3.8 *Then Γ has a model whose elements have names in B, if and only if there exists a set $\Gamma^* \supseteq \Gamma$ of $F(\Gamma, B)$-sentences such that*
  (a) *if $\exists x\, \psi(x) \in \Gamma^*$, then for some $d \in B$, $\psi(d) \in \Gamma^*$,*
  (b) *if $\forall x\, \psi(x) \in \varphi \in \Gamma^*$, then for all $d \in B$, $\psi(d) \in \Gamma^*$,*
  (c) *for any open sentence φ (in the language $L \cup B$), one of φ, $\neg \varphi$ is in Γ^*,*
  (d) *the set of open sentences in Γ^* is consistent.*

Exercise: Prove Proposition 3.8. [Hint: For a set Γ^* satisfying (a)-(d), if $\mathcal{B}$ is the structure whose atomic diagram is the set of basic sentences in Γ^*, to show that $\mathcal{B}$ is a model of Γ, show by induction on n that for a sentence $\varphi(\bar{b})$ that is Σ_n or Π_n, $\mathcal{B} \models \varphi(\bar{b}) \Leftrightarrow \varphi(\bar{b}) \in \Gamma^*$.]

3.11 Compactness

The theorem below is probably more widely used than any other result in model theory.

Theorem 3.9 (Compactness) *If Γ is a set of sentences such that every finite subset of Γ is consistent (i.e., has a non-empty model), then Γ is consistent.*

The Compactness Theorem is usually proved using facts about a formal system of proof. Following this approach, we would first prove the *Completeness Theorem*, saying that if Γ is a set of sentences, then Γ is consistent if and only if there is no proof of $\bot$ from Γ. Then we would argue that if Γ has no model, there is a proof of $\bot$ from Γ, and since a proof is finite, there is actually a proof of $\bot$ from a finite subset of Γ, so this subset has no model.

The standard proof of the Completeness Theorem, for the direction saying that if there is no proof of $\bot$ from Γ, then Γ has a model, involves adding an infinite set of constants B to the language of Γ, and, in an infinite sequence of steps, building the complete diagram of the desired model. Below, we prove the

Compactness Theorem without mentioning formal proofs, but in other respects, following the standard proof of Completeness rather closely.

Proof: We may suppose that the language of Γ, say L, is relational. Let B be an infinite set of new constants. Let $(\varphi_n)_{n \in \omega}$ be a list of all $(L \cup B)$-sentences. We form a sequence $(S_n)_{n \in \omega}$ of finite sets of sentences such that
  (1) for all finite subsets Γ' of Γ, $S_n \cup \Gamma'$ is consistent,
  (2) one of φ_n, $\neg \varphi_n$ is in S_{n+1},
  (3) if φ_n is in S_{n+1} and φ_n has the form $\exists x\, \psi(x)$, then $\psi(b)$ is in S_{n+1} for some $b \in$ B.

It is not difficult to show that the sequence exists. We let $S_0 = \emptyset$. Given S_n satisfying (1), we can add one of φ_n, $\neg \varphi_n$, for if each is inconsistent with some finite subset of Γ, we could take the union of the two subsets, say Γ', and there is a non-empty model of $S_n \cup \Gamma'$ satisfying one of φ_n, $\neg \varphi_n$. Suppose that we have added φ_n, and it has the form $\exists x\, \psi(x)$. For each finite subset Γ' of Γ, there is a model of $\Gamma' \cup S_n \cup \{\varphi_n\}$. Moreover, we may suppose that the elements have names in B, and $\psi(x)$ is satisfied by the element named by the first b not appearing in S_n or φ_n.

To show that Γ has a model, we use the Consistency Criterion.

Claim: The set $\Gamma^* = \cup_n S_n$ has the properties from Theorem 3.7.

Proof of Claim: First of all, $\Gamma \subseteq \Gamma^*$, for if $\varphi \in \Gamma$, and $\neg \varphi \in S_n$, then S_n would not be consistent with certain finite subsets of Γ. Properties (a), (b), (c), (d), and (e) are easy to check. If (f) failed, there would be a finite sub-language of $L \cup$B such that the restriction of Γ^* to the basic sentences in this sub-language is not the diagram of a structure. This bad set is finite, so it is a subset of some S_n. However, S_n has a model, and this model has a substructure with diagram exactly equal to the bad set.

The usual proof of the Completeness Theorem is due to Henkin. A *Henkin construction* for a model of Γ is a sequence $(S_n)_{n \in \omega}$ of finite sets of sentences with the features (1), (2), and (3) above.

3.12 Some special kinds of structures

The notions of "saturation" and "homogeneity" are usually defined for structures of arbitrary cardinality. Here we consider only only countable structures and countable languages. A structure $\mathcal{A}$ is *saturated* if for all tuples $\bar{a}$ in $\mathcal{A}$ and all sets $\Gamma(\bar{a}, x)$ of formulas with parameters $\bar{a}$, if every finite subset of $\Gamma(\bar{a}, x)$ is satisfied in $\mathcal{A}$, then the whole set is satisfied. We say that $\mathcal{A}$ is *homogeneous* if for any tuples $\bar{a}$ and $\bar{b}$ satisfying the same formulas, there is an automorphism of $\mathcal{A}$ that takes $\bar{a}$ to $\bar{b}$. For example, all vector spaces over the rationals are homogeneous. The one of infinite dimension is saturated.

Remark: If $\mathcal{A}$ is a finite structure, then $\mathcal{A}$ is saturated. Let $\Gamma(\bar{a}, x)$ be a set of formulas in the language of $\mathcal{A}$, and suppose that every finite subset is satisfied in $\mathcal{A}$. For each element c of $\mathcal{A}$ if there is some formula of $\Gamma(\bar{a}, x)$ not satisfied by c, choose one. The set of chosen formulas is finite, so it is satisfied by some element c. Then c satisfies all of $\Gamma(\bar{a}, x)$.

Exercise: Let T be a complete theory (in a countable language). Show that T has a (countable) homogeneous model. Starting with an arbitrary (countable) model $\mathcal{A}_0$, form a sequence of countable models $(\mathcal{A}_n)_{n \in \omega}$ such that $\mathcal{A}_n \prec \mathcal{A}_{n+1}$ and for any tuples $\bar{a}$ and $\bar{b}$ satisfying the same formulas in $\mathcal{A}_n$, and any further element c in $\mathcal{A}_n$, there exists d in $\mathcal{A}_{n+1}$ such that $\bar{a}, c$ and $\bar{b}, d$ satisfy the same formulas in $\mathcal{A}_{n+1}$.

A *complete type* is the set of all formulas (in some fixed finite tuple of variables $\bar{x}$) satisfied by some tuple in a structure. We say that a tuple $\bar{a}$ *realizes* the type $\Gamma(\bar{x})$ if $\bar{a}$ satisfies all of the formulas in $\Gamma(\bar{x})$. For a structure $\mathcal{A}$ and a type $\Gamma(\bar{c}, x)$, with parameters $\bar{c}$ in $\mathcal{A}$, $\Gamma(\bar{c}, x)$ is realized in an elementary extension of $\mathcal{A}$ if and only if each finite subset of $\Gamma(\bar{c},x)$ is realized in $\mathcal{A}$.

We may re-phrase the definitions of saturation and homogeneity in terms of types. The structure $\mathcal{A}$ is saturated if and only if all types with parameters in $\mathcal{A}$ realized in elementary extensions of $\mathcal{A}$ are already realized in $\mathcal{A}$. The structure $\mathcal{A}$ is homogeneous if and only if for any tuples $\bar{a}$ and $\bar{b}$ realizing the same complete type in $\mathcal{A}$, and any further element c, there is an element d such that $\bar{a}, c$ and $\bar{b}, d$ realize the same complete type.

Saturated structures have a feature of expandability, sometimes called "resplendence".

Theorem 3.10 *Let $\mathcal{A}$ be a saturated structure. Let Γ be a countable set of sentences, in a language with new symbols, but with at most finitely many parameters from $\mathcal{A}$. If $D^c(\mathcal{A}) \cup \Gamma$ is consistent, then $\mathcal{A}$ can be expanded to a model of Γ.*

Proof sketch: We are assuming that $\mathcal{A}$ and its language are countable. Let $(\varphi_n)_{n \in \omega}$ be a list of all sentences in the language of $\mathcal{A}$ augmented by the new symbols of Γ and constants naming all elements of $\mathcal{A}$. We produce an expansion $\mathcal{B}$, taking $D^c(\mathcal{B})$ to be the union of a chain of finite sets S_n such that

$$D^c(\mathcal{A}) \cup \Gamma \cup S_n$$

is consistent.

Let $S_0 = \emptyset$. Given S_n, we add φ_n or $\neg\varphi_n$, maintaining consistency with $D^c(\mathcal{A}) \cup \Gamma \cup S_n$. This is all of S_{n+1} except in the case where we have added φ_n and this has form $\exists x\, \psi$. In this case, let $\theta(\bar{a}, x)$ be the conjunction of the sentences in S_n and the formula $\psi(x)$. Let $\Lambda(\bar{a}, x)$ be the set of formulas $\gamma(\bar{a}, x)$, in the language of $\mathcal{A}$ with parameters $\bar{a}$, which are *logical consequences* of

$$D^c(\mathcal{A}) \cup \Gamma \cup \{\theta(\bar{a}, x)\}\,.$$

3.12. SOME SPECIAL KINDS OF STRUCTURES

This means that in any model of $D^c(\mathcal{A}) \cup \Gamma$, $\gamma(\bar{a}, x)$ is true of all elements satisfying $\theta(\bar{a}, x)$. Every finite subset of $\Lambda(\bar{a}, x)$ is satisfied in $\mathcal{A}$, so, by saturation, the whole set is satisfied by some c. We add $\theta(\bar{a}, c)$ to S_{n+1}.

Saturation implies homogeneity.

Theorem 3.11 *If a $\mathcal{A}$ is saturated, then it is homogeneous.*

Proof: Let $\mathcal{F}$ be the set of finite partial functions p from $\mathcal{A}$ to $\mathcal{A}$, preserving satisfaction. To show homogeneity, we must show that every element of $\mathcal{F}$ extends to an automorphism of $\mathcal{A}$. By Proposition 3.4, it is enough to show that $\mathcal{F}$ has the back-and-forth property. The arguments for going "back" and "forth" are essentially the same. Here is the argument for "forth". Let $p \in \mathcal{F}$, where p takes $\bar{a}$ to $\bar{b}$, and let c be a further element of $\mathcal{A}$. Let $\Gamma(\bar{u}, x)$ be the complete type realized by $\bar{a}, c$. If $\psi(\bar{u}, x)$ is the conjunction of a finite subset of $\Gamma(\bar{u}, x)$, then $\exists x\, \psi(\bar{u}, x)$ is true of $\bar{a}$, so it is true of $\bar{b}$. By saturation, $\Gamma(\bar{b}, x)$ must be realized in $\mathcal{A}$, say by d. Then $p \cup \{(c, d)\} \in \mathcal{F}$.

Saturated structures realize all possible types. A homogeneous structure is determined, up to isomorphism, by its types (and its cardinality).

Theorem 3.12 *If $\mathcal{A}$ and $\mathcal{B}$ are (countable) homogeneous structures realizing the same complete types, then $\mathcal{A} \cong \mathcal{B}$.*

Proof sketch: Let $\mathcal{F}$ be the set of finite partial functions p from $\mathcal{A}$ to $\mathcal{B}$ preserving satisfaction. As in the proof of Theorem 3.11, we can show that $\mathcal{F}$ has the back-and-forth property.

Remark: We have said that we are only interested in countable structures. With appropriate extensions of our definitions, Theorems 3.10, 3.11, also hold for uncountable structures, and Theorem 3.12 holds for pairs of structures of the same uncountable cardinality.

We mention one more special kind of structure. Let T be a complete theory (in a countable language). A model $\mathcal{A}$ of T is said to be *prime* if for every model $\mathcal{B}$ of T, there is an "elementary" embedding of $\mathcal{A}$ into $\mathcal{B}$; i.e., a mapping from $\mathcal{A}$ into $\mathcal{B}$ that preserves satisfaction of arbitrary formulas. For example, for the theory of algebraically closed fields of characteristic 0, the field of algebraic numbers is a prime model.

A saturated model of a theory realizes all possible types. As we shall see, a prime model realizes only those types that *must* be realized. We need a definition to describe such types. Let T be a theory (not necessarily complete). A type $\Gamma(\bar{x})$ is *principal* (over T) if there is a formula $\gamma(\bar{x})$ such that
   (i) $\exists \bar{x}\, \gamma(\bar{x})$ is true in some model,
   (ii) for all formulas $\psi(\bar{x}) \in \Gamma(\bar{x})$, the sentence $\forall \bar{x}\, (\gamma(\bar{x}) \to \psi(\bar{x}))$ is true in all models.
The formula $\gamma(\bar{x})$ is called a *generator* for $\Gamma(\bar{x})$.

If a type is not realized in a structure $\mathcal{A}$, it is said to be *omitted*. The theorem below is called the "Omitting Types Theorem". Henkin provided the proof, but it was Keisler who realized the importance of the result and attached the descriptive name.

Theorem 3.13 (Henkin) *Let T be a consistent theory, and let $\Gamma(\bar{x})$ be a non-principal type. Then T has a model in which $\Gamma(\bar{x})$ is omitted.*

Sketch of proof: Let L be the language of T and let B be a countably infinite set of new constants. We form an increasing sequence of finite sets of sentences $(S_n)_{n \in \omega}$, in the language $L \cup B$, satisfying the following conditions:
(1) $T \cup S_n$ is consistent,
(2) for each sentence φ, some n contains φ or $\neg\varphi$,
(3) if $\varphi \in S_n$, where φ has the form $\exists x\,\psi(x)$, then for some constant $b \in B$, $\psi(b) \in S_{n+1}$,
(4) for each tuple $\bar{b}$ appropriate for $\bar{x}$, there exist $\gamma(\bar{x}) \in \Gamma(\bar{x})$ and n such that $\neg\gamma(\bar{b}) \in S_n$.

The fact that $\Gamma(\bar{x})$ is non-principal is used in satisfying condition (4). Suppose we have determined S_n, with conjunction $\chi(\bar{b},\bar{d})$, where $\bar{b}$ is a tuple of constants appropriate to substitute for $\bar{x}$. Then $\exists \bar{x}\, \exists \bar{u}\, \chi(\bar{x},\bar{u})$ is true in some model of T. Since $\Gamma(\bar{x})$ is non-principal, there is some $\psi(\bar{x}) \in \Gamma(\bar{x})$ such that $\exists \bar{x}\,(\exists \bar{u}\, \chi(\bar{x},\bar{u})\ \&\ \neg\psi(\bar{x}))$ is true in some model of T. We can add $\neg\psi(\bar{b})$ to S_n.

Now, for the sequence $(S_n)_{n \in \omega}$ described above, it is not difficult to see that $\cup_n S_n$ is the complete diagram of a model of T omitting $\Gamma(\bar{x})$.

For a complete theory T, an *atomic model* is one that realizes only principal types.

Theorem 3.14 *For a complete theory T, a model $\mathcal{A}$ is prime if and only if it is atomic.*

Proof: It follows from the Omitting Types Theorem that if $\mathcal{A}$ is a prime model of T, then $\mathcal{A}$ is atomic. Supposing that $\mathcal{A}$ is an atomic model of T, we show that it is prime. Let $\mathcal{B}$ be another model. To show that $\mathcal{A}$ can be elementarily embedded in $\mathcal{B}$, it is enough to show that the family of finite partial elementary embeddings has the "forth" property. Suppose $\bar{a}$ in $\mathcal{A}$ and $\bar{b}$ in $\mathcal{B}$ realize the same complete type, and let c be a further element of $\mathcal{A}$. Say $\gamma(\bar{u},x)$ generates the type of $\bar{a},c$. Since $\exists x\,\gamma(\bar{u},x)$ is true of $\bar{a}$ in $\mathcal{A}$, it is true of $\bar{b}$ in $\mathcal{B}$, so we have d such that the type of $\bar{b},d$ in $\mathcal{B}$ matches that of $\bar{a},c$ in $\mathcal{A}$.

We have seen that saturated models are homogeneous. The same is true of prime models.

Exercise: Show that a prime model is homogeneous.

3.13 Computable sets of sentences

We restrict our attention to structures whose underlying language is computable. By this, we mean that the set of non-logical symbols is computable, and, in addition, there is a computable function which, for each symbol, tells the type (relation, operation, or constant), and for a relation or operation symbol, the arity. Having assigned codes to the symbols in an effective way, we obtain codes for formulas, since these are finite sequences of symbols. For historical reasons, these codes are called *Gödel numbers*.

For a given computable language L, we can effectively determine whether a given sequence of symbols is an L-formula—the process involves counting parentheses, and breaking the sequence into successively smaller parts to check. For a given formula, we can effectively determine such things as the set of free variables. The process of determining consistency of a given sentence, or a computable set of sentences is more complicated.

Let Γ be a computable set of sentences. We may assume that the language of Γ is relational. We describe a computable tree P whose paths yield all possible complete diagrams of models of Γ with a fixed infinite computable set B as the set of names for the elements. First, let $(\alpha_n)_{n \in \omega}$ be a computable list of the sentences of Γ. Second, let $(\varphi_n)_{n \in \omega}$ be a computable list of all $(L \cup B)$-sentences. Finally, let $\mathcal{C}$ be the family of finite sets S of $(L \cup B)$-sentences, such that the set of basic sentences in S is consistent.

Now, the tree P consists of the finite sequences $S_0, \ldots, S_n$, where $S_0 = \{\top\}$, $S_i \in \mathcal{C}$, and S_{i+1} is an extension of S_i obtained by adding the following (if it is not already present):

(a) α_i,
(b) one of φ_i, $\neg\varphi_i$,
(c) both of ψ, χ, if $(\psi \, \& \, \chi) \in S_i$, and one of $\neg\psi$, $\neg\chi$, if $\neg(\psi \, \& \, \chi) \in S_i$,
(d) one of ψ, χ, if $(\psi \vee \chi) \in S_i$, and both of $\neg\psi$, $\neg\chi$, if $(\neg(\psi \vee \chi)) \in S_i$,
(e) $\psi(b)$, for all b among the first i constants, if $\forall x \, \psi \in S_i$, and $\neg\psi(b)$, for some $b \in B$, if $\neg\forall x \, \psi \in S_i$,
(f) $\psi(b)$, for some $b \in B$, if $\exists x \, \psi \in S_i$, and $\neg\psi(b)$, for all b among the first i constants, if $\neg\exists x \, \psi \in S_i$.

If P has a path, then the union is a set satisfying the conditions of Theorem 3.7, so Γ is consistent. Conversely, if Γ is consistent, there is a model with B as the set of names of elements, and then P has a path. Moreover, for any model $\mathcal{B}$ with B as the set of names of the elements, there is a path whose union is $D^c(\mathcal{B})$. The tree P is infinitely branching, but we can form a subtree P^* which is only finitely branching.

We let P^* consist of the sequences $S_0, \ldots, S_n$ from P such that for $i < n$, if $\neg \forall x \, \varphi$ is in S_i, then S_{i+1} includes $\neg\varphi(b)$ for some b such that either b appears in $\varphi(x)$ or some sentence of S_i, or else b is the first new constant. Note that in P^*, we can compute a bound on the number of successors of any given node, and we can determine the set of nodes at any given level. The paths through P^* do not yield all *complete diagrams* of models of Γ with B as the set of names for the elements, but all *isomorphism types* are represented.

Next, we pass to a computable subtree P^{**} of $2^{<\omega}$ whose paths represent completions of Γ. Let $(\psi_n)_{n\in\omega}$ be a computable list of all sentences in the language L. For $\sigma \in 2^n$, we let $\sigma \in P^{**}$ just in case

$$\{\psi_k : \sigma(k) = 1\} \cup \{\neg\psi_k : \sigma(k) = 0\}$$

is contained in some set S at level n of P^* (or P). Again Γ is consistent if and only if P^{**} has a path. For any path π,

$$\{\psi_n : \pi(n) = 1\} \cup \{\neg\psi_n : \pi(n) = 0\}$$

is a completion of Γ. Moreover, we obtain all possible completions this way.

We have established the following.

Theorem 3.15 *Given an index for a computable set Γ of sentences in a relational language L, we can find an index for a computable subtree of $2^{<\omega}$ such that Γ has a (non-empty) model if and only if the tree has a path. Moreover, the paths correspond to the completions of Γ.*

Combining Theorem 3.15 with the Low Basis Theorem (Theorem 2.15), we obtain the following.

Corollary 3.16 (Jockusch-Soare) *If Γ is a computable, consistent set of axioms, then Γ has a low completion.*

Proof: Let P be as in Theorem 3.15. By the Low Basis Theorem, there is a low path π through P, and the corresponding completion of Γ is Turing equivalent to π, so it is low.

Corollary 3.17 *If Γ is a computable, consistent set of axioms, then the only sets computable in all completions of Γ are the computable sets.*

Proof: Let P be as in Theorem 3.15. If a set X is computable in all completions of Γ, then it is computable in all paths through P. Then by Theorem 2.17, X is computable.

3.14 Complexity of structures

To measure the complexity of a structure $\mathcal{A}$, we must identify the structure with a set of natural numbers. Already, we are assuming that the language of $\mathcal{A}$, say L, is computable. We shall usually also suppose that the universe $|\mathcal{A}|$ is a computable set of constants. We then have codes, or Gödel numbers, for the sentences in the language $L \cup |\mathcal{A}|$ The set of all sentences in this language is a computable subset of ω, and $D^c(\mathcal{A})$ and $D(\mathcal{A})$ are subsets.

We could adopt a less restrictive convention, saying that for a structure $\mathcal{A}$, the universe, $|\mathcal{A}|$, is a *subset* of some computable set of constants. Again $D^c(\mathcal{A})$

and $D(\mathcal{A})$ would be subsets of a computable subset of sentences. There is no real advantage in this less restrictive convention, since $\mathcal{A}$ is isomorphic to a structure with a computable universe, under an isomorphism computable in $|\mathcal{A}|$.

There are several different sets of sentences associated with a structure $\mathcal{A}$. We have the atomic diagram $D(\mathcal{A})$, the complete diagram $D^c(\mathcal{A})$, and the theory $Th(\mathcal{A})$. These are all identified with subsets of ω, which satisfy the relations

$$D(\mathcal{A}) \leq_T D^c(\mathcal{A}) \text{ and } Th(\mathcal{A}) \leq_T D^c(\mathcal{A}).$$

We could consider other sets, such as the open diagram, which is Turing equivalent to $D(\mathcal{A})$, and the existential diagram, whose Turing degree lies somewhere between those of $D(\mathcal{A})$ and $D^c(\mathcal{A})$.

For a complete theory T, the "Henkin" construction yields some model $\mathcal{A}$ such that $D^c(\mathcal{A}) \equiv_T T$. It is possible to have $D(\mathcal{A})$ less complicated than $Th(\mathcal{A})$. For example, consider $\mathcal{N} = (\omega, +, \cdot, S, 0)$, the standard model of arithmetic. While $D(\mathcal{N})$ is computable, the theory $Th(\mathcal{N})$, which is called "true arithmetic", or "TA", is not even arithmetical. It is also possible for $D(\mathcal{A})$ to be much more complicated than $Th(\mathcal{A})$. As we shall soon see, $\mathcal{N}$ has copies of all Turing degrees.

The atomic diagram of a structure tells what the structure looks like, how it is constructed. The complete diagram of a structure tells how the structure might have been obtained as a model of a finitary theory. In much of what follows, we consider non-elementary properties of structures, and we ignore finitary theories. In this context, the complete diagram is not a particularly natural object to consider. Even when we do focus on finitary theories and their models, questions about the atomic diagram may be more interesting than the corresponding questions about the complete diagram. This will be illustrated in Chapter 19.

For the most part, we shall measure the complexity of a structure $\mathcal{A}$ by its atomic diagram. In fact, we may identify $\mathcal{A}$ with $D(\mathcal{A})$. Thus, we say that $\mathcal{A}$ is *computable*, Δ^0_2, or *computable relative to* X, if this is true of $D(\mathcal{A})$. In some settings, the terms "decidable" and "computable" mean the same thing. However, when we are talking about structures, there is a difference. As we indicated in the introduction, a structure $\mathcal{A}$ is *decidable* if $D^c(\mathcal{A})$ is computable, and it is *computable* if $D(\mathcal{A})$ is computable. There have been investigations of finitary theories and their decidable models. While this is not our main focus, we shall mention one or two of these results later.

3.15 Complexity of definable relations

A relation $R(\overline{x})$ is *definable* in a structure $\mathcal{A}$, if there is a formula $\varphi(\overline{x})$, possibly with a tuple of *parameters*, such that

$$R(\overline{a}) \Leftrightarrow \mathcal{A} \models \varphi(\overline{a}).$$

An element a is *definable in* $\mathcal{A}$ if the singleton set $\{a\}$ is a definable relation; i.e., a is the unique element satisfying some formula.

Example: In the standard model of arithmetic $\mathcal{N}$, every element is definable, without parameters. The number n is defined by the atomic formula

$$x = S^{(n)}(0) .$$

It follows that any definable relation is definable by a formula with no parameters—we can incorporate definitions of parameters into the formula defining the relation.

It turns out that the relations definable in $\mathcal{N}$ are precisely the arithmetical relations. This is the explanation of the name.

Theorem 3.18 *A relation is arithmetical if and only if it is definable in $\mathcal{N}$. Moreover, the relation is Σ_n^0 (or Π_n^0) if and only if it is definable in $\mathcal{N}$ by a formula that is Σ_n, Π_n (or Π_n).*

Proof sketch: Since addition, multiplication, and successor are computable, it is clear that any relation definable in $\mathcal{N}$ by an open formula is computable. It follows that a relation definable by a Σ_n formula, must be Σ_n^0, and a relation definable in $\mathcal{N}$ by a Π_n formula must be Π_n^0. To show that all arithmetical relations are definable in $\mathcal{N}$, it is enough to show that all computable relations are definable. The usual proof of Gödel's Incompleteness Theorem establishes this along the way. In fact, for any computable relation, there are two natural definitions: one consisting of a block of existential quantifiers followed by a formula with only *bounded* quantifiers (($\forall x < u$) and ($\exists x < u$))), and the other consisting of a block of universal quantifiers followed by a formula with only bounded quantifiers.

As for the more precise claim that any Σ_n^0 (or Π_n^0) relation is definable in $\mathcal{N}$ by a Σ_n (or Π_n) formula, it is enough to know that any relation definable in $\mathcal{N}$ by a formula β with only bounded quantifiers is definable by one formula σ that is genuinely Σ_1, and by another formula τ that is genuinely Π_1. This is the main lemma in Matijasevic's proof of the unsolvability of Hilbert's 10th Problem (see the Appendix). Using the lemma, we see that a block of existential quantifiers followed by a formula with only bounded quantifiers can be replaced by a Σ_1 formula, and a block of universal quantifiers followed by a formula with only bounded quantifiers can be replaced by a Π_1 formula.

Above, we said that in $\mathcal{N}$, any relation that is definable by a Σ_n formula is Σ_n^0, and any relation definable by a Π_n formula is Π_n^0. The same is true in any computable structure, with a great deal of uniformity.

Theorem 3.19 *For any structure $\mathcal{A}$ and any n, for all Σ_n formulas $\varphi(\bar{x})$, $\varphi^{\mathcal{A}}$ is $\Sigma_n^0(\mathcal{A})$, and for all Π_n formula $\varphi(\bar{x})$, $\varphi^{\mathcal{A}}$ is $\Pi_n^0(\mathcal{A})$. Moreover, from the Gödel number for φ, we can determine a uniform index for $\varphi^{\mathcal{A}}$ relative to $\mathcal{A}$, for all $\mathcal{A}$.*

The *satisfaction predicate* on a structure $\mathcal{A}$ for a class of formulas Γ is the relation which holds between a formula in Γ, or a code for the formula, and a tuple satisfying the formula in $\mathcal{A}$. The uniformity in Theorem 3.19 is such

3.16 Copies of a given structure

The standard model of arithmetic, $\mathcal{N}$, is computable. Isomorphic copies of $\mathcal{N}$, which are also called *standard*, need not be computable. There is a natural set of axioms generating the theory elementary first order Peano arithmetic, or PA. See the Appendix for the list of axioms. The *nonstandard* models of PA are never computable, by the following result of Tennenbaum [149].

Proposition 3.20 (Tennenbaum) *If $\mathcal{A}$ is a non-standard model of PA, then $\mathcal{A}$ is not computable.*

Proof sketch: Let X and Y be a disjoint pair of computably enumerable sets with no computable separator. (See the example after Theorem 1.8.) There are natural formulas with the meanings $y \in X_s$ and $y \in Y_s$, where X_s and Y_s denote the stage s approximations for X and Y, respectively (see Chapter 2, first paragraph of 2.6). There is also a natural formula with the meaning $p_y | u$ (the yth prime divides u). Using these formulas, we let $\psi(x, u)$ say

$$((\exists s \leq x \ y \in X_s) \rightarrow p_y | u) \ \& \ ((\exists s \leq x \ y \in Y_s) \rightarrow p_y \not| u) .$$

For all finite c, $\mathcal{A} \models \exists u \, \psi(c, u)$. Hence, by "Overspill" (see the Appendix), there is some infinite c' such that

$$\mathcal{A} \models \exists u \, \psi(c', u) .$$

Suppose $\mathcal{A} \models \psi(c', d)$, and let

$$S = \{m \in \omega : \mathcal{A} \models p_m | d\} .$$

It is not difficult to see that S is a separator for X and Y. Moreover, we can show that $S \leq_T \mathcal{A}$. For any $n \in \omega$ and any $r < n$, we have an atomic formula saying (of y and u) that $ny + r = u$, or

$$\overbrace{y + \cdots + y}^{n} + S^{(r)}(0) = u .$$

To decide whether $m \in S$, we enumerate the diagram of $\mathcal{A}$, searching for an atomic sentence saying, for d as above, some $r < p_m$ and some $a \in \mathcal{A}$, that $p_m a + r = d$. The search will halt, since the division algorithm holds in models of PA. It follows that $\mathcal{A}$ is not computable.

While a non-standard model of PA cannot be computable, it can be low. To produce such a model, we apply Corollary 3.16 to obtain a low completion

T of PA. Since T is complete, there is a model $\mathcal{A}$ such that $D^c(\mathcal{A}) \leq_T$ T. Since T is low, T $\neq TA$, so the model is necessarily non-standard.

We have said that isomorphic copies of a given structure may have different complexity. For almost all structures $\mathcal{A}$, the set of degrees of copies of $\mathcal{A}$ is closed upwards. The exceptions are the "trivial" structures. A structure $\mathcal{A}$ is *trivial* if there is a tuple of elements $\bar{a}$ such that every permutation of $|\mathcal{A}|$ that fixes $\bar{a}$ pointwise is an automorphism. As an example of a trivial structure, consider ω, with finitely many named elements, and with a family of unary relations, each of which is either $\emptyset$ or ω. It is not difficult to see that for a trivial structure, all copies, with computable universe, have the same degree. None of the structures that we really care about are trivial.

Theorem 3.21 (Solovay, Marker, Knight) *Suppose that $\mathcal{A}$ is a non-trivial structure. If*
$$\mathcal{A} <_T X,$$
then there is a copy $\mathcal{B}$ with an isomorphism F such that
$$\mathcal{B} \equiv_T X .$$
In fact, we have
$$X \leq_T \mathcal{B} \leq_T F \oplus \mathcal{A} \leq_T X.$$

Idea of proof: We give the proof in the case where the structure $\mathcal{A}$ includes a linear ordering. (For the general case, see [81]). Suppose $\mathcal{A}$ has computable universe $A = \{a_n : n \in \omega\}$. We shall take $\mathcal{B}$ to have computable universe $B = \{b_n : n \in \omega\}$. We define a $1-1$ function F from A onto B, taking a_{2n} and a_{2n+1} to the elements b_{2n} and b_{2n+1} (possibly switched) so that
$$n \in X \Leftrightarrow \mathcal{A} \models F^{-1}(b_{2n}) < F^{-1}(b_{2n+1}).$$
We let $\mathcal{B}$ be the induced structure such that $\mathcal{A} \cong_F \mathcal{B}$.

Using Theorem 3.21, we obtain the fact that $\mathcal{N}$ has copies of all Turing degrees.

Exercise: Prove Theorem 3.21 for an arbitrary finite relational language.

Use the following outline. First, show that if $\mathcal{A}$ is a non-trivial structure for the language, then there is an atomic formula $\varphi(\bar{c}, \bar{x})$ with parameters $\bar{c}$, such that for any finite $S \subseteq \mathcal{A}$, there exist tuples $\bar{a}, \bar{b}$, lying entirely outside S, such that
$$\mathcal{A} \models \varphi(\bar{c}, \bar{a}) \ \& \ \neg\varphi(\bar{c}, \bar{b}).$$
Let B be as above. Define a $1-1$ function F from $\mathcal{A}$ onto B using
$$\varphi(\bar{c}, \bar{x}) \ \& \ \neg\varphi(\bar{c}, \bar{y})$$
instead of $x < y$ to code X into $\mathcal{B}$.

Remark: Theorem 3.21 is also true for infinite computable languages. It may be necessary to use a *sequence* of atomic formulas, instead of just one, to code the set X.

3.17 Complexity of quotient structures

Normally, the universe of a structure is a computable set of constants. For a quotient structure $\mathcal{B} = \mathcal{A}/\sim$, it may be convenient *not* to enforce this convention. Suppose that the universe of $\mathcal{A}$ is a computable set of constants. According to the definition, the universe of $\mathcal{B}$ is the set of $\sim$-equivalence classes of constants. Thus, a single element of $\mathcal{B}$ may have many names.

We could replace the quotient structure $\mathcal{B} = \mathcal{A}/\sim$ by an isomorphic copy $\mathcal{B}^*$ whose universe is a computable set of constants. Since $|\mathcal{A}|$ is a computable subset of ω, it inherits the usual ordering. Taking the first element of each $\sim$-class, we obtain an appropriate universe for $\mathcal{B}^* \cong \mathcal{A}/\sim$. Now, $|\mathcal{B}^*|$ is a subset of $|\mathcal{A}|$ computable in $\sim$, and $D(\mathcal{B}^*)$ is computable in the pair $\sim, D(\mathcal{A})$. We could replace $\mathcal{B}^*$ by a further copy $\mathcal{C}$, with *computable* universe, such that $D(\mathcal{C}) \leq_T D(\mathcal{B}^*)$.

For a quotient structure $\mathcal{A}/\sim$, we may measure the complexity of $\mathcal{A}$ and $\sim$ separately. If $\mathcal{A}$ is computable, then we take the complexity of the quotient to be that of $\sim$. Thus, we say that $\mathcal{A}/\sim$ is a *c.e. quotient structure* if $\mathcal{A}$ is computable and $\sim$ is c.e. Similarly, we say that $\mathcal{A}/\sim$ is a Σ^0_n, or Π^0_n quotient structure if $\mathcal{A}$ is computable and $\sim$ is Σ^0_n, or Π^0_n.

We give some examples to illustrate this way of measuring complexity. The first two are from pure computability theory.

Example 1: Let $\mathcal{A} = (\omega, \vee, \wedge)$, where $\vee, \wedge$ are computable binary operations on ω such that $W_a \cup W_b = W_{a \vee b}$ and $W_a \cap W_b = W_{a \wedge b}$. Let

$$a \sim b \Leftrightarrow W_a = W_b \ .$$

Then $\sim$ is a Π^0_2 congruence relation on $\mathcal{A}$. Therefore, $\mathcal{A}/\sim$ is a Π^0_2 quotient structure, the lattice of c.e. sets, often denoted by $\mathcal{E}$.

Example 2: Let $\mathcal{A}$ be as in Example 1. We write $X \Delta Y$ for the *symmetric difference* $(X - Y) \cup (Y - X)$. Let

$$a \sim^* b \Leftrightarrow W_a \Delta W_b \text{ is finite} \ .$$

Then $\sim^*$ is a Σ^0_3 congruence relation on $\mathcal{A}$. Therefore, $\mathcal{A}/\sim^*$ is a Σ^0_3 quotient structure, often denoted by $\mathcal{E}^*$.

Example 3: Consider the group generated by elements $a_1, a_2, \ldots$, under a set of relations R. If R is computable, then the group is said to be *computably* (or *recursively*) presented. If R is finite, then the group is *finitely* presented. Every computably presented group is a c.e. quotient structure. For the group to be computable, the congruence relation, which is equality in the quotient structure, would have to be computable. When this is so, the group is said to have *solvable word problem*. A group with a computable presentation need not have a computable copy. In fact, by a result of Boone [25] and Novikov [128], there is a *finitely* presented group with no computable copy.

Example 4: Let T be the theory of $(Z, +, 0, 1)$ (the integers under addition, with constants 0 and 1). This theory is a trivial variant of Presburger arithmetic, the theory of $(\omega, +, 0, 1)$. The theory T is decidable. Let $\mathcal{A}$ be a model of T, and let D be the subgroup consisting of the elements that are divisible by all natural numbers. Let
$$a \sim b \Leftrightarrow a - b \in D \ .$$
This is a Π_2^0 congruence relation. The resulting Π_2^0 quotient structure is the quotient group $\mathcal{A}/D$.

Chapter 4

Ordinals

Without giving set-theoretic definitions, we state some basic facts about the class of ordinals and the natural relation $<$ on this class. For precise definitions, and proofs of the most basic facts, consult [93], or any other basic treatment of set theory. In the remainder of the chapter, we discuss computable ordinals and ordinal notation, defining Kleene's system O, and proving various results that we shall need later. For further information of this kind, see [134].

4.1 Set theoretic facts

The result below says that the ordinals under $<$ are irreflexive, transitive, comparable, and, in general, behave like a well ordering, except for the fact that elements do not form a set. (See the Appendix for basic information about orderings, and well orderings, in particular.)

Theorem 4.1 *(a) For all ordinals α, $\alpha \not< \alpha$.*

(b) For all ordinals α, β, and γ, if $\alpha < \beta$ and $\beta < \gamma$, then $\alpha < \gamma$.

(c) For all ordinals α and β, either $\alpha < \beta$ or $\beta < \alpha$ or $\alpha = \beta$.

(d) Any non-empty set of ordinals has a $<$-least element; more generally, for any property P satisfied by some ordinal, there is a $<$-least ordinal α such that $P(\alpha)$.

It follows from Theorem 4.1 that for any ordinal α, the set $\{\beta : \beta < \alpha\}$ is well ordered under $<$. We may write $pred(\alpha)$ for either the set or the ordered structure.

Theorem 4.2 *For every well ordering $(A, <_A)$, there is an ordinal α such that $(A, <_A) \cong pred(\alpha)$. Moreover, the ordinal α and the isomorphism are both unique.*

Let $(A, <_A)$ be a well ordering, and suppose that h is an isomorphism from $(A, <_A)$ onto $pred(\alpha)$, as in Theorem 4.2. For $b \in A$, the ordinal $h(b)$ is called the *height* of b, and α is called the *order type* of $(A, <_A)$. Thus, $pred(\alpha)$ is the standard well ordering of order type α.

The first few ordinals are the natural numbers $0, 1, 2, \ldots$, used for counting. There is a next ordinal beyond the natural numbers, denoted by ω, then one beyond this, denoted by $\omega + 1$, and so on, forming a list

$$0, 1, 2, \ldots, \omega, \omega+1, \omega+2, \ldots, \omega+\omega, \omega+\omega+1, \omega+\omega+2, \ldots .$$

The list goes on "forever", in a sense made precise in the next result.

Proposition 4.3 *For any initial segment of the ordinals that forms a set S, there is a least ordinal greater than all elements of S.*

Proof: Since S is a set of ordinals, S is well ordered under $<$, by Theorem 4.1. By Theorem 4.2, S is isomorphic to $pred(\alpha)$, for some ordinal α. Moreover, since S is an initial segment, we can see that the unique isomorphism is the identity function. Then $S = pred(\alpha)$, and α is the first ordinal not in S.

A *countable* ordinal is one for which the set of predecessors can be matched up with natural numbers. We shall deal almost exclusively with countable ordinals. However, uncountable ordinals do exist. To see this, note that the countable ordinals form a set which is an initial segment of the ordinals. Therefore, by Proposition 4.3, there is a first ordinal greater than all of these. This ordinal, denoted by ω_1, must be uncountable.

We classify ordinals α in terms of some simple properties of $pred(\alpha)$. The unique ordinal α for which $pred(\alpha) = \emptyset$ is 0. If $pred(\alpha)$ has a greatest element β, so that α is the first ordinal greater than β, then α is called a *successor* ordinal. We write $\alpha = \beta + 1$. If $pred(\alpha)$ has no greatest element, and is not empty, then α is called a *limit* ordinal. For example, ω is a limit ordinal.

We have been stating properties of the ordinals and the relation $<$ without giving a precise definition of either. Under the usual set theoretic definitions, an ordinal is a *transitive* set (i.e., it contains all elements of its elements) that is linearly ordered by $\in$, and $\beta < \alpha$ if and only if $\beta \in \alpha$. This implies that for each ordinal α, $pred(\alpha)$ is actually equal to α. Thus, $0 = \emptyset$, $1 = \{0\}$, $2 = \{0, 1\}$, ... Also, ω is the set of natural numbers, and ω_1 is the set of countable ordinals.

4.2 Inductive proofs and definitions

Theorem 4.1 provides a method for proving that property P holds for all ordinals α. We prove, for an arbitrary ordinal α, that if $P(\beta)$ holds for all $\beta < \alpha$, then $P(\alpha)$ holds. This is clearly sufficient, for if there is some ordinal α for which $P(\alpha)$ fails, consider the first. An argument of this kind is known as a *proof by transfinite induction* (the ordinals beyond the finite numbers are sometimes called *transfinite* numbers).

4.3. OPERATIONS ON ORDINALS

Similar considerations justify the principle of *definition by transfinite recursion*, in which for each ordinal α, we associate with α an object that is described in terms of objects already associated with the ordinals $\beta < \alpha$. For example, given a set S of real numbers, we may associate with each ordinal α a set S_α such that

(a) $S_0 = S$,

(b) if $\alpha = \beta + 1$, then S_α is the result of removing from S_β, the first element, if any,

(c) if α is a limit ordinal, then $S_\alpha = \cap_{\beta < \alpha} S_\beta$.

We must eventually reach an α for which $S_\alpha = S_{\alpha+1}$, indicating that S_α has no first element. Intuitively, this follows from the principal that the ordinals go on for ever. The reals which for some α witness that $S_\alpha \neq S_{\alpha+1}$ form a set, and the corresponding ordinals form a set S—an initial segment of the ordinals, so by Proposition 4.3, there is some first ordinal α greater than all ordinals in S. For this α, we can prove, by transfinite induction, that for all ordinals $\beta \geq \alpha$, $S_\beta = S_\alpha$.

Proofs by transfinite induction and definitions by transfinite recursion are often subdivided as above into the three cases, where the ordinal is 0, a successor ordinal, or a limit ordinal. On the other hand, it is sometimes taken as a point of elegance to arrange matters so that these distinctions need not be made.

4.3 Operations on ordinals

We have operations of addition and multiplication of linear orderings, where the order types of $\mathcal{A} + \mathcal{B}$ and $\mathcal{A} \cdot \mathcal{B}$ are determined by the order types of $\mathcal{A}$ and $\mathcal{B}$. We also have an operation of exponentiation, with the order type of $\mathcal{A}^\mathcal{B}$ depending not just on the order types of $\mathcal{A}$ and $\mathcal{B}$, but also on the choice of a particular element of $\mathcal{A}$. (These operations are all described in the Appendix.)

If $\mathcal{A}$ and $\mathcal{B}$ are well orderings, then $\mathcal{A} + \mathcal{B}$ and $\mathcal{A} \cdot \mathcal{B}$ turn out to be well orderings. If in forming $\mathcal{A}^\mathcal{B}$, we choose the first element of $\mathcal{A}$, then again we get a well ordering, with order type determined by those of $\mathcal{A}$ and $\mathcal{B}$. Thus, we obtain well-defined operations on the ordinals. Let α and β be ordinals, and let $\mathcal{A}$ and $\mathcal{B}$ be orderings of type α and β, respectively. Then

(a) $\alpha + \beta$ is the order type of $\mathcal{A} + \mathcal{B}$,

(b) $\alpha \cdot \beta$ is the order type of $\mathcal{A} \cdot \mathcal{B}$,

(c) if $\alpha \neq 0$, then α^β is the order type of $\mathcal{A}^\mathcal{B}$, where the chosen element is the first in $\mathcal{A}$.

Remark: These operations on the ordinals could also be defined by transfinite recursion. In particular, we could define ordinal addition, $\alpha + \beta$, by recursion on β, saying that $\alpha + 0 = \alpha$, $\alpha + (\beta + 1) = (\alpha + \beta) + 1$, and for limit ordinals β, $\alpha + \beta$ is the least upper bound of $\{\alpha + \gamma : \gamma < \beta\}$.

4.4 Cantor normal form

The operations on the ordinals—addition, multiplication, and exponentiation — extend the usual operations on the natural numbers. Some of the properties true of the operations on the natural numbers hold for the extended operations, while others fail. For example, associativity holds for both addition and multiplication of ordinals, while commutativity fails for both. A version of the division algorithm holds, saying that for any ordinals α and β, there exist unique ordinals γ and ρ such that $\rho < \beta$ and $\alpha = \beta \cdot \gamma + \rho$. This yields the result below, saying that we can represent any ordinal as a kind of polynomial in powers of ω, with natural number coefficients.

Theorem 4.4 (Cantor normal form) *Each ordinal $\alpha \neq 0$ can be uniquely represented in the form*

$$\omega^{\beta_1} \cdot n_1 + \cdots + \omega^{\beta_k} \cdot n_k,$$

where $\beta_1, \ldots, \beta_k$ are ordinals with $\beta_1 > \ldots > \beta_k$, and $n_1, \ldots, n_k$ are non-zero natural numbers.

For an ordinal α, the *Cantor normal form* is the unique expression given in Theorem 4.4, if $\alpha \neq 0$, and 0 if $\alpha = 0$.

We have already defined one notion of addition for ordinals. Cantor normal form gives rise to a different notion. The *commutative sum* of ordinals α and β is the ordinal γ whose Cantor normal form is obtained by taking the Cantor normal forms for α and β, and adding the coefficients of like terms (with the understanding that the coefficient is 0 if the term does not appear). As the name implies, the operation of commutative sum is commutative. In addition, this operation has the following useful property.

Proposition 4.5 *Suppose that $\alpha \leq \alpha'$ and $\beta \leq \beta'$, where at least one of the inequalities is strict. Then the commutative sum of α and β is strictly less than the commutative sum of α' and β'.*

4.5 Constructive and computable ordinals

Above, we attached names to some of the ordinals, and we can continue to do so for quite a long way. However, there can only be countably many names (involving symbols that we could recognize and code as numbers). There are uncountably many countable ordinals, so we cannot invent distinct names for all of these. In principle, for any countable ordinal α, we can assign numbers (as names) to all ordinals less than α, but we may not be able to do so in a "useful" manner. We would wish, at least, to know whether the ordinal represented by one number is less than that represented by another.

4.6. KLEENE'S O

A *computable well ordering* is a computable structure $(A, <_A)$, where $<_A$ is a well ordering of A. For example, the natural numbers can be computably ordered as
$$0, 2, 4, \ldots; 1, 3, 5, \ldots$$
to give a computable well ordering of type $\omega + \omega$. A *computable ordinal* is an ordinal which is the order type of some computable well ordering.

Proposition 4.6 *The computable ordinals form an initial segment of the ordinals.*

Proof: We must show that if α is a computable ordinal and $\beta < \alpha$, then β is also a computable ordinal. Since α is computable, it is the order type of some computable well ordering $(A, <_A)$. Let d be the element of A of height β, let $B = \{b \in A : b <_A d\}$, and let $<_B$ be the restriction of $<_A$ to B. Then $(B, <_B)$ is a computable well ordering of order type β.

Clearly, each computable ordinal is countable. Therefore, the computable ordinals form a set which is a subset of $pred(\omega_1)$. Note that by Proposition 4.3, there is a first non-computable ordinal.

Theorem 4.7 *The first non-computable ordinal is strictly less than ω_1.*

Proof: For each computable ordinal α, there is a computable ordering of order type α. We may associate with α the least computable index for the atomic diagram of an ordering of order type α. Thus, we have assigned numbers (in a non-effective way) to all of the computable ordinals, while as we remarked earlier, we cannot do this for all the ordinals less than ω_1.

A different, and more useful, way of assigning numbers to what turns out to be the same set of ordinals was given by Kleene [79].

4.6 Kleene's O

In this system of *notations* for ordinals, we define, simultaneously, a set O, a function $|\ |_O$ taking each $a \in O$ to an ordinal $\alpha = |a|_O$, and a strict partial ordering $<_O$ on O. The elements of O are the notations, and $|a|_O$ is the ordinal represented by the notation a. Some of the conventions are rather arbitrary, but nobody so far has had a good enough reason to change them.

We let 1 be the notation for 0; that is, $|1|_O = 0$. If a is a notation for α, then 2^a is a notation for $\alpha + 1$. In the partial ordering, we let $b <_O 2^a$ if either $b <_O a$ or $b = a$. So, the system begins with
$$1, 2, 2^2, 2^{2^2}, \ldots$$
as notations for
$$0, 1, 2, 3, \ldots.$$

For a limit ordinal α, the notations are the numbers $3 \cdot 5^e$ such that φ_e is a total computable function, with values in O, where

$$\varphi_e(0) <_O \varphi_e(1) <_O \varphi_e(2) <_O \ldots,$$

and α is the least upper bound of the sequence of ordinals $\alpha_n = |\varphi_e(n)|_O$. In the partial ordering, we let $b <_O 3 \cdot 5^e$ if there exists n such that $b <_O \varphi_e(n)$. We have now described the whole system.

To be more precise, we would define, by transfinite recursion on α, the partial system O_α, made up of the set of notations for ordinals $\beta \leq \alpha$, together with the restriction of the relation $<_O$ and the operation $|\ |_O$ to this set.

Note: For each finite ordinal $0, 1, 2, 3, \ldots$, there is a unique notation, while for an infinite ordinal, if there is one notation, then there are infinitely many.

Since we cannot number all the ordinals less than ω_1, there is a countable ordinal for which there is no notation. The first such ordinal will necessarily be a limit ordinal, with no φ_e satisfying the required conditions. The ordinals having notations in O are called *constructive* ordinals, and the least non-constructive ordinal is denoted by ω_1^{CK} (CK stands for Church-Kleene).

4.7 Constructive and computable ordinals

Spector [147] showed that the constructive ordinals are exactly the computable ones. We shall prove this, establishing some other important facts about O along the way.

Lemma 4.8 *If $|a| = \alpha$, then the set $\{b : b <_O a\}$ consists of exactly one notation for each $\beta < \alpha$. Moreover, this set, under the restriction of $<_O$, is isomorphic to $\mathrm{pred}(\alpha)$ via the isomorphism $|\ |_O$.*

This is easily shown by induction on α.

As we shall see, the set O is extremely complicated, and as a consequence, the relation $<_O$ is not computably enumerable. However, $<_O$ is the restriction to O of a c.e. relation $<_O^+$, which we define next. Let $a <_O^+ b$ if and only if there is a finite sequence $a_0, \ldots, a_m$ such that $m > 0$, $a_0 = b$, $a_m = a$, and for each $i < m$, we have one of the following:

(i) $a_i = 2^{a_{i+1}}$,

(ii) $a_i = 3 \cdot 5^e$, for some e, and for some n, $a_{i+1} = \varphi_e(n)$.

For example, if e is an index for the partial computable function that has value 7 at $x = 0$ and is undefined elsewhere, then $7 <_O^+ 2^{3 \cdot 5^e}$, as witnessed by the sequence $a_0 = 2^{3 \cdot 5^e}$, $a_1 = 3 \cdot 5^e$, $a_2 = 7$. Obviously, the numbers in this example are not in O. In fact, for any $a \in \omega$, there exists b such that $a <_O^+ b$—let $b = 2^a$. It is easy to see that the relation $<_O^+$ is c.e.

Lemma 4.9 *For $a \in O$, $b <_O^+ a$ if and only if $b <_O a$.*

4.7. CONSTRUCTIVE AND COMPUTABLE ORDINALS

This is easily shown by induction on $|a|_O$.

From the lemma, together with the fact that $<_O^+$ is c.e., we get the following.

Proposition 4.10 *There is a computable function p such that for all $b \in O$, $\{a : a <_O b\} = W_{p(e)}$.*

The next lemma says that every constructive ordinal is computable.

Lemma 4.11 *For all $a \in O$, if $|a| = \alpha$, then there is a computable well ordering of order type α.*

Proof: If $a \in O$, where $|a| = \alpha$, then $\{b : b <_O^+ a\}$ is c.e. Let us computably enumerate this set as $b_0, b_1, b_2, \ldots$. Now, we determine a c.e. binary relation $<^*$ on ω, letting $i <^* j$ if and only if $b_i <_O^+ b_j$. By Lemma 4.9, $b_i <_O^+ b_j$ if and only if $b_i <_O b_j$. Then $(\omega, <^*)$ has order type α. We know that $<^*$ is c.e., and the comparability property guarantees that a c.e. linear ordering on a computable set is actually computable. This completes the proof.

We must show that every computable ordinal is constructive. The proof involves two lemmas.

Lemma 4.12 *Given a computable index for a linear ordering $\mathcal{A}$, we can find a computable index for a linear ordering $\mathcal{B}$ of type $\omega \cdot (1 + \mathcal{A}) + 1$, in which we can apply uniform effective procedures to*

(a) determine whether a given element b is first, last, a successor, or a limit point,

(b) for any a successor element b, find the immediate predecessor, and for any limit element b, determine an increasing sequence $\ell(b,i)$ with limit b.

Furthermore, we can effectively determine a computable index for the embedding that takes a, in $\mathcal{A}$, to the first element of the corresponding copy of ω, in $\mathcal{B}$.

Note: If $\mathcal{A}$ is a linear ordering, then $\omega \cdot (1 + \mathcal{A}) + 1$ is well ordered if and only if $\mathcal{A}$ itself is well ordered. Moreover, if $\mathcal{A}$ is a well ordering of order type α, then the order type of $\omega \cdot (1 + \mathcal{A}) + 1$ is an ordinal β strictly greater than α.

Proof of Lemma 4.12: We may suppose that the universe of $\mathcal{A}$ is $\omega - \{0\}$ (or a finite initial segment). Let $\mathcal{A}^*$ be the result of adding 0 at the front of $\mathcal{A}$. Let B consist of the elements $2^x 3^m$ for $x \in \mathcal{A}^*$ and $m \in \omega$, plus the single element 5. We define the ordering $<_B$ on B such that $2^x 3^m <_B 5$ for all x and m, and $2^x 3^m <_B 2^y 3^n$ if either $\mathcal{A}^* \models x < y$, or else $x = y$ and $m < n$. There is a natural embedding that takes $x \in \mathcal{A}$ to $(x, 0)$. It is easy to check that $(B, <_B)$ is a computable ordering satisfying (a) and the first part of (b) (finding the immediate predecessor, if any).

We turn to the other part of (b)—involving limit elements. Let $b_0, b_1, b_2, \ldots$ be the elements of B, listed according to the usual ordering on ω. If b is a limit element, then we let $\ell(b,i) = b_n$, for the first n such that $b_n <_B b$ and for all $j < i$, we do not have $b_n \leq_B \ell(b,j)$. It is easy to see that for any $b' <_B b$, there exists i such that $b' <_B \ell(b,i) < b$.

The next lemma is very important. We shall use it right away in showing that every computable ordinal is constructive. We shall use it again, in Chapter 5, in showing that O is a sufficiently powerful oracle to recognize well orderings. The proof of the lemma involves the Recursion Theorem.

Lemma 4.13 *Given an index for a computable linear ordering B with the special properties in Lemma 4.12 (there are first and last elements, the sets of successor elements and limit elements are computable, with computable witnessing functions, and the index for B tells us how to compute all of these things), we can find a computable index for a function g, defined on all of B, such that*
    (a) if $pred(x)$ is well ordered, then $g(x)$ is a notation in O for the height of x (the order type of $pred(x)$),
    (b) if $pred(x)$ is not well ordered, then $g(x) \notin O$.

Proof: We begin by defining a partial computable function f such that if e is a computable index for a function with the behavior we want for g on $pred(x)$ (thinking of B as well ordered), then $f(e,x)$ has the value we want for $g(x)$.

We consider three cases.

Case 1: If x is the first element of B, then $f(e,x) = 1$.

Case 2: If x is the successor of y in B, then $f(e,x) = 2^{f(e,y)}$.

Case 3: If x is a limit element of B, and $\ell(x,i)$ is the computable increasing sequence with limit b defined in the proof of Lemma 4.12, let $h(e,b)$ be the computable function given by the $s-m-n$ Theorem, such that

$$\varphi_{h(e,x)}(i) = \varphi_e(\ell(x,i)) \ .$$

Then $f(e,x) = 3 \cdot 5^{h(e,x)}$.

Note that $f(e,x)$ is defined for all e and all x in B, even when B is not well ordered or the function with index e is badly behaved. If x is first in B, then the value of $f(e,x)$ is given by Case 1. If x is a limit element, then the value of $f(e,x)$ is given by Case 3, since $h(e,x)$ is total. If x is neither first nor a limit element in B, then the nth predecessor of x is, for some n, and we can determine the value of $f(e,x)$ by n applications of Case 2.

There is a computable function k such that

$$f(e,x) = \varphi_{k(e)}(x) \ ,$$

using the $s-m-n$ Theorem again. Then by the Recursion Theorem, there exists n for which $\varphi_{k(n)} = \varphi_n$. This means that for all $x \in B$,

$$\varphi_n(x) = \varphi_{k(n)}(x) = f(n,x) \ ,$$

4.7. CONSTRUCTIVE AND COMPUTABLE ORDINALS

where f(n,x) is defined as above.

Claim 1: For all $x \in \mathcal{B}$, $\varphi_n(x)$ is defined.

This is clear from the fact that $f(n,x)$ is defined.

Claim 2: If $pred(x)$ is well ordered and x has height α, then $|\varphi_n(x)|_O = \alpha$.

This is easy to show, by transfinite induction on α.

Claim 3: If $\varphi_n(x) \in O$, then $pred(x)$ is well ordered in $\mathcal{B}$.

We show by transfinite induction on α that if $|a|_O = \alpha$ and $\varphi_n(x) = a$, then x has height α in $\mathcal{B}$. If $\alpha = 0$, then $a = 1$, and $\varphi_n(x) = 1$ only if x is the first element of $\mathcal{B}$. If $\alpha = \beta + 1$, then a has form 2^b, where $|b|_O = \beta$. We have $\varphi_n(x) = a$ only if x is the successor of some y such that $\varphi_n(y) = b$. By H.I., $pred(y)$ is well ordered, so $pred(x)$ is well ordered. Finally, if α is a limit ordinal, then a has form $3 \cdot 5^e$, where $\beta_i = |\varphi_e(i)|_O$ is an increasing sequence of ordinals with least upper bound α. We have $\varphi_n(x) = a$ only if x is a limit element of $\mathcal{B}$ and $e = h(n,x)$, where for all i,

$$\varphi_{h(n,x)}(i) = \varphi_n(\ell(x,i)).$$

By H.I., $pred(\ell(x,i))$ is well ordered. It follows that $pred(x)$ is well ordered.
Thus, φ_n is the desired function g, completing the proof of the lemma.

We are now prepared to show that every computable ordinal is constructive, to finish Spector's proof of equivalence.

Theorem 4.14 *The least non-computable ordinal is the least non-constructive ordinal—both are equal to ω_1^{CK}.*

Proof: By Lemma 4.11, if α has a notation in O, then α is a computable ordinal. Suppose α is a computable ordinal, so there is a computable ordering $\mathcal{A}$ of order type α. Let $\mathcal{B}$ be the computable well ordering obtained from $\mathcal{A}$ as in Lemma 4.12, of order type strictly greater than α. Now, take the function g as in Lemma 4.13, such that for $b \in \mathcal{B}$, $g(b)$ is a notation in O for the height of b. One of the elements of $\mathcal{B}$ has height α.

In addition to its part in the proof of Theorem 4.14, Lemma 4.13 yields the following result.

Proposition 4.15 *There is a partial computable function g such that if x is an index for a computable linear ordering $\mathcal{A}$, then $g(x)$ is defined, and $g(x) \in O$ if and only if $\mathcal{A}$ is well ordered.*

We shall abbreviate $|a|_O$ by $|a|$. According to our conventions, O is not a tree—the elements are not finite sequences, and there are elements (notations for infinite ordinals) with infinitely many $<_O$-predecessors. Nonetheless, we may speak of a *path* through O, by which we mean a maximal subset of O that is linearly ordered by $<_O$. Often we shall focus on a single path in O and consider only ordinals that have notations on this path. Since each ordinal has at most one notation on the path, we may identify the notations with the ordinals. This allows us to forget about ordinal notation entirely. Occasionally, we are forced to consider *incomparable* notations a and b, such that $a \not\leq_O b$ and $b \not\leq_O a$. The following technical result helps us pass from notations which may be incomparable to notations along a single path in O.

Proposition 4.16 *Suppose that* $(a_n)_{n \in \omega}$ *is a computable sequence of elements of O. Then there exist*

$a^* \in O$ *and a partial computable function k such that*

(a) *for all n and all $b \leq_O a_n$, $k(n,b) <_O a^*$, and*

(b) *if $d <_O b \leq_O a_n$, then $k(n,d) <_O k(n,b)$.*

Moreover, from an index for $(a_n)_{n \in \omega}$, we can effectively determine a^* and an index for k.

Proof: Uniformly in n, we determine a computable linear ordering $\mathcal{A}_n$ of order type $|a_n| + 1$, as in Lemma 4.11, such that the partial function taking $b \leq_O a_n$ to the element of height $|b|$ in $\mathcal{A}_n$ is computable, uniformly in n. Then we form a computable ordering $\mathcal{A}$ of type $\Sigma_n \mathcal{A}_n$, together with a partial computable function f such that for $b \leq_O a_n$, $f(n,b)$ is the element which, in the restriction of $\mathcal{A}$ to $\mathcal{A}_n$, has height $|b|_O$. From $\mathcal{A}$, we obtain a computable ordering

$$\mathcal{B} \cong \omega \cdot (1 + \mathcal{A}) + 1$$

and a natural computable embedding h of $\mathcal{A}$ into $\mathcal{B}$, as in Lemma 4.12. By Lemma 4.13, there is a computable function g from $\mathcal{B}$ into O such that $g(b)$ is a notation for the height of b in $\mathcal{B}$, and

$$\mathcal{B} \models d < b \Rightarrow g(d) <_O g(b) \ .$$

Then $k(n,b) = g(h(f(n,b)))$ is the desired function. We let $a^* = g(b^*)$, where b^* is the last element of $\mathcal{B}$.

4.8 Transfinite induction on ordinal notation

To show that all elements of O have some property P, it is enough to prove the following:

(a) $P(1)$,

(b) $P(b) \Rightarrow P(2^b)$, and

(c) $(\forall n\, P(\varphi_e(n))) \Rightarrow P(3 \cdot 5^e)$.

Or, we may prove the single statement that for all $b \in O$,

$$(\forall d <_O b\, P(d)) \Rightarrow P(b) \ .$$

4.8. TRANSFINITE INDUCTION ON ORDINAL NOTATION

Such an argument may be called a proof by *transfinite induction on ordinal notation*.

To define a partial computable function f with domain including all of O, it is enough to
 (a) specify $f(1)$,
 (b) say how $f(2^d)$ is computed from $f(d)$, and
 (c) say how $f(3 \cdot 5^e)$ is computed from the sequence of values $f(\varphi_e(n))$

Or, we may define f by specifying, in a single statement, how $f(b)$ is computed from the restriction of f to the set $<_O -pred(b) = \{d : d <_O b\}$. A definition of this form is called a definition by *computable transfinite recursion on ordinal notation*. The result below justifies these definitions, under some strong conditions.

Theorem 4.17 *(a) Let g be a partial computable function defined (at least) on the pairs (b, u) where $b \in O$ and φ_u is a function with domain $<_O -pred(b)$. Then there is a partial computable function f such that for $b \in O$, $f(b) = g(b, u)$, where φ_u is the restriction of f to $<_O -pred(b)$.*

(b) Let g be a partial computable function defined on the triples (a, b, u), where $a, b \in O$ and φ_u is a function with domain $\{(a, d) : d <_O b\}$. Then there is a partial computable function f such that for all a and b in O,

$$f(a, b) = g(a, b, u) ,$$

where φ_u is the restriction of f to $\{(a, d) : d <_O b\}$.

Proof of (a): Let r be a partial computable function such that if $b \in O$, then for all n, $r(n, b)$ is an index for the restriction of φ_n to $<_O -pred(b)$. Let $h(n, b) = g(b, r(n, b))$. Thus, h is a partial computable function such that if φ_n has the values desired for f on $<_O -pred(b)$, then $h(n, b)$ has a value appropriate for $f(b)$. By the $s-m-n$ Theorem, there is a computable function k such that $\varphi_{k(n)}(b) = h(n, b)$. By the Recursion Theorem (Theorem 1.11), there exists n such that $\varphi_n = \varphi_{k(n)}$. We let $f = \varphi_n$. Then $f(b) = h(n, b) = g(b, u)$, where $u = r(n, b)$ is an index for the restriction of f to $<_O -pred(b)$.

Remark: The hypothesis of Theorem 4.17 can be relaxed. We may obtain f from g provided that g is a partial computable function which is defined, and has an appropriate value, for each pair (b, u) such that $b \in O$ and φ_u is the restriction to $<_O -pred(b)$ of a function with the features we want for f.

From its statement, the Recursion Theorem (Theorem 1.11) might have been called the "Fixed-Point Theorem". The name "Recursion Theorem" is justified by applications such as Theorem 4.17 and Lemma 4.13, and the name "Recursion Theorem" is sometimes applied to various results resembling Theorem 4.17.

The following result, one of Kleene's original applications of the Recursion Theorem, says that addition on ordinals translates into a computable operation on ordinal notations.

Proposition 4.18 *There is a partial computable function p such that for all $a, b \in O$,*

(a) $|a| + |b| = |p(a,b)|$, and

(b) for all $c \in O$, $a \leq_O c \leq_O p(a,b) \Leftrightarrow (\exists d \leq_O b) \, p(a,d) = c$.

Proof: We define p inductively, letting $p(a, 1) = a$, $p(a, 2^b) = 2^{p(a,b)}$, and $p(a, 3 \cdot 5^e) = 3 \cdot 5^{e'}$, where $\varphi_{e'}(n) = p(a, \varphi_e(n))$. More formally, we may define an appropriate function $g(a, b, u)$ and apply Theorem 4.17.

The following result is another example of a definition by computable transfinite recursion on ordinal notation. This will be important in Chapter 15.

Theorem 4.19 *Given $b \in O$, where $|b| = \beta$, we can find a computable linear ordering $\mathcal{A}_b$ of type ω^β such that for any interval with endpoints in $\mathcal{A}_b \cup \{\infty\}$, we can effectively express the order type in Cantor normal form, using notations $c \leq_O b$ for the ordinal exponents.*

Proof: Formally, what we wish to produce is a partial computable function f such that for $b \in O$, $f(b)$ is a pair consisting of an index for a computable linear ordering $\mathcal{A}_b$ of type ω^{w^β}, and a computable function assigning to each interval in $\mathcal{A}_b$ a code for the order type expressed in Cantor normal form. If an interval has order type $\omega^{\beta_1} \cdot n_1 + \ldots + \omega^{\beta_k} \cdot n_k$, in Cantor normal form, we code this by the Gödel number of the finite sequence $(\langle b_1, n_1 \rangle, \ldots, \langle b_k, n_k \rangle)$, where b_i is the notation for β_i such that $b_i \leq_O b$.

We shall be informal in defining the function f. We describe $\mathcal{A}_1$, and we indicate, for $b = 2^d$, how to obtain $\mathcal{A}_b$ from $\mathcal{A}_d$, and for $b = 3 \cdot 5^e$, where $\varphi_e(n) = b_n$, how to obtain $\mathcal{A}_b$ from the sequence $(\mathcal{A}_{b_n})_{n \in \omega}$. We also explain how the new function giving Cantor normal forms can be gotten from the old one(s).

First, consider $b = 1$. Then $|b| = 0$, and $\omega^0 = 1$. Let $\mathcal{A}_1$ be a 1-element ordering. The intervals in $\mathcal{A}_1$ are trivial, so the function giving the Cantor normal forms is trivial.

Next, consider $b = 2^d$, where $|d| = \gamma$, so $|b| = \gamma + 1$. Let $\mathcal{A}_d = (A_d, <_d)$ be a computable linear ordering of type ω^γ, with a computable function giving the order types for intervals, in Cantor normal form. Let x_0 be the first element distinguished by the Cantor normal form of the interval on its left. Let

$$\mathcal{A}_b = (A_b, <_b) \, ,$$

where

$$A_b = \{(x, n) : n \in \omega \, \& \, x \in A_d\}$$

and

$$(x, m) <_b (y, n) \Leftrightarrow [m < n \vee (m = n \, \& \, x <_d y)] \, .$$

It is clear that $\mathcal{A}_b$ has the desired order type $\omega^{\gamma+1}$. For intervals in $\mathcal{A}_b$, we compute the order types, in Cantor normal form, as follows. An interval

4.8. TRANSFINITE INDUCTION ON ORDINAL NOTATION

$[(x,n), \infty)$ has order type $\omega^{\gamma+1}$. An interval $[(x,n),(y,n))$, lying entirely in the nth copy of $\mathcal{A}_d$, has the same order type as the interval $[x,y)$ in $\mathcal{A}_d$. An interval $[(x,m),(y,n))$, for $m < n$, has order type $\omega^\beta \cdot (n-m) + \xi$, where ξ is the order type of the interval $[x_0, y)$ in $\mathcal{A}_d$.

Finally, consider $b = 3 \cdot 5^e$, where $\varphi_e(n) = b_n$. Say $|b| = \beta$ and $|b_n| = \beta_n$. Note that ω^β can be expressed as $\Sigma_n \omega^{\beta_n}$. Suppose that for each n, we have a computable ordering $\mathcal{A}_{b_n} = (A_{b_n}, <_{b_n})$ of type ω^{β_n}, with a computable function giving the order types for intervals, in Cantor normal form. Let

$$\mathcal{A}_b = (A_b, <_b),$$

where

$$A_b = \{(x,n) : n \in \omega \ \& \ x \in A_{b_n}\},$$

and

$$(x,m) <_b (y,n) \Leftrightarrow [m < n \lor (m = n \ \& \ x <_{b_n} y)].$$

Then $\mathcal{A}_b = \Sigma_n \mathcal{A}_{b_n}$ has order type ω^β, as required. For intervals in $\mathcal{A}_b$, we determine the order types, in Cantor normal form, as follows. An interval $[(x,n), \infty)$ has order type ω^α. An interval $[(x,m),(y,n))$, for $m < n$, has order type

$$\omega^{\beta_m} + \omega^{\beta_{m+1}} + \cdots + \omega^{\beta_{n-1}} + \xi,$$

where ξ is the order type of the interval to the left of y in $\mathcal{A}_{b_n}$. From this, it should be clear that we can compute the order types in Cantor normal form.

Chapter 5

The hyperarithmetical hierarchy

In this chapter, we define the *hyperarithmetical* hierarchy. This will be our main classification tool in later chapters. We also define the *analytical* hierarchy. For the latter, we are chiefly interested in the sets at the lowest level—those which are Δ_1^1, Σ_1^1, or Π_1^1. We prove that the Δ_1^1 relations are exactly the hyperarithmetical relations, and we give other results that are needed for Chapter 8. For further information, see Rogers [134].

5.1 The hyperarithmetical hierarchy

Recall that the arithmetical relations can be characterized in terms of the special sets $\emptyset^{(n)}$. By Theorem 2.10, for all finite $n \geq 1$, a relation is Σ_n^0, Π_n^0, or Δ_n^0 if and only if it is, respectively, c.e., co-c.e., or computable relative to $\emptyset^{(n-1)}$. We take this approach in extending the arithmetical hierarchy to transfinite levels. The first step is to extend the sequence of sets $\emptyset, \emptyset', \emptyset'', \ldots$. Having no single-valued system of notations for all the computable ordinals, we use all possible notations for each ordinal.

We define sets $H(a)$, for $a \in O$, by transfinite recursion on the ordinals $|a|$, as follows:
 (1) $H(1) = \emptyset$,
 (2) $H(2^a) = H(a)'$,
 (3) $H(3 \cdot 5^e) = \{\langle u, v \rangle : u <_O 3 \cdot 5^e \ \& \ v \in H(u)\} = \{\langle u, v \rangle : \exists n \, (u \leq_O \varphi_e(n) \ \& \ v \in H(u))\}$.

Remark: By (3), if a is a notation for ω or some other limit ordinal, then $H(a)$ gives a "uniform" upper bound for the sets $H(u)$ with $u <_O a$.

Each finite ordinal n has a unique notation a, and $H(a) = \emptyset^{(n)}$. For ω, there are infinitely many different notations a, but all give the same set $H(a)$. The same is true for $\omega + n$. For $\alpha \geq \omega \cdot 2$, different notations a for α yield different sets $H(a)$, but all are Turing equivalent. The proof of this is more complicated than might be expected, involving several steps.

Lemma 5.1 *There is a partial computable function which assigns to each pair (x, y) of elements of O such that $x \leq_O y$, an index for $H(x)$ as a set computable in $H(y)$.*

Proof: First, note that if $x, y \in O$, with $x \leq_O y$, then by shortening the stack of 2's in y, we arrive at $z \leq_O y$ and $n < \omega$ such that $|z| + n = y$ and we are in one of two cases.

Case 1: $x = z$,

Case 2: $|z|$ is a limit ordinal and $x <_O z$.

By Corollary 2.2, for any $n \in \omega$, we can find a uniform index for X as a set computable in $X^{(n)}$. This takes care of Case 1. In Case 2, we have $u \in H(x)$ if and only if $\langle x, u \rangle \in H(z)$, so we can find an index for $H(x)$ as a set computable in $H(z)$. Then, using Corollary 2.2 again, we obtain an index for $H(x)$ as a set computable in $H(y)$.

Lemma 5.2 (Spector) *There is a partial computable function f such that for each $a \in O$, $f(a)$ is an index for $\{b \in O : |b| < |a|\}$ as a set computable in $H(a)'$.*

Proof: The function is defined by computable transfinite recursion on ordinal notation.

Case 1: Suppose $a = 1$.

Let $f(1)$ be an index for $\emptyset$ relative to $\emptyset'$.

Case 2: Suppose $a = 2^b$.

We say how $f(2^b)$ is defined in terms of $f(b)$. Note that $(d \in O \ \& \ |d| < |2^b|)$ if and only if one of the following holds:
    (a) $d = 1$,
    (b) d has the form 2^c, where $c \in O$ and $|c| < |b|$, or
    (c) d has the form $3 \cdot 5^e$, where for all n, $\varphi_e(n) \in O$, $|\varphi_e(n)| < |b|$, and $\varphi_e(n) <_O \varphi_e(n+1)$.

Determining whether d satisfies (a) is trivial.
Assuming that $f(b)$ is an index for

$$\{c \in O : |c| < |b|\}$$

5.1. THE HYPERARITHMETICAL HIERARCHY

relative to $H(b)'$, we can determine whether $d = 2^c$ satisfies (b), using $H(2^b)$, or, by Lemma 5.1, using $H(2^b)'$.

We can determine whether $d = 3 \cdot 5^e$ satisfies (c), using $H(2^b)'$, since for each n, we can determine whether $\varphi_e(n) \in O$ and $|\varphi_e(n)| < |b|$, using $H(2^b)$, and we can determine whether $\varphi_e(n) <_O \varphi_e(n+1)$, using $H(2)$ or, by Lemma 5.1, using $H(2^b)$. Now, we can find an index for

$$\{d \in O : |d| < |2^b|\}$$

as a set computable in $H(2^b)'$, and this is $f(2^b)$.

Case 3: Let $a = 3 \cdot 5^e$.

Given the sequence $f(\varphi_e(n))_{n<\omega}$, we define $f(3 \cdot 5^e)$ as follows. We have

$$d \in O \ \& \ |d| < |3 \cdot 5^e| \ \Leftrightarrow \ \exists n \, (d \in O \ \& \ |d| < |\varphi_e(n)|) \, .$$

Assuming that $f(\varphi_e(n))$ is an index for

$$\{d \in O : |d| < |\varphi_e(n)|\}$$

as a set computable in $H(\varphi_e(n))'$, we can determine whether d is a notation for an ordinal less than $|\varphi_e(n)|$, using $H(\varphi_e(n))'$ or, by Lemma 5.1, $H(3 \cdot 5^e)$. The procedure is uniform in n. Then we can determine whether d is in the union, using $H(3 \cdot 5^e)'$. It follows that we can compute an index for

$$\{d \in O : |d| < |3 \cdot 5^e|\}$$

as a set computable in $H(3 \cdot 5^e)'$, and this is $f(3 \cdot 5^e)$.

Corollary 5.3 (Spector) *There are partial computable functions that assign, to each $b \in O$, indices for*

$$\{a \in O : |a| \leq |b|\} \, , \ \{a \in O : |a| = |b|\}$$

as sets computable in $H(2^{2^b})$.

Proof: For $b \in O$, we have

$$(a \in O \ \& \ |a| \leq |b|) \ \Leftrightarrow \ (a \in O \ \& \ |a| < |2^b|) \, ,$$

and

$$(a \in O \ \& \ |a| = |b|) \ \Leftrightarrow \ (a \in O \ \& \ |a| \leq |b| \ \& \ |a| \not< |b|) \, .$$

Theorem 5.4 (Spector) *There is a partial computable function f such that for $a, b \in O$ with $|a| \leq |b|$, $f(a,b)$ is an index for $H(a)$ as a set computable in $H(b)$.*

Proof: Formally, we define, by computable transfinite recursion on ordinal notation, a partial computable function that assigns to each $b \in O$ an index relative to $H(b)$ for a function with values appropriate for $f(a,b)$.

Let u_0 be a single index for $\emptyset$ as a set computable in all sets X.

(1) Consider $b = 1$. Let $f(1,1) = u_0$. For $x \neq 1$, $f(x,1)$ is undefined.

(2) Consider $b = 2^d$. Let $f(1,2^d) = u_0$. Assuming that $c, d \in O$ and $f(c,d)$ is an index for $H(c)$ relative to $H(d)$, we determine an index for $H(c)'$ relative to $H(d)'$ as in Theorem 2.3, and we let this be $f(2^c, 2^d)$. Assuming that $f(3 \cdot 5^e, d)$ is an index for $H(3 \cdot 5^e)$ relative to $H(d)$, we determine an index for $H(3 \cdot 5^e)$ relative to $H(d)'$, as in Corollary 2.2, and we let this be $f(3 \cdot 5^e, 2^d)$.

(3) Consider $b = 3 \cdot 5^e$.

(a) Let $f(1, 3 \cdot 5^e) = u_0$.

(b) We must say how to compute $f(2^c, 3 \cdot 5^e)$. Given $d <_O 5 \cdot 3^e$, we can find an index for $\{a \in O : |a| = |d|\}$ as a set computable relative to $H(2^{2^d})$, by Corollary 5.3, and we can find an index for the same set relative to $H(3 \cdot 5^e)$, by Lemma 5.1. We can determine whether $x \in H(2^c)$ as follows. We enumerate the set $\{d \in O : d <_O 3 \cdot 5^e\}$, searching for the unique element d such that $|2^c| = |d|$. Next, we find the index $f(c,d)$ for $H(c)$ as a set computable in $H(d)$, and then an index for $H(2^c) = H(c)'$ as a set computable in $H(d)'$. We replace this by an index for H(c)' as a set computable in $H(3 \cdot 5^e)$, say $i(d)$. We check whether x is in the set with index $i(d)$. Now, $f(2^c, 3 \cdot 5^e)$ is not $i(d)$—we cannot compute $i(d)$ without the oracle for $H(3 \cdot 5^e)$, and f is supposed to be computable. Rather, $f(2^c, 3 \cdot 5^e)$ is an index for the whole procedure, which first locates d, and then determines and simulates $i(d)$.

(c) Finally, we must say how to compute $f(3 \cdot 5^{e'}, 3 \cdot 5^e)$. We can determine whether $\langle u, v \rangle \in H(3 \cdot 5^{e'})$, using $H(3 \cdot 5^e)$ as follows. First, check that $u <_O 3 \cdot 5^{e'}$ (we can do this using $H(2)$, so we can also use $H(3 \cdot 5^e)$). Next, find $d <_O 3 \cdot 5^e$ such that $|u| = |d|$, as above. Assuming that $f(u, d)$ is an index for $H(u)$ relative to $H(d)$, find an index $i(d)$ for $H(u)$ relative to $H(3 \cdot 5^e)$, as in Lemma 5.1. Finally, we apply the procedure with this index to see if $v \in H(u)$.

Corollary 5.5 *If $x, y \in O$ and $|x| = |y|$, then $H(x) \equiv_T H(y)$.*

We have already defined the classes Σ_n^0, Π_n^0, and Δ_n^0, for $1 \leq n < \omega$. Now, following Kleene, we define classes Σ_α^0, Π_α^0, and Δ_α^0, for all computable ordinals $\alpha \geq \omega$. For infinite α, a relation is said to be Σ_α^0, Π_α^0, or Δ_α^0 if it is, respectively, c.e., co-c.e., or computable relative to $H(a)$ for *some* $a \in O$ with $|a| = \alpha$. By Corollary 5.5, such a relation will be c.e., co-c.e., or computable relative to $H(a)$ for *every* $a \in O$ with $|a| = \alpha$. A total function is said to be Δ_α^0 if it is computable relative to $H(a)$ for some, or every $a \in O$ with $|a| = \alpha$, and similarly for a partial function. By the relativization of Theorem 1.5 to $H(a)$, a (partial) function $f(\overline{x})$ is Δ_α^0 if and only if the relation $f(\overline{x}) = y$ is Σ_α^0. Finally, a relation or function is said to be *hyperarithmetical* if it is Δ_α^0 for some computable ordinal α.

Remark: There is a lack of uniformity in the definition above when we pass from finite to infinite computable ordinals. For finite α, say $\alpha = n$, the

5.2. THE ANALYTICAL HIERARCHY

Σ^0_n relations are the ones that are c.e. relative to $\emptyset^{(n-1)}$, and $\emptyset^{(n-1)} \equiv_T H(a)$, where $|a|$ is $n-1$, *not* n. (There is the same lack of uniformity for Π^0_α and Δ^0_α relations.) We might be tempted to change the definition of the arithmetical hierarchy, and say that for finite as well as infinite α, the Σ^0_α relations are those which are c.e. relative to $H(a)$, where $|a| = \alpha$ (with similar changes in the definition for Π^0_α and Δ^0_α relations). However, the reasons for leaving things as they are are more compelling. The Σ^0_n and Π^0_n relations are expressible with n alternations of quantifiers, and we shall see that *this* pleasing match extends naturally through all levels of the hyperarithmetical hierarchy.

It is convenient to have a uniform way to refer to a canonical complete Δ^0_α set associated with a given notation a for a computable ordinal α, whether α is finite or infinite. For $a \in O$, we let

$$\Delta^0_a = \begin{cases} H(a) & \text{if } |a| \geq \omega \,, \\ H(b), \text{ where } |b|+1 = a, & \text{if } 0 < |a| < \omega \,. \end{cases}$$

If we are working along a particular path in O, then for an ordinal α with a notation a on this path, we may identify α with a. In this spirit, we may write Δ^0_α for the complete Δ^0_α set Δ^0_a.

The fact that the hyperarithmetical hierarchy is a proper hierarchy should be clear from the definition and the arguments we gave for the arithmetical hierarchy. The importance and usefulness of the hyperarithmetical hierarchy will become clear as we continue. Now, however, we turn to a hierarchy which includes still more complicated sets and relations.

5.2 The analytical hierarchy

We defined the *arithmetical* relations to be the members of the least class which includes the computable relations and is closed under complements and projections. The relations that we had in mind involved only *individual* variables, ranging over numbers. The class of *analytical* relations is defined in the same way except that we consider *function* variables, ranging over total functions from ω to ω, in addition to the individual variables. We use $f_0, f_1, f_2, \ldots$, or f, g, h, etc. as function variables.

We begin by broadening the class of computable relations. A relation

$$R(x_1, \ldots, x_n; f_1, \ldots, f_m) \,,$$

on *numbers and functions*, is said to be *computable* if there is some e such that for all appropriate tuples $(x_1, \ldots, x_n; f_1, \ldots, f_m)$,

$$\varphi_e^{f_1, \ldots, f_m}(x_1, \ldots, x_n) = \begin{cases} 1 & \text{if } R(x_1, \ldots, x_n; f_1, \ldots, f_m) \\ 0 & \text{otherwise} \end{cases}$$

In thinking of the oracles here, we identify each total function with its graph, and we have already described how to replace a relation by a set, and how to replace a tuple of sets by a single set.

The *analytical* relations, on numbers and functions, are the members of the least class that includes this broader class of computable relations and is closed under complements and projections on both individual and function variables.

Remark: We could define a more general class of *arithmetical* relations, on numbers and functions, by starting with the class of computable relations defined above and closing under complements and projections on *individual* variables (see [134]). However, we are primarily interested in relations $S(\overline{x})$ just on numbers, and we would not obtain any further arithmetical relations of this sort by adopting the more general definiton.

It is not difficult to see that a relation is analytical just in case it has a definition consisting of a string of quantifiers, over individual and function variables, followed by a computable relation, of the new, more general kind. Such a definition can be re-written in various ways without changing the meaning. We can insert extra quantifiers, and we can collapse like quantifiers. Less trivially, we can bring a number quantifier across a function quantifier. In particular, we can replace

$$\forall x \, \exists f \, R(x, \ldots, f \ldots)$$

by

$$\exists g \, \forall x \, R'(x, \ldots, g \ldots) \, ,$$

if we think of the function $f(y)$ corresponding to x as $g(\langle x, y \rangle)$ and let

$$R'(x \ldots, g, \ldots) \Leftrightarrow R(x, \ldots, f, \ldots) \, .$$

From this, it follows (by considering negations) that we can replace

$$\exists x \, \forall f \, R(x, \ldots, f, \ldots)$$

by

$$\forall g \, \exists x \, R'(x, \ldots, g, \ldots) \, .$$

We can replace a string of number quantifiers

$$Q_1 y_1 \cdots Q_n y_n$$

by either $\exists f \, \forall x$ or $\forall f \, \exists x$ (we add a number quantifier of the desired kind at the end and take care of the rest by absorption and bringing number quantifiers across function quantifiers).

We can see that a relation is analytical if it has a definition of the form

$$Q_1 f_1 \cdots Q_n f_n Q_{n+1} y \, R \, ,$$

where R is computable and the quantifiers Q_i alternate between $\exists$ and $\forall$, with all but the last ranging over functions. In terms of these definitions, in "normal" form, we can now describe the *analytical hierarchy*.

We say that S is Σ_n^1 if it has a definition with $n+1$ alternating quantifiers, where all but the last ranges over functions, and the first is $\exists$. We say that S

5.2. THE ANALYTICAL HIERARCHY

is Π_n^1 if $\neg S$ is Σ_n^1, or, equivalently, if S has a definition with $n + 1$ alternating quantifiers, where all but the last ranges over functions and the first is $\forall$. A relation is Δ_n^1 if it is both Σ_n^1 and Π_n^1. (The "1" in the notation indicates that there variables ranging over functions, which, as we mentioned in Chapter 2, are sometimes called objects of *type* 1).

In view of the manipulations of quantifiers described above, we have the following.

Remarks: (1) If $S(\overline{x})$ is arithmetical, then it is Δ_1^1.
(2) If $S(\overline{x}) \Leftrightarrow \forall f \, \exists y \, R(\overline{x}, y, f)$, where $R(\overline{x}, y, f)$ is Π_1^1, then $S(\overline{x})$ is also Π_1^1.

Consider an analytic relation $S(\overline{x})$ defined by

$$Q_1 f_1 \cdots Q_n f_n Q_{n+1} y \, R(\overline{x}, y, f_1, \ldots, f_n) ,$$

where $R(\overline{x}, y, f_1, \ldots, f_n)$ is a computable relation on numbers and functions. We shall alter the definition further, replacing the relation

$$Q_{n+1} y \, R(\overline{x}, y, f_1, \ldots, f_n) ,$$

on numbers and functions, by a relation just on numbers, that is either c.e. or co-c.e. For a function f and a number t, we use $f|t$ to denote either the finite sequence $(f(0), \ldots, f(t-1))$, or its code. Suppose

$$\chi_R(\overline{x}, y, f_1, \ldots, f_n) = \varphi_e^{f_1, \ldots, f_n}(\overline{x}, y) .$$

If $Q_{n+1} = \exists$, then we have

$$\exists y \, R(\overline{x}, y, f_1, \ldots, f_n) \Leftrightarrow \exists t \, R^*(\overline{x}, f_1|t, \ldots, f_n|t) ,$$

where $R^*(\overline{x}, u_1, \ldots, u_n)$ is a c.e. relation on numbers saying that the u_i are (codes for) finite sequences and

$$\exists y \, \exists s \, \varphi_{e,s}^{u_1, \ldots, u_n}(\overline{x}, y) \downarrow = 1 .$$

If $Q_{n+1} = \forall$, then

$$\forall y \, R(\overline{x}, y, f_1, \ldots, f_n) \Leftrightarrow \forall t \, R^*(\overline{x}, f_1|t, \ldots, f_n|t) ,$$

where $R^*(\overline{x}, u_1, \ldots, u_n)$ is a co-c.e. relation on numbers saying that the u_i are (codes for) finite sequences, and

$$\forall y \, \forall s \, (\varphi_{e,s}^{u_1, \ldots, u_n}(\overline{x}, y) \downarrow = v \to v = 0) .$$

To arrive at a nice system of indices for the Σ_n^1 relations, in which every natural number is an index, we reverse the process above. For an arbitrary c.e. relation $P(\overline{x}, u_1, \ldots, u_n)$ on numbers, we derive a computable relation $R(\overline{x}, y, f_1, \ldots, f_n)$, on numbers and functions, such that

$$\exists y \, R(\overline{x}, y, f_1, \ldots, f_n) \Leftrightarrow \exists t \, P(\overline{x}, f_1|t, \ldots, f_n|t) .$$

We let $R(\bar{x}, y, f_1, \ldots, f_n)$ hold if there exists $t \leq y$ such that the tuple of numbers $(\bar{x}, f_1|t, \ldots, f_n|t)$ appears in P by stage y (we compute $f_i|t$ using the oracle for f_i). Clearly, R is a computable relation on numbers and functions with the desired property. Similarly, for an arbitrary co-c.e. relation $P(\bar{x}, u_1, \ldots, u_n)$ on numbers, we can derive a computable relation $R(\bar{x}, y, f_1, \ldots, f_n)$, on numbers and functions, such that

$$\forall y\, R(\bar{x}, y, f_1, \ldots, f_n) \Leftrightarrow \forall t\, P(\bar{x}, f_1|t, \ldots, f_n|t) \ .$$

To show that the analytical hierarchy does not collapse, we proceed as we did for the arithmetical hierarchy. That is, we first show the existence of enumeration relations at all levels.

Theorem 5.6 (Kleene) *For each $n \geq 1$ and each tuple of variables $\bar{x}$, there is a Σ_n^1 enumeration relation $E^1(e, \bar{x})$ that, as e varies, yields all of the Σ_n^1 relations in variables $\bar{x}$. Similarly, there is a Π_n^1 enumeration relation that yields all of the Π_n^1 relations in variables $\bar{x}$.*

Proof: As in the proof of the analogous Theorem 2.4, we need only consider the last part of the defining expression—the part of form

$$Q_{n+1} y\, R(\bar{x}, y, f_1, \ldots, f_n) \ ,$$

where $R(\bar{x}, y, f_1, \ldots, f_n)$ is computable. We deal with the case where Q_{n+1} is $\exists$, the other case being the same except for the insertion of a negation. Above, we saw that

$$\exists y\, R(\bar{x}, y, f_1, \ldots, f_n) \Leftrightarrow \exists t\, R^*(\bar{x}, f_1|t, \ldots, f_n|t) \ ,$$

where $R^*(\bar{x}, u_1, \ldots, u_n)$ is a c.e. relation on numbers. We have

$$R^*(\bar{x}, u_1, \ldots, u_n) \Leftrightarrow \exists z\, T_{m+n}(e, u_1, \ldots, u_n, z) \ ,$$

for some e. Therefore,

$$\exists t\, R(\bar{x}, f_1|t, \ldots, f_n|t) \Leftrightarrow \exists t\, (\exists z \leq t)\, T_{m+n}(e, \bar{x}, f_1|t, \ldots, f_n|t, z) \ .$$

We define $E^1(e, \bar{x})$ to be

$$Q_1 f_1 Q_2 f_2 \cdots Q_n f_n \exists t\, (\exists z \leq t)\, T_{m+n}(e, \bar{x}, f_1|t, \ldots, f_n|t, z) \ .$$

Corollary 5.7 (Kleene) *(a) For each n, there is a Σ_n^1 relation that is not Π_n^1. (b) For each n, there is a Π_n^1 relation that is not Σ_n^1.*

Proof: (a) Consider the Σ_n^1 $(m+1)$-ary enumerating relation $E^1(e, x_1, \ldots, x_m)$. The m-ary relation $E^1(x_1, x_1, \ldots, x_m)$ is Σ_n^1. Suppose that it is also Π_n^1. Then

5.2. THE ANALYTICAL HIERARCHY

$\neg E^1(x_1, x_1, \ldots, x_m)$ is Σ_n^1, so it can be expressed as $E^1(e, x_1, \ldots, x_m)$, for some e. Letting x_1 be this e, we have

$$E^1(e, e, x_2, \ldots, x_m) \Leftrightarrow \neg E^1(e, e, x_2, \ldots, x_m) ,$$

a contradiction.

We can prove (b) from (a), by taking negations.

Our interest from now on is in the Σ_1^1, Π_1^1, or Δ_1^1 relations on numbers. It will be convenient to have special names for the Σ_1^1 and Π_1^1 enumeration relations from Theorem 5.6, in the case where $n = 1$. We use E_Σ^1 and E_Π^1. A Σ_1^1 *index* for a Σ_1^1 relation $S(\overline{x})$ is a number e such that

$$S(\overline{x}) \Leftrightarrow E_\Sigma^1(e, \overline{x}) .$$

We note (following the proof of Theorem 5.6) that a Σ_1^1 index for $S(\overline{x})$ can be effectively obtained from a computable index for a relation R such that

$$S(\overline{x}) \Leftrightarrow \exists f \, \forall y \, R(\overline{x}, y, f) ,$$

and vice versa. We define Π_1^1 indices for Π_1^1 relations similarly. Finally, a Δ_1^1 *index* for a Δ_1^1 relation is a number $e = \langle e_1, e_2 \rangle$ such that e_1 is a Σ_1^1 index and e_2 is a Π_1^1 index for the relation.

Theorem 5.8 *(a) If $S(y_1, \ldots, y_m)$ is Σ_1^1, then so is $S(g_1(\overline{x}), \ldots, g_m(\overline{x}))$, for any sequence of computable functions $(g_1, \ldots, g_m)$.*
(b) If $S_1(\overline{x})$, $S_2(\overline{x})$ are Σ_1^1, then so are $(S_1(\overline{x}) \,\&\, S_2(\overline{x}))$, $(S_1(\overline{x}) \vee S_2(\overline{x}))$.
(c) If $S(\overline{x}, y)$ is Σ_1^1, then so are $\exists y \, S(\overline{x}, y)$, $\forall y \, S(\overline{x}, y)$.
The same is true when Σ_1^1 is replaced everywhere by Π_1^1.

Remarks: In each case, a Σ_1^1 index for the new relation can be obtained effectively from Σ_1^1 indices for the old relations, plus, for (a), computable indices for the computable functions. In (a), we could take arbitrary arithmetical functions, not just computable ones.

Proof of Theorem 5.8: (a) Suppose

$$S(y_1, \ldots, y_m) \Leftrightarrow \exists f \, \forall z \, R(y_1, \ldots, y_m, z, f) ,$$

where R is a computable relation with a function variable. Then

$$S(g_1(\overline{x}), \ldots, g_m(\overline{x})) \Leftrightarrow$$
$$\exists f \, \forall z \, R(g_1(\overline{x}), \ldots, g_m(\overline{x}), z, f) \Leftrightarrow$$
$$\exists f \, \forall z \, R^*(\overline{x}, z, f) ,$$

where R^* is the computable relation defined by

$$R^*(\overline{x}, z, f) \Leftrightarrow R(g_1(\overline{x}), \ldots, g_m(\overline{x}), z, f) .$$

(b) Suppose
$$S_i(\overline{x}) \Leftrightarrow \exists f \, \forall y \, R_i(\overline{x}, y, f) ,$$
where R_i is computable. Then we have
$$(S_1(\overline{x}) \,\&\, S_2(\overline{x})) \Leftrightarrow \exists f_1 \, \exists f_2 \, \forall y_1 \, \forall y_2 \, (R_1(\overline{x}, y_1, f_1) \,\&\, R_2(\overline{x}, y_2, f_2)) ,$$
and
$$(S_1(\overline{x}) \vee S_2(\overline{x})) \Leftrightarrow \exists f \, \forall y_1 \, \forall y_2 \, (R_1(\overline{x}, y_1, f) \vee R_2(\overline{x}, y_2, f)) .$$
Using the quantifier manipulations discussed earlier, we can replace each of these definitions by one of the form $\exists g \, \forall z \, R^*(\overline{x}, z, g)$, where R^* is computable.

(c) This follows from the comments immediately preceding our definition of Σ_1^1 and Π_1^1 relations.

In each case, we can pass effectively from indices for the computable relations appearing in the definitions of the old Σ_1^1 relations to a computable index for the computable relation appearing in the definition of the new Σ_1^1 relation. From this, it follows that we can pass effectively from Σ_1^1 indices of the old relations to a Σ_1^1 index for the new relation.

The results for Π_1^1 relations follow from those for Σ_1^1 relations by taking complements.

Using Theorem 5.8, we obtain the following useful result on Π_1^1 functions.

Corollary 5.9 *Let f be a partial function whose graph is Π_1^1. If $dom(f)$ is Δ_1^1, then the graph of f and $ran(f)$ are both Δ_1^1.*

Proof: To show that the graph of f is Δ_1^1, it is enough to show that it is Σ_1^1, or that the complement is Π_1^1. We have $f(x) \neq y$ if and only if
$$x \notin dom(f) \vee (x \in dom(f) \,\&\, \exists z \, (z \neq y \,\&\, f(x) = z)) ,$$
and using Theorem 5.8, we can see that this is Π_1^1. To show that $ran(f)$ is Δ_1^1, we note that
$$y \in ran(f) \Leftrightarrow \exists x \, f(x) = y .$$
The graph of f is Σ_1^1 and Π_1^1, so by Theorem 5.8, $ran(f)$ is also Σ_1^1 and Π_1^1.

Earlier, we established that every arithmetical relation is Δ_1^1. In the remainder of this section, we extend this, showing that every hyperarithmetical relation is Δ_1^1. We use the following.

Proposition 5.10 *If X is a Δ_1^1 set, S is is a relation c.e. relative to X, then S is Δ_1^1. Furthermore, we can compute a Δ_1^1 index for S from a Δ_1^1 index for X and a c.e. index for S relative to X.*

Proof: Recall that D_n is the canonical index for the nth finite set on a fixed computable list. From a c.e. index for S relative to X, we obtain an index for a c.e. relation W such that
$$S(\overline{x}) \Leftrightarrow \exists u \, \exists v \, (W(\overline{x}, u, v) \,\&\, D_u \subseteq X \,\&\, D_v \subseteq \neg X) .$$

5.2. THE ANALYTICAL HIERARCHY

From a Δ_1^1 index for X, we obtain Δ_1^1 indices for the relations $D_u \subseteq X$ and $D_v \subseteq \neg X$. We note that

$$D_u \subseteq X \Leftrightarrow \forall y \, (y \notin D_u \vee y \in X),$$

and

$$D_v \subseteq \neg X \Leftrightarrow \forall y \, (y \notin D_v \vee y \notin X),$$

and we apply Theorem 5.8. By further applications of Theorem 5.8, we obtain Σ_1^1 and Π_1^1 indices for S. We combine these to produce the required Δ_1^1 index for S.

Proposition 5.10 yields, incidentally, the following.

Corollary 5.11 *A relation R computable relative to a Δ_1^1 set X is itself Δ_1^1. Moreover a Δ_1^1 index for R can be determined effectively from a computable index for R relative to X and a Δ_1^1 index for X.*

To show that every hyperarithmetical relation is Δ_1^1, it is sufficient to prove the following.

Proposition 5.12 *For each $a \in O$, the set $H(a)$ is Δ_1^1. In fact, there is a partial computable function f such that for all $a \in O$, $f(a)$ is a Δ_1^1 index for $H(a)$.*

Proof: We prove the first statement by proving the second. The function f is defined by computable transfinite recursion on ordinal notation.

(a) Let f(1) be a Δ_1^1 index for $\emptyset$.

(b) Suppose that $f(b)$ is a Δ_1^1 index for $H(b)$, and recall the partial computable function g from Proposition 5.10, which, given a Δ_1^1 index for a set X, yields a Δ_1^1 index for X'. Then let $f(2^b) = g(f(b))$.

(c) Suppose that $3 \cdot 5^e \in O$, and for all $u <_O 3 \cdot 5^e$, $f(u)$ is a Δ_1^1 index for $H(u)$. Letting E_Σ^1, E_Π^1 be the enumeration relations for Σ_1^1 and Π_1^1 sets, we have

$$\langle u, v \rangle \in H(3 \cdot 5^e) \Leftrightarrow$$
$$u <_O^+ 3 \cdot 5^e \ \& \ E_\Sigma^1(f(u)_1, v) \Leftrightarrow$$
$$u <_O^+ 3 \cdot 5^e \ \& \ E_\Pi^1(f(u)_2, v).$$

Applying Theorem 5.8, we see that $H(3 \cdot 5^e)$ is both Σ_1^1 and Π_1^1, and we can compute the indices. Let $f(3 \cdot 5^e)$ be this pair of indices.

We have shown that every hyperarithmetical relation is Δ_1^1. In the next section, we shall prove the converse.

5.3 The Kleene-Brouwer ordering

In this section, we prove several results simultaneously. In particular, we show that every Δ^1_1 set is hyperarithmetical, and we show that the set O is Π^1_1 complete. We begin by linearly ordering the set $\omega^{<\omega}$, consisting of finite sequences of natural numbers, as follows. Let $\sigma < \tau$ if either τ is a proper initial segment of σ, or else neither is an initial segment of the other, and at the first place where they differ, the entry in σ is smaller than that in τ. Thus, if $\omega^{<\omega}$ is visualized as a tree growing downwards, with $\sigma(n)$ to the left of $\sigma(n+1)$, then the nodes are ordered from left to right, with the added stipulation that if one node is below another, then it is smaller. The specific ordering that we have described is the *Kleene-Brouwer ordering*.

Lemma 5.13 *If P is a subtree of $\omega^{<\omega}$, then P is well ordered under the Kleene-Brouwer ordering (or its restriction) if and only if P has no path.*

Proof: If P is well ordered under the Kleene-Brouwer ordering, then there is no infinite decreasing sequence in P, and this clearly implies that the tree P has no infinite path. If P is not well ordered, then there is an infinite decreasing sequence, say

$$\sigma_0 > \sigma_1 > \sigma_2 > \dots .$$

We can show, by induction on x, that the value of $\sigma_k(x)$ is constant for all sufficiently large k. If π is the function such that for all x, $\pi(x) = \lim_{k \to \infty} \sigma_k(x)$, then π is a path.

Theorem 5.14 (Kleene, Spector) *Let S be a Π^1_1 set. There is a computable function which assigns to each x in ω an index for a computable linear ordering $\mathcal{A}_x$, such that*

$$x \in S \Leftrightarrow \mathcal{A}_x \text{ is a well ordering} .$$

Proof: We have

$$x \in S \Leftrightarrow \forall f\, \exists t\, R(x, f|t) ,$$

where $R(x, u)$ is a c.e. relation on numbers. Let A_x be the subtree of ω^ω consisting of those σ such that if $\sigma)$ has length s, then for all $t \leq s$, the pair $(x, \sigma|t)$ is not in our stage s approximation of $R(x, u)$. Clearly, A_x is a computable subtree of $\omega^{<\omega}$. Moreover,

$$x \in S \Leftrightarrow A_x \text{ has no path} .$$

Let $\mathcal{A}_x = (A_x, <_x)$, where $<_x$ is the restriction of the Kleene-Brouwer ordering to A_x. By Lemma 5.13,

$$A_x \text{ has no path} \Leftrightarrow \mathcal{A}_x \text{ is a well ordering} .$$

Given x, we can find a computable index for $\mathcal{A}_x$. This completes the proof of Theorem 5.14.

5.3. THE KLEENE-BROUWER ORDERING

In Chapter 4, we proved an important lemma (Lemma 4.13), saying that there is a partial computable function g such that if x is a computable index for a linear ordering $\mathcal{A}$, then $g(x) \in O$ if and only if $\mathcal{A}$ is a well ordering. We combine this with what we have just done. Let S be an arbitrary Π_1^1 set. Given a number x, we can pass effectively to an index for a computable linear ordering $\mathcal{A}$, as in Theorem 5.14, where

$$x \in S \Leftrightarrow \mathcal{A} \text{ is a well ordering}.$$

We can pass effectively from the index, say y, for $\mathcal{A}$ to a number $z = g(y)$, as in Lemma 4.13, such that

$$\mathcal{A} \text{ is a well ordering} \Leftrightarrow z \in O.$$

We have established the following.

Corollary 5.15 *For any Π_1^1 set S, $S \leq_m O$.*

We cannot conclude from Corollary 5.15 that O is a complete Π_1^1 set, for we have not yet shown that O is Π_1^1. In the remainder of the section, for $\Lambda \subseteq O$, we let Λ^* denote the set $\{a \in O : (\exists x \in \Lambda)(|a| < |x|)\}$.

Lemma 5.16 *Suppose $\Lambda \subseteq O$. If Λ is Π_1^1, or Σ_1^1, then so is Λ^*.*

Proof: By Lemma 5.2, Corollary 5.11, and Proposition 5.12, there is a partial computable function g such that for $x \in O$, $g(x)$ is a Π_1^1 index for the set $\{a \in O : |a| < |x|\}$. We have

$$a \in \Lambda^* \Leftrightarrow \exists x\, (\, x \in \Lambda \,\&\, E_\Pi^1(g(x), a)\,)\ .$$

If Λ is Π_1^1, then by Theorem 5.8, Λ^* is Π_1^1. Similarly, if Λ is Σ_1^1, then so is Λ^*.

Let $\Lambda \subseteq O$. Note that either $\Lambda^* = O$ or else $\Lambda^* = \{a \in O : |a| < \beta\}$, for some computable ordinal β. If $\Lambda^* = \{a \in O : |a| < \beta\}$, then by Lemma 5.2, it is hyperarithmetical, and therefore Δ_1^1. Let S be Π_1^1, say $S \leq_m O$ via f. Note that $S \leq_m ran(f|S)^*$. By Theorem 5.8, $ran(f|S)$ is Π_1^1, so by Lemma 5.16, $ran(f|S)^*$ is Π_1^1. By Corollary 5.7, there is some Π_1^1 set S which is not Δ_1^1. For such S, if $S \leq_m O$ via f, then $(ran(f|S))^*$ cannot be Δ_1^1, so it must be O. Therefore, O is Π_1^1.

Proposition 5.17 *The set O is Π_1^1 complete.*

Proof: This follows from Corollary 5.15, together with the fact that O is Π_1^1, which we have just established.

At last, we can complete the result saying that the Δ_1^1 and hyperarithmetical sets are the same.

Theorem 5.18 (Kleene) *For any set S, the following are equivalent:*
(a) S is Δ_1^1,
(b) $S \leq_m \{a \in O : |a| < \beta\}$, *for some computable ordinal β,*
(c) S *is hyperarithmetical.*

Proof: It follows from Proposition 5.12 that (c) $\Rightarrow$ (a). The argument above shows that (a) $\Rightarrow$ (b) and (b) $\Rightarrow$ (c).

The next result says that Π_1^1 relations have Π_1^1 "Skolem" functions, or "choice" functions.

Theorem 5.19 (Single-Valuedness) *Let Q be a Π_1^1 binary relation. Then there is a Π_1^1 function F such that $F \subseteq Q$ and $\forall x\, [\exists y\, Q(x,y) \to Q(x, F(x))]$.*

Proof: Say $Q \leq_m O$ via g. For each x, we define $F(x) = y$ so that the ordinal $|g(x,y)| = \beta$ is as small as possible, and y is least such that $|g(x,y)| = \beta$. We give the definition so as to make clear the fact that the function is Π_1^1. By Corollary 5.3, Proposition 5.12, and Corollary 5.11, there is a partial computable function h such that for $u \in O$, $h(u)$ is a Σ_1^1 index for $\{v \in O : |v| \leq |u|\}$. Let $F(x) = y$ if and only if
(1) $Q(x,y)$
(2) $\forall z\, [Q(x,z) \to E_\Sigma^1(h(g(x,z)), g(x,y)]$
(3) $\forall z\, [Q(x,z)\ \&\ E_\Sigma^1(h(g(x,y)), g(x,z)) \to y \leq z)]$.
It is clear from the definition that F has all of the desired properties.

Proposition 5.20 *Let f be a partial computable function defined on all of O. Let $\Gamma = ran(f|O)$, and let $\Gamma_\beta = \{f(y) : y \in O\ \&\ |y| < \beta\}$. Then Γ is Π_1^1, and for any Δ_1^1 set $\Lambda \subseteq \Gamma$, there is a computable ordinal β such that $\Lambda \subseteq \Gamma_\beta$.*

Proof: Using Theorem 5.8, we can see that Γ is Π_1^1. By applying Theorem 5.19 to the Π_1^1 relation
$$Q = \{(x,y) : y \in O\ \&\ f(y) = x\},$$
we obtain a Π_1^1 function g such that for $x \in \Gamma$, $f(g(x)) = x$. Now, $g|\Lambda$ is a Π_1^1 function with Δ_1^1 domain. By Corollary 5.9, $ran(g|\Lambda)$ is Δ_1^1. By Lemma 5.16,
$$ran(g|\Lambda) \subseteq \{y \in O : |y| < \beta\},$$
for some computable ordinal β. For this β, we have $\Lambda \subseteq \Gamma_\beta$.

5.4 Relativizing

Here we indicate how to relativize the hyperarithmetical hierarchy to an arbitrary set X. We start by defining sets $H(a)(X)$, for all $a \in O$.
(1) $H(1)(X) = X$,

5.4. RELATIVIZING

(2) $H(2^b)(X) = (H(b)(X))'$, and

(3) $H(3 \cdot 5^e)(X) = \{\langle u, v \rangle : \exists n \, (u \leq_O \varphi_n(e) \, \& \, v \in H(u)(X)\}$.

For finite n, a relation is $\Delta_n^0(X)$, $\Sigma_n^0(X)$, or $\Pi_n^0(X)$ if it is, respectively, computable, c.e., or co-c.e. relative to $H(a)(X)$, where $|a| = n - 1$. For an infinite computable ordinal α, a relation is $\Delta_\alpha^0(X)$, $\Sigma_\alpha^0(X)$, or $\Pi_\alpha^0(X)$ if it is, respectively, computable, c.e., or co-c.e. relative to $H(a)(X)$, where $|a| = \alpha$.

Note: There may be well orderings that are computable relative to X but not computable. Here, we are ignoring the ordinals representing the types of such well orderings. When we relativize, it is to learn more about the hyperarithmetical hierarchy itself. We would proceed differently if we wished to study generalizations of the family of hyperarithmetical sets. For $a \in O$, where $|a| = \alpha$, we may write $\Delta_a^0(X)$ for the canonical complete $\Delta_\alpha^0(X)$ set associated with a. This is $H(a)(X)$ if $|a|$ is infinite, and $H(b)(X)$ if $|a|$ is finite and $|b| + 1 = |a|$. If we have in mind a fixed path in O, where the ordinal α has notation a, then we may write $\Delta_\alpha^0(X)$ for $\Delta_a^0(X)$.

The following result is intuitively clear.

Proposition 5.21 *If X is Δ_α^0 and Y is $\Delta_\beta^0(X)$, then Y is $\Delta_{\alpha+\beta}^0$.*

Proof: Let p be the function from Proposition 4.18. Given $a, b \in O$, and an index for X as a set computable in $H(a)$, we can find an index for $H(b)(X)$ as a set is computable in $H(p(a,b))$. We proceed by transfinite induction on b.

First let $b = 1$. We have $H(1)(X) = X$ and $H(p(a,1)) = H(a)$, so the statement holds.

Next, suppose $b = 2^d$, where the statement holds for d. We have $H(2^d)(X) = (H(d)(X))'$ and $H(p(a,2^d)) = H(2^{p(a,d)}) = H(p(a,d))'$, so the statement holds for b.

Finally, suppose $b = 3 \cdot 5^e$, where the statement holds for $d <_O b$. We have $H(3 \cdot 5^e)(X) = \{\langle u, v \rangle : u \leq_O 3 \cdot 5^e \, \& \, v \in H(u)(X)\}$. Since $p(a, 3 \cdot 5^e) = 3 \cdot 5^{e'}$, where $\varphi_{e'}(n) = p(a, \varphi_e(n))$, we have

$$H(p(a, 3 \cdot 5^e)) = \{\langle u', v \rangle : u' \leq_O 3 \cdot 5^{e'} \, \& \, v \in H(u')\}$$
$$= \{\langle p(a, u), v \rangle : u \leq_O 3 \cdot 5^e \, \& \, v \in H(p(a, u))\}.$$

From this, it should be clear that the statement holds for $3 \cdot 5^e$.

Proposition 5.21 is not sharp, for finite β. In particular, for $\beta = 1$, if X is Δ_α^0 and Y is $\Delta_1^0(X)$, then Y is Δ_α^0.

Exercise: Give the sharp version of Proposition 5.21, for finite $\beta = n \geq 1$.

5.5 Ershov's hierarchy

We mention one more hierarchy. Ershov defined a hierarchy of Δ^0_2 sets, based on differences of c.e. sets (see [44], [45], and [46], or [47]). For simplicity, we identify the tuples in a relation with their codes, and consider only sets. A set X is 1-c.e. if it is c.e. It is d-c.e., or 2-c.e., if it can be expressed in the form $Y - Z$, where Y and Z are c.e. More generally, for $n > 1$, X is n-c.e. if it can be expressed in the form $Y - Z$, where Y is c.e. and Z is $(n-1)$-c.e.

Exercise: Show that X is n-c.e. if and only if there is a total computable function $g : \omega \times \omega \to \{0,1\}$ such that for all x, $\lim_{s\to\infty} g(x,s) = \chi_X(x)$, $g(x,0) = 0$, and the set $\{s : g(x,s) \neq g(x,s+1)\}$ has cardinality at most n.

Still more generally, we define what it means for X to be α-c.e.. We fix a path through O with a notation for α, and we identify ordinals with their notations on this path. Then X is α-c.e. if there exist a total computable *guessing function* $g : \omega \times \omega \to \{0,1\}$ and a total computable *ordinal function* $o : \omega \times \omega \to \{\beta : \beta \leq \alpha\}$ with the following properties:

(1) $g(x,0) = 0$,
(2) $o(x,0) = \alpha$,
(3) for all s, $o(x,s+1) \leq o(x,s)$,
(4) if $g(x,s+1) \neq g(x,s)$, then $o(x,s+1) < o(x,s)$,
(5) $\lim_{s\to\infty} g(x,s) = \chi_X(x)$.

The following is closer to Ershov's original definition.

Proposition 5.22 *Let α be a computable ordinal. A set X is α-c.e. if and only if there exist uniformly c.e. families $(A_\beta)_{\beta<\alpha}$, $(B_\beta)_{\beta<\alpha}$ such that*
  (a) if $x \in A_\beta \cap B_\beta$, then $x \in A_\gamma \cup B_\gamma$, for some $\gamma < \beta$, and
  (b) $X = \cup_{\beta<\alpha}(A_\beta - \cup_{\gamma<\beta} B_\gamma)$.

Proof: First suppose that X is α-c.e., as witnessed by the functions g and o. Let
$$A_\beta = \{x : g(x,s) = 1 \ \& \ o(x,s) = \beta\},$$
and let
$$B_\beta = \{x : g(x,s) = 0 \ \& \ o(x,s) = \beta\}.$$
Now, suppose that there are uniformly c.e. families $(A_\beta)_{\beta<\alpha}$ and $(B_\beta)_{\beta<\alpha}$ satisfying (a) and (b). We define a guessing function and an ordinal function as follows. Let $g(x,0) = 0$, $o(x,0) = \alpha$. At stage $s+1$, if $g(x,s) = 1$, we search $s+1$ steps for $\gamma < o(x,s)$ such that $x \in B_\gamma$. If we find such a γ, then $g(x,s+1) = 0$ and $o(x,s+1) = \gamma$; otherwise, $g(x,s+1) = g(x,s)$ and $o(x,s+1) = o(x,s)$. Similarly, if $g(x,s) = 0$, we search $s+1$ steps for $\gamma < o(x,s)$ such that $x \in A_\gamma$. If we find such a γ, then $g(x,s+1) = 1$ and $o(x,s+1) = \gamma$; otherwise, $g(x,s+1) = g(x,s)$ and $o(x,s+1) = o(x,s)$.

Remark: We can relativize Ershov's hierarchy to an arbitrary set C, in the same way that we did the hyperarithmetical hierarchy. The change is that the guessing function g and the ordinal function o are computable relative to C.

5.5. ERSHOV'S HIERARCHY

We shall use the following relativization of Proposition 5.22. Again we fix a path in O, with notations for the ordinals that we are considering.

Proposition 5.23 *Let α be a computable ordinal. A set X is α-c.e. relative to C if and only if there exist families $(A_\beta)_{\beta<\alpha}$ and $(B_\beta)_{\beta<\alpha}$, uniformly c.e. relative to C (by which we mean that there are computable functions giving the indices for A_β and B_β as sets c.e. relative to C), such that*
  (a) if $x \in A_\beta \cap B_\beta$, then $x \in A_\gamma \cup B_\gamma$, for some $\gamma < \beta$, and
  (b) $X = \cup_{\beta<\alpha}[A_\beta - \cup_{\gamma<\beta}B_\gamma]$.

In the definition of the hyperarithmetical hierarchy, there was some apparent dependence on notation, but we were able to show that the class of Σ_α^0 sets and relations is independent of the notation for α. Ershov's hierarchy is truly dependent on notation. Along any path in O, we have a proper hierarchy. However, with different notations for ω^2, all Δ_2^0 sets are ω^2-c.e. (see [47]).

Recall that a set X is co-c.e. if the complement is c.e. By Theorem 1.4, X is computable if and only if it is both c.e. and co-c.e. By analogy, we say that X is *co-α-c.e.* if the complement is α-c.e., and it is *α-computable* if it is both α-c.e. and co-α-c.e. Just as there are non-computable c.e. sets, there are, when we fix a notation a for α, α-c.e. sets that are not α-computable. To show this, we proceed as we did for the hyperarithmetical hierarchy. We define an α-c.e. enumeration relation $E(a, e, x)$, and form the set

$$\{e : E(a, e, e)\},$$

which is α-c.e., while the complement is not.

Remark: The term ω-c.e. is sometimes applied to a Δ_2^0 set X having a computable guessing function $g : \omega \times \omega \to \{0, 1\}$ such that for each x,

$$lim_{s\to\infty} g(x, s) = \chi_S(x)$$

and for each n, $g(n, s)$ changes value at most n times, or $b(n)$ times, for some fixed computable function b. We would prefer to say that such a set is $\langle n \rangle$-c.e., or $\langle b(n) \rangle$-c.e., as this suggests the sequence of bounds. The relation between this notion and Ershov's hierarchy is given in the following result.

Proposition 5.24 *The following are equivalent:*
  (1) X is ω-computable
  (2) there is a computable function b such that X is $\langle b(n) \rangle$-c.e.

Proof: (1) $\Rightarrow$ (2) Suppose that X is ω-computable. Let g and o witness the fact that X is ω-c.e., and let g^* and o^* witness the fact that $\neg X$ is ω-c.e. We define a computable guessing function h and a computable bounding funtion b as follows. For each n, there is some first s such that one of $o(n, s)$, $o^*(n, s)$ is finite. For this s, if $o(n, s)$ is finite, then we let $b(n) = o(n, s)$, and we let $h(n, t) = g(n, t)$ for all t. If, for this s, $o(n, s) = \omega$, then we let $b(n) = o^*(n, s)$, and we let $h(n, t) = 1 - g^*(n, t)$ for all t.

(2) ⇒ (1) Suppose that X is $\langle b(n) \rangle$-c.e., witnessed by the computable guessing function g and the computable bounding function b. Then let $o(n,0) = \omega$, and let $o(n,1) = b(n)$. For $s \geq 1$, let $o(n, s+1) = o(n,s)$ if $g(n, s+1) = g(n,s)$ and $o(n,s) - 1$, otherwise.

Chapter 6

Infinitary formulas

Some mathematical properties, such as the Archimedean property (true of subfields of the ordered field of reals), are expressed in a natural way by an infinitely long formula, not by a finitary formula of the kind considered in Chapter 3. There are various kinds of infinitary formulas have been considered. The language may be propositional or predicate. Arbitrary disjunctions and conjunctions may be allowed, or only restricted ones. In a predicate language, there may or may not be infinitely nested quantifiers. The individual variables may be taken from a set which is countably infinite, uncountable, or finite.

We shall use both propositional and predicate languages. For a predicate language L, we consider formulas of $L_{\omega_1\omega}$. The basic reference is Keisler's beautiful book [75]. In the notation $L_{\omega_1\omega}$, the "ω_1" indicates that the disjunctions and conjunctions are over only *countable* sets. In [75], it also implies that the set of variables has size $\aleph_1$, but we shall use only countably many variables. The "ω" indicates that there is only *finite* nesting of quantifiers. The formulas of $L_{\omega_1\omega}$ are not elementary, but they are first order. For a propositional language P, we consider formulas of what we call P_{ω_1}, with disjunctions and conjunctions over only countable sets.

6.1 Predicate formulas

As logical symbols, we use the connectives $\bigvee$ (countable disjunction), $\bigwedge$ (countable conjunction), $\neg$ (negation), the logical constants $\top$ (truth) and $\bot$ (falsity), the quantifiers $\exists$ and $\forall$, $=$ (equality), countably many variables, and parentheses. Informally, we may write &, $\vee$ for disjunctions and conjunctions of two formulas, and we may use $\to$, $\leftrightarrow$, thinking of them as abbreviations.

Let L be a countable language (i.e., a countable set of non-logical symbols). The *formulas* of $L_{\omega_1\omega}$ are the members of of the least class F satisfying the following:

(1) if φ is an atomic L-formula, then $\varphi \in F$ ($\top$ and $\bot$ are included),
(2) if φ is in F, then so are $\neg\varphi$, $\exists x\, \varphi$, and $\forall x\, \varphi$,

(3) if R is a countable subset of F, then $\bigvee R$ and $\bigwedge R$ are in F (R may be finite, even empty).

Notation: We may write $\bigvee_{\varphi \in R} \varphi$ for $\bigvee R$, and $\bigwedge_{\varphi \in R} \varphi$ for $\bigwedge R$. Or, listing the elements of R as $\varphi_0, \varphi_1, \ldots$, we may write $\bigvee_i \varphi_i$ and $\bigwedge_i \varphi_i$.

We could represent formulas of $L_{\omega_1 \omega}$ in a set-theoretic way, as is done in [75]. However, starting in Chapter 7, we shall concentrate on *computable* infinitary formulas, which we code by natural numbers. To give a different representation now would be more distracting than helpful.

Satisfaction of formulas of $L_{\omega_1 \omega}$ is defined in the natural way. We note that $\top$ and the empty conjunction are always true, while $\bot$ and the empty disjunction are always false. As usual, formulas are said to be *logically equivalent* if they are satisfied by the same tuples in the same structures.

6.2 Sample formulas

Here are some sample formulas illustrating the expressive power of $L_{\omega_1 \omega}$.

Example 1: In the language of arithmetic, there is a sentence φ whose models are exactly the copies of $\mathcal{N}$.

We take φ to be the conjunction of the first order Peano axioms (actually, a finite subset would suffice), together with the sentence

$$\forall x \, (\bigvee_n x = S^{(n)}(0)) \, .$$

(See the Appendix for the axioms of PA.)

Example 2: In the language of ordered fields, there is a sentence whose models are exactly the Archimedean ordered fields—these are subfields of the ordered field of reals, or isomorphic copies.

We take the conjunction of the usual axioms for ordered fields (a finite set of finitary sentences), plus the sentence

$$\forall x \, (\bigvee_n x < \tau_n) \, , \text{ where } \tau_n = \overbrace{1 + \ldots + 1}^{n} \, .$$

Example 3: In the language of vector spaces over the field of rationals, for each n, there is a formula $\varphi_n(x_1, \ldots, x_n)$ saying that $x_1, \ldots, x_n$ are linearly independent.

6.2. SAMPLE FORMULAS

Let $\varphi_n(x_1, \ldots, x_n)$ be the conjunction of the natural formulas saying

$$q_1 x_1 + \ldots + q_n x_n \neq 0 ,$$

for tuples $(q_1, \ldots, q_n)$ from Q with some $q_i \neq 0$.

Example 4: In the language of linear orderings, for each countable ordinal α, there are formulas $\lambda_\alpha(x)$, $\mu_\alpha(x,y)$, and $\rho_\alpha(x)$, saying, respectively, that the interval to the left of x, between x and y (where $x < y$), or to the right of x, has order type α.

We let $\lambda_0(x) = \forall y \, \neg y < x$. The formulas $\mu_0(x,y)$ and $\rho_0(x)$ are similar to $\lambda_0(x)$. Having determined formulas λ_{ω^β}, μ_{ω^β}, and ρ_{ω^β}, we let

$$\lambda_{\omega^\beta \cdot n}(x) = \exists y_1 \ldots y_{n-1} \, (\lambda_{\omega^\beta}(y_1) \, \& \, \mu_{\omega^\beta}(y_1, y_2) \, \& \, \ldots \, \& \, \mu_{\omega^\beta}(y_{n-1}, x))$$

and we let $\lambda_{\omega^{\beta+1}}(x) =$

$$\bigwedge_n \exists y \, (y < x \, \& \, \lambda_{\omega^\beta \cdot n}(y)) \, \& \, \forall y \, (y < x \to \exists z \, (y < z < x \, \& \, \bigvee_n \lambda_{\omega^\beta \cdot n}(z))) .$$

The formulas $\mu_{\omega^{\beta+1}}(x,y)$ and $\rho_{\omega^{\beta+1}}(x)$ are defined similarly.

For a limit ordinal β, given formulas λ_{ω^γ}, μ_{ω^γ}, and λ_{ω^γ}, for $\gamma < \beta$, we let $\lambda_{\omega^\beta}(x) =$

$$\bigwedge_{\gamma < \beta} \exists y \, (y < x \, \& \, \lambda_{\omega^\gamma}(y)) \, \& \, \forall y \, (y < x \to \exists z \, (y < z < x \, \& \, \bigvee_{\gamma < \beta} \lambda_{\omega^\gamma}(z))) .$$

The formulas μ_{ω^β} and ρ_{ω^β} are defined similarly.

Having determined the formulas corresponding to 0 and the ordinals of the form ω^β, for all β, we can easily write formulas corresponding to an arbitrary countable ordinal α. We express α in Cantor normal form. To say that an interval has order type α, we say that it is a sum of sub-intervals of type ω^β for the appropriate ordinals β.

Example 5: In the language of Boolean algebras, for each countable ordinal α, there is a formula α-atom(x) saying that x is an α-atom. (See the Appendix for basic material on Boolean algebras.)

First, let 0-atom(x) be the finitary formula

$$(\forall y \leq x)(y = 0 \vee y = z) ,$$

saying that x is an atom. Now, let $\alpha > 0$, and suppose that we have determined the formulas β-atom(x), for $\beta < \alpha$. In terms of these, we can write formulas

$$a_{n,\beta}(x) ,$$

saying that x is the join of n disjoint β-atoms. Taking the disjunction of these formulas over $n \in \omega$, we have

$$a_{<\omega,\beta}(x) \; ,$$

saying that x is the join of finitely many β-atoms. Taking the disjunction over $\gamma < \beta$, we have

$$a_{<\omega,<\beta}(x)$$

saying that for some $\gamma < \beta$, x is the join of finitely many γ-atoms. If $\alpha = \beta + 1$, then we let α-atom(x) say

$$(\forall y \leq x)\,[a_{<\omega,\beta}(y) \vee a_{<\omega,\beta}(x-y)] \; \& \; \bigwedge_n (\exists y \leq x)\, a_{n,\beta}(y) \; .$$

If α is a limit ordinal, then we let α-atom(x) say

$$(\forall y \leq x)\,[\,a_{<\omega,<\alpha}(y) \vee a_{<\omega,<\alpha}(x-y)\,] \; \& \; \bigwedge_{\beta<\alpha} \exists y \, \beta\text{-atom}(y)\,] \; .$$

Example 6: For each countable ordinal α, there is a formula $height_\alpha(x)$ saying, in reduced Abelian p-groups, that x has height at least α. (See the Appendix for some basic material on Abelian groups.)

First, let $height_0(x)$ be $x = x$. Let $\alpha > 0$, and suppose that we have determined the formulas $height_\beta(x)$, for $\beta < \alpha$. If $\alpha = \beta + 1$, let $height_\alpha(x)$ say

$$\exists y \, (\overbrace{y + \ldots + y}^{p} = x \; \& \; height_\beta(y)) \; .$$

Finally, if α is a limit ordinal, let $height_\alpha(x)$ say

$$\bigwedge_{\beta<\alpha} \exists y \, (\overbrace{y + \ldots + y}^{p} = x \; \& \; height_\beta(y)) \; .$$

6.3 Subformulas and free variables

We define the set of *subformulas* of the formula φ, denoted by $sub(\varphi)$, inductively, as follows.
   (a) If φ is atomic, $sub(\varphi) = \{\varphi\}$,
   (b) if φ is $\neg\psi$, $\exists x\, \psi$, or $\forall x\, \psi$, then $sub(\varphi) = \{\varphi\} \cup sub(\psi)$,
   (c) if φ is $\bigvee_{\psi \in R} \psi$ or $\bigwedge_{\psi \in R} \psi$, then $sub(\varphi) = \{\varphi\} \cup \bigcup_{\psi \in R} sub(\psi)$.
Note that the set of subformulas of the disjunction (or conjunction) of R, does *not* include the disjunctions (or conjunctions) of arbitrary subsets of R.

We also define the set of *free variables* of the formula φ, denoted by $fv(\varphi)$, inductively:
   (a) if φ is atomic, $fv(\varphi)$ is the set of all variables occurring in φ,

(b) $fv(\neg\varphi) = fv(\varphi)$,
(c) $fv(\exists u\, \psi) = fv(\forall u\, \psi) = fv(\psi) - \{u\}$,
(d) $fv(\bigvee_{\varphi \in R} \varphi) = fv(\bigwedge_{\varphi \in R} \varphi) = \cup_{\varphi \in R} fv(\varphi)$.

A *sentence* is a formula with no free variables. As in the case of finitary formulas, we write $\varphi(\bar{x})$ to indicate that the free variables of φ are included in the tuple $\bar{x}$. At the moment, our definition does not preclude writing formulas with infinitely many free variables. However, the formulas that we are interested in will have only finitely many free variables.

Remark: It is easily shown by induction that if φ has only finitely many free variables, and ψ is a subformula of φ, then ψ has only finitely many free variables. In particular, if φ is a sentence, and ψ is a subformula of φ, then ψ has only finitely many free variables.

6.4 Normal form

For infinitary formulas, we do not have *prenex normal form*, but we can define a useful substitute. A formula of $L_{\omega_1 \omega}$ is said to be in *normal form!for infinitary formulas* if it is Σ_α or Π_α, for some countable ordinal α, where these classes are defined by transfinite induction as follows.

The Σ_0 and Π_0 formulas are the finitary open (i.e., quantifier-free) formulas. For $\alpha > 0$, a Σ_α formula φ is a countable disjunction of formulas of the form $\exists u\, \psi$, where ψ is Π_β for some $\beta < \alpha$. Similarly, a Π_α formula φ is a countable conjunction of formulas of the form $\forall u\, \psi$, where ψ is Σ_β for some $\beta < \alpha$. If φ is Σ_α or Π_α, and it is not Σ_β or Π_β for any $\beta < \alpha$, we may say that φ has *complexity* α.

We could work exclusively with formulas in normal form, since it turns out that every formula of $L_{\omega_1 \omega}$ is logically equivalent to one in normal form. For a formula φ in normal form, we let $NS(\varphi)$ be the set of subformulas of φ in normal form, with finitary open formulas added. The inductive definition is as follows:

(a) if φ is a finitary open formula, then $NS(\varphi) = sub(\varphi)$,

(b) if φ is a Σ_α formula, the disjunction of formulas $\exists u\, \psi$, where ψ is Π_β for some $\beta < \alpha$, or if φ is a Π_α formula, the conjunction of formulas $\forall u\, \psi$, where ψ is Σ_β for some $\beta < \alpha$, then $NS(\varphi)$ consists of φ and the elements of $NS(\psi)$ for the appropriate ψ.

Lemma 6.1 *(a) For any Σ_α formula φ, there is a Π_α formula $neg(\varphi)$ that is logically equivalent to $\neg\varphi$. Similarly, for any Π_α formula φ, there is a Σ_α formula $neg(\varphi)$ that is logically equivalent to $\neg(\varphi)$.*

(b) For any Σ_α formulas φ and ψ, there are Σ_α formulas logically equivalent to $(\varphi \vee \psi)$ and $(\varphi \,\&\, \psi)$. The corresponding statement holds for Π_α formulas.

(c) For any formula φ which is a finite Boolean combination of Σ_β and Π_β formulas, for various $\beta < \alpha$, there exist a Σ_α formula and a Π_α formula, each logically equivalent to φ.

(d) For any Σ_α formula φ, there is a Σ_α formula that is logically equivalent to $\exists x\, \varphi$, and there is a $\Pi_{\alpha+1}$ formula logically equivalent to $\forall x\, \varphi$. Similarly, for any Π_α formula φ, there is a Π_α formula logically equivalent to $\forall x\, \varphi$, and there is a $\Sigma_{\alpha+1}$ formula logically equivalent to $\exists x\, \varphi$.

Theorem 6.2 (Normal Form Theorem) *Every formula of $L_{\omega_1\omega}$ is logically equivalent to a Σ_α formula, for some countable α, and to a Π_α formula, for some countable α.*

The proof is inductive. For negations of formulas that are not open, we use part (a) of Lemma 6.1. If φ is a countable disjunction of formulas in normal form, we may think of each disjunct as the result of attaching a trivial string of existential quantifiers to a formula that is Π_β for some β. If $\alpha > \beta$, for all of these β, then φ is Σ_α. The argument for a conjunction is similar.

Note: In Section 6.2, we gave sample formulas illustrating the expressive power of $L_{\omega_1\omega}$. It may be useful to look back at some of these formulas and determine the complexity. Example 4, in particular, gave formulas $\lambda_\alpha(x)$, $\mu_\alpha(x,y)$ and $\rho_\alpha(x)$ saying, respectively that the interval to the left of x, between x and y, and to the right of x, has order type α. If $\alpha = 0$, we may take these formulas to be Π_1. If α has the form ω^β, we may take the formulas to be $\Pi_{2\beta+1}$. For an ordinal α whose Cantor normal form has leading term $\omega^\beta \cdot n$, where either $n > 1$ or else the leading term is not the only term, we may take the formulas to be $\Sigma_{2\beta+2}$.

Exercise: For the other examples from Section 6.4, put the formulas into normal form and classify as Σ_α or Π_α, aiming for the the least possible α.

6.5 Model existence

Makkai [109], [110], [111] gave a consistency criterion for countable sets of infinitary sentences. (Makkai indicated that he was using a general idea of Smullyan.) We give a version of Makkai's result for the case where the sentences are in normal form. Makkai did not restrict himself to formulas in normal form, but later on, we shall deal exclusively with these formulas, and we want to be sure that we do not need other formulas in constructing a model.

If Γ is a set of formulas of $L_{\omega_1\omega}$, all in normal form, let $NS(\Gamma) = \cup_{\varphi \in \Gamma} NS(\varphi)$. For each formula φ, $NS(\varphi)$ is countable, so if Γ is a countable set of formulas, then $NS(\Gamma)$ is countable. Suppose Γ is a countable set of sentences, all in normal form. Generalizing the notion of elementary substructure to $NS(\Gamma)$, we could prove a version of the Downward Löwenheim-Skolem Theorem. This implies that if Γ is a countable set of sentences, and B is a countably infinite set of new constants, then Γ is consistent if and only if it has a model whose elements are named by constants from B.

Let Γ be a set of sentences in normal form, and let B be a set of new constants. Let $F(\Gamma, B)$ be the set of sentences $\psi(\bar{b})$ formed by substituting a

6.5. MODEL EXISTENCE

tuple of constants from B for the tuple of free variables of ψ, where ψ is either a formula in $NS(\Gamma)$ or a finitary open formula. The following is an analogue of Theorem 3.7, or Proposition 3.8

Theorem 6.3 (Consistency Criterion) *Let L be a relational language, and let B be a set of new constants. Suppose that Γ is a set of sentences of $L_{\omega_1\omega}$, all in normal form. Then Γ has a model with universe B, or a model whose elements have names in B, if and only if there exists a set $\Gamma^* \supseteq \Gamma$ of $F(\Gamma, B)$-sentences such that*

(a) if $\varphi \in \Gamma^*$, where $\varphi = \bigvee_i \exists \overline{u}_i \, \psi_i(\overline{u}_i)$, then for some i and some tuple $\overline{d}$ in B, $\psi_i(\overline{d}) \in \Gamma^*$,

(b) if $\varphi \in \Gamma^*$, where $\varphi = \bigwedge_i \forall \overline{u}_i \, \psi_i(\overline{u}_i)$, then for all i and all appropriate tuples $\overline{d}$ in B, $\psi_i(\overline{d}) \in \Gamma^*$,

(c) for any finitary open sentence φ (in the language $L \cup B$), one of φ, $\neg\varphi$ is in Γ^*,

(d) the set of finitary open sentences in Γ^* is consistent.

Proof: First, suppose that Γ has a model $\mathcal{B}$ with universe B, or whose elements have names in B. If Γ^* is the set of $F(\Gamma, B)$-sentences true in $\mathcal{B}$, then Γ^* clearly has properties (a)–(d). Now, suppose that $\Gamma^* \supseteq \Gamma$ is a set of $F(\Gamma, B)$-sentences with these properties. Let $\mathcal{B}$ be the model of the set of finitary open sentences in Γ^* such that that all elements of $\mathcal{B}$ have names in B. We can show, by induction on β, that if φ is a Σ_β or Π_β sentence in Γ^*, then φ is true in $\mathcal{B}$.

Makkai [110] expressed the consistency criterion for a countable set of sentences Γ in terms of finite approximations for a set Γ^* with essentially the properties in Theorem 6.3. Let L be a relational language, and Γ be a countable set of $L_{\omega_1\omega}$-sentences, all in normal form. Let B be a countably infinite set of new constants. Let $\mathcal{C}$ be a set of finite subsets S of $F(\Gamma, B)$ such that the set of finitary open sentences in S is consistent. We call $\mathcal{C}$ a *consistency property* for Γ if the following properties are satisfied:

(1) for all $S \in \mathcal{C}$ and all $\varphi \in \Gamma$, there exists $S' \in \mathcal{C}$ such that $S' \supseteq S$ and $\varphi \in S'$,

(2) for all $S \in \mathcal{C}$ and all $\varphi \in S$ having the form $\bigvee_i \exists \overline{u}_i \, \psi(\overline{u}_i)$, there exists $S' \in \mathcal{C}$ such that $S' \supseteq S$ and for some i and some tuple $\overline{d}$ in B, $\psi_i(\overline{d}) \in S'$,

(3) for all $S \in \mathcal{C}$ and all $\varphi \in S$ having the form $\bigwedge_i \forall \overline{u}_i \, \psi_i(\overline{u}_i)$, for any i and any appropriate tuple $\overline{d}$ in B, there exists $S' \in \mathcal{C}$ such that $S' \supseteq S$ and $\psi_i(\overline{d}) \in S'$.

We may re-phrase Theorem 6.3 as follows.

Theorem 6.4 (Makkai) *Let L be a countable relational language, and let Γ be a countable set of $L_{\omega_1\omega}$ sentences, in normal form. Then Γ has a model if and only if it has a consistency property.*

6.6 Scott's Isomorphism Theorem

We shall see that $L_{\omega_1\omega}$ has all of the expressive power we could wish for in working with countable structures. One piece of evidence is the Scott Isomorphism Theorem[140], which we prove in this section. Let $\mathcal{A}$ be a countable structure. A *Scott sentence* for $\mathcal{A}$ is a sentence whose countable models are just the structures isomorphic to $\mathcal{A}$.

Theorem 6.5 (Scott Isomorphism Theorem) *Every countable L-structure $\mathcal{A}$ has a Scott sentence.*

Before proving Theorem 6.5, we give an important definition. A *Scott family* for a structure $\mathcal{A}$ is a countable family Φ of formulas (possibly with parameters in some fixed finite set) satisfying the following conditions:
 (a) for each tuple $\bar{a}$ in $\mathcal{A}$, there exists $\varphi \in \Phi$ such that $\mathcal{A} \models \varphi(\bar{a})$,
 (b) if two tuples $\bar{a}$ and $\bar{b}$ satisfy the same formula $\varphi \in \Phi$, then there is an automorphism of $\mathcal{A}$ taking $\bar{a}$ to $\bar{b}$.

Remark: In the definition of a Scott family, we could replace condition (b) by the following:
 (b') the family

$$\mathcal{F} = \{p : p \text{ is a finite partial 1-1 function taking one tuple} \\ \text{to another satisfying the same } \varphi \in \Phi\}$$

has the back-and-forth property.

To prove Theorem 6.5, we first show that any countable structure has a Scott family with some special features. The lemma below implies that for a countable structure, there is a countable family of formulas such that each tuple satisfies one of the formulas, and if two tuples satisfy the same formula, then they satisfy exactly the same formulas of $L_{\omega_1\omega}$. We give a more general statement.

Lemma 6.6 *Let $\mathcal{K}$ be a countable family of countable L-structures.*
 (a) For any $\mathcal{A}$ in $\mathcal{K}$, any tuple $\bar{a}$ in $\mathcal{A}$, and any countable ordinal α, there is a Π_α formula $\varphi^\alpha_{\mathcal{A},\bar{a}}(\bar{x})$ such that for all $\mathcal{B}$ in $\mathcal{K}$, and all tuples $\bar{b}$ in $\mathcal{B}$ such that $length(\bar{b}) = length(\bar{a})$, we have $\mathcal{B} \models \varphi^\alpha_{\mathcal{A},\bar{a}}(\bar{b})$ if and only if all Π_α formulas $\psi(\bar{x})$ true of $\bar{a}$ in $\mathcal{A}$ are true of $\bar{b}$ in $\mathcal{B}$.
 (b) There is a countable ordinal α such that for all $\mathcal{A}$ and $\mathcal{B}$ in $\mathcal{K}$, and all $\bar{a} \in \mathcal{A}$ and all $\bar{b} \in \mathcal{B}$ with $length(\bar{a}) = length(\bar{b})$, if there is an $L_{\omega_1\omega}$ formula $\psi(\bar{x})$ such that $\mathcal{A} \models \psi(\bar{a})$ and $\mathcal{B} \not\models \psi(\bar{b})$, then there is a Π_α formula with this property.

Proof: (a) Let $\bar{x}$ be a tuple of variables corresponding to $\bar{a}$. For each $\mathcal{B} \in \mathcal{K}$ and $\bar{b}$ in $\mathcal{B}$ with $length(\bar{b}) = length(\bar{a})$, if there is a Π_α formula $\psi(\bar{x})$ such that $\mathcal{A} \models \psi(\bar{a})$ and $\mathcal{B} \not\models \psi(\bar{b})$, we choose one. We let $\varphi^\alpha_{\mathcal{A},\bar{a}}(\bar{x})$ be the conjunction of the chosen formulas.

6.6. SCOTT'S ISOMORPHISM THEOREM

(b) For each pair of structures $\mathcal{A}$, $\mathcal{B}$ in K (not necessarily distinct), and for each pair of tuples $\bar{a} \in \mathcal{A}$, $\bar{b} \in \mathcal{B}$, of the same length, if there is a formula ψ true of $\bar{a}$ in $\mathcal{A}$ and not true of $\bar{b}$ in $\mathcal{B}$, choose β such that some Π_β formula ψ has this property. Let α be the least upper bound of the chosen β's. Since each β is countable and there are only countably many, α must be countable.

The next lemma gives the existence of a Scott family that we can easily convert into a Scott sentence.

Lemma 6.7 *If $\mathcal{A}$ is a countable structure, then it has a Scott family Φ with no parameters. Moreover, we may suppose that if $\mathcal{A} \models \varphi(\bar{a})$, where $\varphi(\bar{x}) \in \Phi$, then $\varphi(\bar{x})$ logically implies all basic formulas true of $\bar{a}$ (so that these are true of any $\bar{b}$ satisfying $\varphi(\bar{x})$, in any structure $\mathcal{B}$ for the same language).*

Proof: Let $\mathcal{K} = \{\mathcal{A}\}$, and take the ordinal α as in Lemma 6.6 (b). For each tuple $\bar{a} \in \mathcal{A}$, let $\delta_{\bar{a}}(\bar{x})$ be the conjunction of the basic formulas true of $\bar{a}$, and let $\varphi_{\bar{a}}(\bar{x})$ be the conjunction of $\delta_{\bar{a}}(\bar{x})$ and the formula $\varphi^\alpha_{(\mathcal{A},\bar{a})}$ from Lemma 6.6 (a). Let Φ be the set of formulas $\varphi_{\bar{a}}(\bar{x})$.

We should verify that this is Scott family with the desired properties. In particular, we should check Condition (b'). Let $\mathcal{F}$ be the family of finite partial $1-1$ functions p on $\mathcal{A}$ such that the tuple $\bar{a}$ forming $dom(p)$ and the corresponding tuple $\bar{b}$ forming $ran(p)$ both satisfy the same formula from Φ. This implies that $\bar{a}$ and $\bar{b}$ satisfy exactly the same infinitary formulas. For any c, since $\exists y\, \varphi_{\bar{a},c}(\bar{x}, y)$ is satisfied by $\bar{a}$, it is satisfied by $\bar{b}$. Then for some d, $\bar{a}, c$ and $\bar{b}, d$ both satisfy $\varphi_{\bar{a},c}(\bar{x}, y)$, so $f \cup \{(c, d)\}$ is in $\mathcal{F}$. This completes the proof of Lemma 6.7

We now return to the proof of Theorem 6.5. Let Φ be a Scott family as in Lemma 6.7. It is clear that the following sentences are all true in $\mathcal{A}$:

$$\varphi_{\bar{a}} = \forall \bar{x} \left(\varphi_{\bar{a}}(\bar{x}) \rightarrow \bigwedge_{b \in \mathcal{A}} \exists u\, \varphi_{\bar{a},\bar{b}}(\bar{x}, u) \right),$$

$$\rho_\emptyset = \bigwedge_{b \in \mathcal{A}} \exists u\, \varphi_b(u),$$

$$\sigma_{\bar{a}} = \forall \bar{x} \left(\varphi_{\bar{a}}(\bar{x}) \rightarrow \forall u \bigvee_{b \in \mathcal{A}} \varphi_{\bar{a},b}(\bar{x}, u) \right)$$

$$\sigma_\emptyset = \forall u \bigvee_{b \in \mathcal{A}} \varphi_b(\bar{v}, u).$$

Now, let

$$\sigma = \bigwedge_{\bar{a} \in \mathcal{A}} (\rho_{\bar{a}}\ \&\ \sigma_{\bar{a}}).$$

We show that σ is the required Scott sentence. Clearly, $\mathcal{A} \models \sigma$. Let $\mathcal{B}$ be a countable model of σ. Let $\mathcal{F}$ be the family of finite partial 1-1 functions p from

$\mathcal{A}$ to $\mathcal{B}$ such that if p takes $\bar{a}$ to $\bar{b}$ (where $dom(p) = \bar{a}$), then $\mathcal{B} \models \varphi_{\bar{a}}(\bar{b})$. By our choice of Φ, the functions in $\mathcal{F}$ are partial isomorphisms (preserving satisfaction of basic formulas). To show that

$$\mathcal{A} \cong \mathcal{B},$$

it is enough to show that $\mathcal{F}$ has the back-and-forth property. Suppose $p \in \mathcal{F}$ maps $\bar{a}$ to $\bar{b}$. The fact that $\mathcal{B} \models \rho_{\bar{a}}$ guarantees that for any $c \in \mathcal{A}$, there is an appropriate image $d \in \mathcal{B}$. The fact that $\mathcal{B} \models \sigma_{\bar{a}}$ guarantees that for any $d \in \mathcal{B}$, there is an appropriate pre-image $c \in \mathcal{A}$. Therefore, σ is a Scott sentence for $\mathcal{A}$.

6.7 Ranks and special Scott families

It follows from Lemma 6.6 (b) that for any countable structure $\mathcal{A}$, there is some countable ordinal α such that for all $\bar{a}$ and $\bar{b}$ in $\mathcal{A}$, if the Π_α formulas true of $\bar{a}$ are true of $\bar{b}$, then all $L_{\omega_1\omega}$ formulas true of $\bar{a}$ are true of $\bar{b}$. We let $r(\mathcal{A})$ denote the first such α. This is one possible notion of *rank* for $\mathcal{A}$. A second notion of rank is defined as follows. First, for each tuple $\bar{a}$ in $\mathcal{A}$, let $\rho(\bar{a})$ be the least ordinal β such that for all tuples $\bar{b}$, if the Π_β formulas true of $\bar{a}$ are true of $\bar{b}$, then all $L_{\omega_1\omega}$ formulas true of $\bar{a}$ are true of $\bar{b}$. Now, let $R(\mathcal{A})$ be the least ordinal α greater than $\rho(\bar{a})$ for all tuples $\bar{a}$ in $\mathcal{A}$.

The following result says that the values of of the two ranks $r(\mathcal{A})$ and $R(\mathcal{A})$ are not far apart.

Proposition 6.8 *Let $\mathcal{A}$ be a countable structure. If $\rho(\bar{a}) = r(\mathcal{A})$ for some tuple $\bar{a}$, then $R(\mathcal{A}) = r(\mathcal{A}) + 1$; if $\rho(\bar{a}) < r(\mathcal{A})$ for all tuples $\bar{a}$, then $R(\mathcal{A}) = r(\mathcal{A})$.*

We mention a third notion of rank. This one has perhaps a better claim than the others to the name "Scott rank", because it comes from Scott's proof of Theorem 6.5. We begin with a family of equivalence relations $\sim_\beta$ on tuples from $\mathcal{A}$:

(1) $\bar{a} \sim_0 \bar{b}$ if $\bar{a}$ and $\bar{b}$ satisfy the same open formulas,

(2) for $\beta > 0$, $\bar{a} \sim_\beta \bar{b}$ if for all $\gamma < \beta$, for each $\bar{c}$, there exists $\bar{d}$ such that $\bar{a}, \bar{c} \sim_\gamma \bar{b}, \bar{d}$, and for each $\bar{d}$, there exists $\bar{c}$ such that $\bar{a}, \bar{c} \sim_\gamma \bar{b}, \bar{d}$.

From this point, the definition of $SR(\mathcal{A})$ is like that of $R(\mathcal{A})$, with $\sim_\beta$ playing the role of the relation "all Π_β formulas true of $\bar{a}$ are true of $\bar{b}$". For each tuple $\bar{a}$, let $S\rho(\bar{a})$ be the least ordinal β such that for all tuples $\bar{b}$, if $\bar{a} \sim_\beta \bar{b}$, then $\bar{a}$ and $\bar{b}$ satisfy the same $L_{\omega_1\omega}$ formulas. Then $SR(\mathcal{A})$ is the least ordinal greater than $S\rho(\bar{a})$ for all tuples $\bar{a}$ in $\mathcal{A}$.

Clearly, $SR(\mathcal{A}) \leq R(\mathcal{A})$. Again, the values are not very different. Neither SR nor R can take value 0. If one of these ranks has value 1, or a limit ordinal, or the successor of a limit ordinal, then the other has the same value, and the difference in value is never infinite.

Among the three notions of rank, SR has been studied extensively (see [127]). We prefer r and R because they are more closely to the complexity of formulas. The ranks r and R are based on asymmetric relations "the Π_α formulas true of

6.7. RANKS AND SPECIAL SCOTT FAMILIES

$\bar{x}$ are true of $\bar{y}$. These turn out to be the same as the "back-and-forth" relations that will form the object of study in Chapter 15. The rank SR is based on symmetric relations $\bar{x} \sim_\alpha \bar{y}$, defined as follows. We have $\bar{x} \sim_1 \bar{y}$ if and only if the two tuples satisfy the same Π_1 formulas. For larger successor ordinals α, there is no β such that $\bar{x} \sim_\alpha \bar{y}$ if and only if $\bar{x}$ and $\bar{y}$ satisfy the same Π_β formulas.[1]

The result below says that a structure of small rank has a Scott family made up of simple formulas.

Proposition 6.9 *Let $\mathcal{A}$ be a countable structure.*

(a) If $r(\mathcal{A}) = \alpha$, then $\mathcal{A}$ has a Scott family consisting of Π_α formulas with no parameters.

(b) If $R(\mathcal{A}) = \alpha$, then $\mathcal{A}$ has a Scott family in which each formula is Π_β for some $\beta < \alpha$.

Exercise: Prove Proposition 6.9

In what follows, Scott families will be important, while Scott sentences will not. Proposition 6.9 gave conditions guaranteeing the existence of Scott families of special kinds. The next two results also do this. The one below gives conditions under which a family of finitary existential formulas is a Scott family.

Proposition 6.10 *Let $\mathcal{A}$ be a countable structure. Suppose that Φ is a family consisting of finitary existential formulas with parameters from a fixed finite tuple $\bar{c}$. Suppose that*

(a) each $\bar{a}$ in $\mathcal{A}$ satisfies some formula in Φ, and

(b) if there is some formula φ in Φ satisfied by tuples $\bar{a}$ and $\bar{b}$, then the two tuples satisfy exactly the same existential formulas with parameters $\bar{c}$.

Then Φ is a Scott family.

Exercise: Prove Proposition 6.10

The result below gives conditions for the existence of a Scott family made up of Σ_α formulas, where α is an arbitrary countable ordinal.

Proposition 6.11 *Suppose $\mathcal{A}$ is a countable structure, $\bar{c}$ is a tuple from $\mathcal{A}$, and α is a countable ordinal. Suppose that for each tuple $\bar{a}$, there exist an ordinal $\beta < \alpha$ and a tuple $\bar{a}_1$ such that for all tuples $\bar{b}$ and $\bar{b}_1$ of the same lengths as $\bar{a}$ and $\bar{a}_1$, if all Π_β formulas $\varphi(\bar{c}, \bar{x}, \bar{u})$ true of $\bar{a}, \bar{a}_1$ are true of $\bar{b}, \bar{b}_1$, then all Π_α formulas $\varphi(\bar{c}, \bar{x})$ true of $\bar{a}$ are true of $\bar{b}$. Then $\mathcal{A}$ has a Scott family consisting of Σ_α formulas with parameters $\bar{c}$.*

---

[1] The asymmetric and symmetric relations represent winning strategies for games with different rules. In both cases, two players add tuples to the domain and range of a partial isomorphism. For the asymmetric "back-and-forth" relations, Player I must add to the domain and range alternately. For the symmetric relations that Scott used, Player I may add to either on any move.

Exercise: Prove Proposition 6.11.

Remark: Let $\mathcal{A}$ be a countable structure. If $\mathcal{A}$ has a Scott family consisting of Σ_α formulas, then it has a Scott family consisting of Π_α formulas. (We can say that $\bar{x}$ does not satisfy a particular formula in the given Scott family by saying that it does *not* satisfy various other, inequivalent formulas.)

Here are some examples of structures with simple Scott families, and low rank.

Example 1: Let $\mathcal{A}$ be the rationals with the usual ordering. There is a Scott family consisting of finitary open formulas with no parameters. Moreover, $r(\mathcal{A}) = 0$, while $R(\mathcal{A}) = SR(\mathcal{A}) = 1$.

Example 2: Let $\mathcal{A}$ be an infinite dimensional vector space over the rationals. There is a Scott family consisting of Π_1 formulas, with no parameters. Each formula in the Scott family is a conjunction of linear equations and inequations. As in Example 1, $r(\mathcal{A}) = 0$, while $R(\mathcal{A}) = SR(\mathcal{A}) = 1$.

Example 3: Let $\mathcal{A}$ be a finite dimensional vector space over the rationals. If $\bar{a}$ is a basis, then there is a Scott family consisting of finitary open formulas with parameters $\bar{a}$. Again, $r(\mathcal{A}) = 0$, and $R(\mathcal{A}) = SR(\mathcal{A}) = 1$.

Example 4: Let $\mathcal{A}$ be an ordering of type ω. There is a Scott family consisting of finitary Σ_2 formulas, with no parameters. As for the ranks, we have $r(\mathcal{A}) = 2$, $R(\mathcal{A}) = 3$, and $SR(\mathcal{A}) = 2$.

6.8 Rigid structures and defining families

We describe a special kind of Scott family, for a special kind of structure. A structure $\mathcal{A}$ is *rigid* if there are no non-trivial automorphisms. It is *almost rigid* if for some tuple $\bar{c}$, $(\mathcal{A}, \bar{c})$ is rigid. For example, the ordering $(\omega, <)$ is rigid. More generally, any well ordering is rigid. The ordering $(Z, <)$ is not rigid, but it is almost rigid, since $(Z, <, 0)$ is rigid. A finite dimensional vector space is almost rigid; it becomes rigid once we name the elements of a basis.

Remark: If $\mathcal{A}$ is rigid, then for any copy $\mathcal{B}$, there is a unique isomorphism from $\mathcal{A}$ onto $\mathcal{B}$.

Theorem 6.12 *Let $\mathcal{A}$ be a countable structure. If $\mathcal{A}$ has fewer than $2^{\aleph_0}$ automorphisms, then it is almost rigid.*

Proof sketch: If, for each tuple $\bar{c}$, there is an automorphism that fixes $\bar{c}$ and moves some other element, then we can produce a tree with $2^{\aleph_0}$ paths representing distinct automorphisms of $\mathcal{A}$.

A *defining family* for $\mathcal{A}$ is a set Φ of formulas (possibly having parameters in a fixed finite set) such that
  (i) each element of $\mathcal{A}$ satisfies some formula in Φ,
  (ii) no formula in Φ is satisfied by two distinct elements of $\mathcal{A}$.

Example 1: The ordering of type ω has a defining family made up of finitary Σ_2 formulas. For each n, we take the formula
$$\varphi_n(x) = \exists y_0 \cdots \exists y_{n-1} \forall z\, (\, y_0 < \ldots < y_{n-1}\ \&\ (z < x \leftrightarrow (z = y_0 \vee \ldots \vee z = y_{n-1}))\,)$$
saying that there are exactly n elements to the left of x.

Example 2: The ordered field $\mathcal{A}$ of real algebraic numbers has a defining family made up of finitary Σ_1 formulas, in which each formula $\varphi(x)$ says, for some polynomial $p(x)$ and some k, that x is the kth root of $p(x)$.

Exercise: Describe a defining family for an ordering of type ω^2. [There is one consisting of finitary Σ_4 formulas with no parameters. There is another consisting of infinitary Π_4 formulas with no parameters.] Calculate the various ranks.

Exercise: Describe a defining family for an ordering of type ω^α, where α is an arbitrary countable ordinal. Again calculate the various ranks.

Theorem 6.13 *If $\mathcal{A}$ is a countable structure, then*
  (a) $\mathcal{A}$ is rigid if and only if it has a defining family Φ, consisting of $L_{\omega_1\omega}$ formulas $\varphi(x)$ with no parameters,
  (b) $\mathcal{A}$ is almost rigid if and only if it has a defining family Φ, consisting of $L_{\omega_1\omega}$ formulas $\varphi(\bar{c}, x)$ with parameters in a fixed finite set $\bar{c}$.

The proof is straightforward. For (a), if $\mathcal{A}$ is rigid, then to obtain the defining family, we take the formulas $\varphi(x)$ in just one variable from a Scott family consisting of formulas with no parameters, as in Proposition 6.9.

6.9 Definability of relations

The Scott Isomorphism Theorem is one piece of evidence that the formulas of $L_{\omega_1\omega}$ say everything we want to say about countable structures. Another piece of evidence is the next result, due to Kueker [92] and Makkai [109]. We shall give a proof using Scott families.

Theorem 6.14 *Let $\mathcal{A}$ be a countable structure, and let R be a further relation on $\mathcal{A}$. Then the following are equivalent:*
  (1) R has fewer than $2^{\aleph_0}$ different images under automorphisms of $\mathcal{A}$,
  (2) R is defined in $\mathcal{A}$ by an $L_{\omega_1\omega}$ formula $\varphi(\bar{c}, \bar{x})$, with a finite tuple of parameters $\bar{c}$.

Proof: First, suppose (2). Then the image of R under an automorphism of $\mathcal{A}$ is determined by the image of $\bar{c}$, so we have (1). Now, suppose (1).

Claim: There exists a tuple $\bar{c}$ such that for each $\bar{a} \in R$, there is a formula $\varphi_{\bar{c},\bar{a}}(\bar{c},\bar{x})$ satisfied by $\bar{a}$ and not by any $\bar{a}' \notin R$.

Proof of claim: Let Φ be a Scott family for $\mathcal{A}$, with no parameters. For each $\bar{a}$ in $\mathcal{A}$, choose a formula $\varphi_{\bar{a}}(\bar{x})$, in Φ, true of $\bar{a}$. Suppose there is no $\bar{c}$ as in the claim. Then for each $\bar{c}$, we have $\bar{a} \in R$ and $\bar{a}' \notin R$, both satisfying $\varphi_{\bar{c},\bar{a}}(\bar{c},\bar{x})$. Using this, we can produce a tree with $2^{\aleph_0}$ paths, yielding automorphisms of $\mathcal{A}$ under which the images of R are distinct—contradicting (1). Each element of the tree represents the restriction of some automorphism to a finite subset of $\mathcal{A}$. Suppose p is a tree element, and let $\bar{c} = ran(p)$. Take $\bar{a} \in R$, and $\bar{a}' \notin R$, where $\mathcal{A} \models \varphi_{\bar{c},\bar{a}}(\bar{c},\bar{a}')$. Let q be an extension of p, extending to an automorphism, where q takes some tuple $\bar{b}$ to $\bar{a}$. Let r be another extension of p, taking $\bar{b}$ to $\bar{a}'$—then r also extends to an automorphism. We put both q and r into the tree, extending, if necessary, so that if q, r are at level n, then the first n elements of $\mathcal{A}$ are included in the domain and range. This completes the proof of the claim.

Now, to complete the proof of the theorem, we take a tuple $\bar{c}$, and formulas $\varphi_{\bar{a}}(\bar{c},\bar{x})$, for all $\bar{a} \in R$, as in the claim. Then R is defined by the formula

$$\bigvee_{\bar{a} \in R} \varphi_{\bar{a}}(\bar{a},\bar{x}) \ ,$$

so we have (2).

Remark: If $\mathcal{A}$ has a Scott family consisting of Π_α formulas (or Σ_α formulas), and R is a relation on $\mathcal{A}$ with fewer than $2^{\aleph_0}$ images under automorphisms of $\mathcal{A}$, then R is defined by a Σ_α formula, and also by a Π_α formula.

6.10 Propositional formulas

Let P be a countable propositional language. We define the class of P_{ω_1} formulas by analogy with the class of $L_{\omega_1 \omega}$ formulas, allowing disjunctions and conjunctions over only countable sets. Satisfaction of P_{ω_1} formulas in structures $A \subseteq P$ is defined in the natural way. We also define, in the natural way, the special classes of P_{ω_1} formulas that are Σ_α and Π_α, for various α. Then we have a Normal Form Theorem, the analogue of Theorem 6.2, saying that every formula of P_{ω_1} is logically equivalent to one in normal form— Σ_α or Π_α, for some α.

Example: For any $A \subseteq P$, there is a formula σ in P_{ω_1} such that A is the unique model of σ. Let

$$\sigma = \bigwedge_{p \in A} p \ \& \ \bigwedge_{p \notin A} \neg p \ .$$

6.10. PROPOSITIONAL FORMULAS

Below we state a further result indicating the expressive power of P_{ω_1}. First, we recall what it means for a set of subsets of P to be Borel. First, for each function σ from a finite subset of P to $\{0,1\}$, we have a basic set B_σ consisting of those $A \subseteq P$ such that $\chi_A \supseteq \sigma$. Then the *Borel* sets are the members of the least class that contains the sets B_σ and is closed under countable union and complement.

Proposition 6.15 *The following are equivalent:*
 (a) B is a Borel set,
 (b) B is the class of models for some formula of P_{ω_1}.

Exercise: Prove Proposition 6.15.

Chapter 7

Computable infinitary formulas

Computable infinitary formulas were introduced in [1]. Roughly speaking, they are infinitary formulas in which the disjunctions and conjunctions are over c.e. sets. We consider computable infinitary formulas in both predicate and propositional languages. In the predicate formulas, we allow only finitely many free variables, and for both predicate and propositional languages, we limit ourselves, at least formally, to formulas in normal form. It turns out that most of what we want to say using infinitary formulas can be said using computable infinitary formulas. Moreover, the complexity of a computable infinitary formula matches the complexity of its interpretations in computable structures.

7.1 Informal definitions

Let L be a fixed computable language. To each formula φ, we associate a tuple of variables $\overline{x}$, including the free variables of φ. We define the class of computable infinitary formulas by induction on the complexity, which is a computable ordinal.

Informal Definition I: The *computable Σ_0 and Π_0 formulas* are the finitary open formulas. For a computable ordinal $\alpha > 0$, a *computable Σ_α formula* $\varphi(\overline{x})$ is the disjunction of a c.e. set of formulas of the form $\exists \overline{u}\, \psi$, where ψ is computable Π_β for some $\beta < \alpha$ and $\overline{u}$ includes the variables of ψ that are *not* in $\overline{x}$ ($\overline{u}$ may also include some variables that *are* in $\overline{x}$). Similarly, a *computable Π_α formula* $\varphi(\overline{x})$ is the conjunction of a c.e. set of formulas of the form $\forall \overline{u}\, \psi$, where ψ is computable Σ_β for some $\beta < \alpha$ and $\overline{u}$ includes the variables of ψ not in $\overline{x}$.

The definition above is useful, but, for $\alpha \geq 2$, it is not precise. In Chapter 1, we said that the elements of a c.e. set are natural numbers, or objects that can be coded by natural numbers. Now, a computable Σ_1 or Π_1 formula is obtained from a c.e. set of finitary formulas, and these have codes. For $\alpha \geq 2$, a computable Σ_α or Π_α formula is obtained from a c.e. set of infinitary formulas. The definition makes no sense unless we assign codes to these formulas.

We shall assign codes only to formulas in normal form. We modify the informal definition above so that instead of deriving a computable Σ_α (or computable Π_α) formula $\varphi(\overline{x})$ from c.e. set of disjuncts $\exists \overline{u}\, \psi$ (or conjuncts $\forall \overline{u}\, \psi$), we use a c.e. set of (codes for) pairs $(\psi, \overline{u})$. The reason for doing this is that ψ will have strictly lower complexity than φ—it is Π_β (or Σ_β), for some $\beta < \alpha$, while $\exists \overline{u}\, \psi$ (or $\forall \overline{u}\, \psi$) is already Σ_α (or Π_α).

Informal Definition II: Let $\alpha > 0$. A *computable Σ_α formula* $\varphi(\overline{x})$ is the disjunction of formulas $\exists \overline{u}\, \psi$, for pairs $(\psi, \overline{u})$ with codes in a c.e. set, where ψ is computable Π_β for some $\beta < \alpha$ and $\overline{u}$ includes the variables specified for ψ not in $\overline{x}$. A *computable Π_α formula* $\varphi(\overline{x})$ is the conjunction of formulas $\forall \overline{u}\, \psi$, for pairs $(\psi, \overline{u})$ with codes in a c.e. set, where ψ is computable Σ_β for some $\beta < \alpha$ and $\overline{u}$ includes the variables of ψ not in $\overline{x}$.

It might seem that in our "computable" infinitary formulas, we should take disjunctions and conjunctions of *computable* rather than *computably enumerable* sets. In fact, by "padding", we could replace the c.e. sets by computable ones. We illustrate this for a computable Σ_1 formula. Let $\varphi(\overline{x})$ be the disjunction of formulas $\exists \overline{u}\, \psi$, where $(\psi, \overline{u})$ is a finitary open formula in a certain c.e. set S. If n is first such that $(\psi, \overline{u}) \in S_n$ (where S_n is the stage n approximation of S), let ψ^* be the conjunction of n copies of ψ. The set

$$S^* = \{(\psi^*, \overline{u}) : (\psi, \overline{u}) \in S\}$$

is computable, and S^* yields a formula $\varphi^*(\overline{x})$ logically equivalent to $\varphi(\overline{x})$. The reason for using c.e. sets rather than computable ones is that this gives us a nice system of codes.

We now describe the codes precisely. The system that we use here was introduced in [8]. For computable Σ_0 and Π_0 formulas (finitary open formulas), the codes are the usual Gödel numbers. For $\alpha > 0$, the codes for computable Σ_α and Π_α formulas are (codes for) quadruples in which the first component is the symbol Σ or Π (we could use 0 and 1), the second is a notation for α, the third is (the code for) a tuple of variables $\overline{x}$, and the fourth is a natural number e. The set of codes having a given pair as first and second components is computable, because we can use *any* tuple of variables $\overline{x}$ and any number e as third and fourth components.

We should explain how an arbitrary number e can be made to yield a c.e. set of pairs of the appropriate kind. We illustrate for the formula φ with code $\langle \Sigma, 2, \overline{x}, e \rangle$. The first three components indicate that φ is a computable Σ_1 formula, with free variables among $\overline{x}$. Let P be the set of pairs $(\psi, \overline{u})$, where ψ is a finitary open formula and $\overline{u}$ is a tuple of variables including the variables of ψ that are not in $\overline{x}$. We identify these pairs with their codes. Associated with e, we have the set $W_e \cap P$, which is c.e. since P is computable. Our formula

$\varphi(\overline{x})$ is the disjunction of the formulas $\exists \overline{u}\, \psi$ corresponding to pairs $(\psi, \overline{u})$ with codes in this set. The set of pairs may be empty—there are many codes for the empty disjunction.

7.2 Formal definition

We have sets of indices S_a^Σ and S_a^Π, for all $a \in O$. If $|a| = \alpha$, then the formulas with indices in S_a^Σ will be Σ_α, and the formulas with indices in S_a^Π will be Π_α. The computable infinitary Σ_0 and Π_0 formulas are the finitary open formulas. The set of Gödel numbers of these formulas is computable, and it serves as both S_1^Σ and S_1^Π (recall that $|1| = 0$). For $|a| > 0$, S_a^Σ is the set of quadruples of the form $\langle \Sigma, a, \overline{x}, e \rangle$, where $\overline{x}$ is a finite sequence of variables, and $e \in \omega$, and S_a^Π is the set of quadruples of this same form, except that Π replaces Σ.

Let $P(\Pi, a, \overline{x})$ be the set of (codes for) pairs $(i, \overline{u})$ such that i has the form $\langle \Pi, b, \overline{v}, e' \rangle$, for some $b <_O a$, and $\overline{u}$ is a finite sequence of variables including the variables of $\overline{v}$ not in $\overline{x}$. Let $P(\Sigma, a, \overline{x})$ be the same except that i has the form $\langle \Sigma, b, \overline{v}, e' \rangle$. Then the index $\langle \Sigma, a, \overline{x}, e \rangle$ represents a computable Σ_α formula $\varphi(\overline{x})$ with disjuncts corresponding to the pairs in $W_e \cap P(\Pi, a, \overline{x})$, where if $(i, \overline{u})$ is one of these pairs, and ψ is the formula with index i, then the corresponding disjunct is $\exists \overline{u}\, \psi$. Similarly, the index $\langle \Pi, a, \overline{x}, e \rangle$ represents a computable Π_α formula $\varphi(\overline{x})$ with a conjunct for each pair in $W_e \cap P(\Sigma, a, \overline{x})$, where if $(i, \overline{u})$ is one of these pairs, and ψ is the formula with index i, then the corresponding conjunct is $\forall \overline{u}\, \psi$.

We have not lost expressive power by assigning indices only to computable infinitary formulas in normal form. The result below says that other formulas which we might wish to think of as being computable can be replaced, effectively, by formulas in normal form. In stating the result, we suppress the ordinal notation. The computable Σ_α or Π_α formulas mentioned should all be thought of as having indices in S_a^Σ or S_a^Π, where a is a notation for α. Furthermore, we have in mind a fixed path through O, containing notations for all of the ordinals mentioned, the notations used in our indices are the unique ones on this path.

Theorem 7.1 *(a) Given an index for a computable Σ_α (or Π_α) formula φ, we can find an index for a computable Π_α (or Σ_α) formula $neg(\varphi)$ that is logically equivalent to $\neg \varphi$.*

(b) Given indices for a pair of computable Σ_α (or Π_α) formulas φ and ψ, we can find indices for computable Σ_α (or Π_α) formulas logically equivalent to $(\varphi \vee \psi)$ and $(\varphi \,\&\, \psi)$.

(c) Given a formula φ that is a finite Boolean combination of computable Σ_β and Π_β formulas, for various $\beta < \alpha$ (with indices in $\cup_{b <_O a} S_b^\Sigma \cup S_b^\Pi$, where a is the notation for α on our fixed path), we can find a computable Σ_α formula and a computable Π_α formula, both logically equivalent to φ.

(d) Given a computable Σ_α formula φ, we can find a computable Σ_α formula logically equivalent to $\exists x\, \varphi$, and a computable $\Pi_{\alpha+1}$ formula logically equivalent to $\forall x\, \varphi$. Similarly, given a computable Π_α formula φ, we can find a computable

Π_α formula logically equivalent to $\forall x\, \varphi$, and a computable $\Sigma_{\alpha+1}$ formula logically equivalent to $\exists x\, \varphi$.

Sketch of proof for (a): Formally, we define, by computable transfinite recursion on ordinal notation, a computable partial function F such that for all $a \in O$, $F(a)$ is a computable index for a function f_a, defined on $S_a^\Sigma \cup S_a^\Pi$, such that if i is an index for φ (in $S_a^\Sigma \cup S_a^{Pi}$), then $f_a(i)$ is an index for the formula $neg(\varphi)$ (also in $S_a^\Sigma \cup S_a^\Pi$). Informally, we describe f_1, and say how to calculate f_a, given f_b for $b <_O a$.

For f_1, if i is the Gödel number of a finitary open formula φ, then $f_1(i)$ is the Gödel number of $\neg\varphi$. For other a, suppose that we have indices for f_b, for $b <_O a$. If $i = \langle \Sigma, a, \overline{x}, e \rangle$, we find a c.e. index e' for

$$\{\langle f_b(j), \overline{u}\rangle : (\exists b <_O a)\, j \in S_b^\Pi \,\&\, (j, \overline{u}) \in W_e \cap P(\Pi, a, \overline{x})\}\ ,$$

and we let $f_a(i) = \langle \Pi, a, \overline{x}, e'\rangle$. Similarly, if $i = \langle \Pi, a, \overline{x}, e \rangle$, we find a c.e. index e' for

$$\{(f_b(j), \overline{u}) : (\exists b <_O a)\, j \in S_b^\Sigma \,\&\, (i, \overline{u}) \in W_e \cap P(\Sigma, a, \overline{x})\}\ ,$$

and we let $f_a(i) = \langle \Sigma, a, \overline{x}, e'\rangle$.

7.3 Sample formulas

In Chapter 6, we described infinitary sentences characterizing the Archimedean ordered fields and the copies of $\mathcal{N}$ (the standard model of arithmetic). Both sentences can be taken to be computable. In fact, the natural formulas expressing general properties that we are interested in usually turn out to be computable. We illustrate this by revisiting some examples from Chapter 6.

First, we consider the formulas describing well-ordered intervals in a linear ordering. In stating the next result, we suppress ordinal notation, identifying ordinals with their notations on a fixed path through O.

Proposition 7.2 *For each computable ordinal α, there is a computable infinitary formula $\lambda_\alpha(x)$ saying, in linear orderings, that the interval to the left of x has order type α, and there are similar computable infinitary formulas $\mu_\alpha(x, y)$, and $\rho_\alpha(x)$ saying, respectively, that the interval between x and y, and to the right of x, have order type α.*

(a) We can take $\lambda_0(x)$, $\mu_0(x, y)$, and $\rho_0(x)$ to be finitary Π_1; for other finite n, we can take $\lambda_n(x)$, $\mu_n(x, y)$, and $\rho_n(x)$ to be finitary Σ_2.

(b) For each computable ordinal $\beta > 0$, we can take λ_{ω^β}, μ_{ω^β}, and ρ_{ω^β} to be computable $\Pi_{2\beta+1}$.

(c) If α is a computable ordinal with $\omega^\beta < \alpha < \omega^{\beta+1}$, then we can take λ_α, μ_α, and ρ_α to be computable $\Sigma_{2\beta+2}$.

Proof: (a) is clear. We would omit the proof of (b) and (c) except that there may be questions about the indices. Here we must consider ordinal notation. First, note that for $b \in O$, $2|b|$ differs only finitely from $|b|$, even when $|b|$ is an infinite ordinal. Then we have unique notations b^* and b^{**}, for $2|b| + 1$ and $2|b| + 2$ such that $b <_O b^*, b^{**}$.

Formally, to prove (b), we define, by computable transfinite recursion on ordinal notation, computable partial functions f, g, and h such that if $|b| = \beta$, then $f(b)$, $g(b)$, and $h(b)$ are indices in S_b^Π for formulas with the meanings we want for λ_{ω^β}, μ_{ω^β}, and ρ_{ω^β}. From the description of the formulas in Chapter 6, together with Theorem 7.1, it should be clear that we can determine $f(b)$, $g(b)$ and $h(b)$, given the restrictions of f, g, and h to the set $\{d : d <_O b\}$. Then by Theorem 4.17, we have the desired functions f, g, and h.

For (c), we begin by expressing α in Cantor normal form. We see that an interval of type α is a sum of sub-intervals of order types ω^γ, for various $\gamma \leq \beta$, and n. If $|b| = \beta$, then (b) (or its proof) yields formulas describing the sub-intervals, with indices in $S_{d^*}^\Pi$, for $d^* \leq_O b$. Combining these formulas in a natural way, we obtain formulas with the meaning desired for λ_α, μ_α, and ρ_α. Using Theorem 7.1, we obtain the desired formulas with indices in $S_{b^{**}}^\Sigma$.

The next two results say that there are computable infinitary formulas defining important relations in Boolean algebras and groups.

Proposition 7.3 *For each computable ordinal α, there is a computable $\Pi_{2\alpha+1}$ formula α-atom(x) saying, in Boolean algebras, that x is an α-atom.*

Proof: We described these formulas in Chapter 6. For any notation a for α, there is a unique notation a^* for $2\alpha + 1$ such that $a <_O a^*$, and we can pass effectively from a to an index in $S_{a^*}^\Pi$ for α-atom(x).

Proposition 7.4 *Let p be a prime. For each computable ordinal α, there is a formula height$_\alpha(x)$ saying, in reduced Abelian p-groups (where all elements have height), that height$(x) \geq \alpha$. If $\alpha = \omega \cdot \beta$, then we can take height$_\alpha(x)$ to be computable $\Pi_{2\beta}$, and if $\alpha = \omega \cdot \beta + n$, for $0 < n < \omega$, then we can take height$_\alpha(x)$ to be computable $\Sigma_{2\beta+1}$.*

Proof: We described these formulas in Chapter 6. For any notation b for β, there are unique notations b^* for 2β and b^{**} for $2\beta + 1$ such that $b <_O b^*, b^{**}$. We can pass effectively from b to an index in $S_{b^*}^\Pi$ for $height_{\omega \cdot \beta}(x)$, and from b and $n > 0$ to an index in $S_{b^{**}}^\Sigma$ for $height_{\omega \cdot \beta + n}(x)$.

7.4 Satisfaction and the hyperarithmetical hierarchy

The next result gives the single most important fact about computable infinitary formulas. It says that the complexity of a computable infinitary formula matches the complexity of its interpretation, at least in computable structures.

Theorem 7.5 *(a) For computable structures $\mathcal{A}$, if $\varphi(\overline{x})$ is a computable Σ_α formula, then $\varphi^\mathcal{A}$ is Σ_α^0, and if $\varphi(\overline{x})$ is a computable Π_α formula, then $\varphi^\mathcal{A}$ is Π_α^0. Moreover, this is true with all conceivable uniformity.*

(b) For arbitrary structures $\mathcal{A}$, if $\varphi(\overline{x})$ is computable Σ_α, then $\varphi^\mathcal{A}$ is $\Sigma_\alpha^0(\mathcal{A})$, and if $\varphi(\overline{x})$ is computable Π_α, then $\varphi^\mathcal{A}$ is $\Pi_\alpha^0(\mathcal{A})$. Moreover, given a notation a for α and an index for φ in S_a^Σ, or S_a^Π, we can find an index for $\varphi^\mathcal{A}$, as a set c.e., or co-c.e., relative to $\Delta_a^0(\mathcal{A})$, $\mathcal{A}$. The index is independent of $\mathcal{A}$.

Proof: We give the proof for (a). Formally, we proceed by induction on ordinal notation. Satisfaction of finitary open formulas is clearly computable. Suppose $\varphi(\overline{x})$ is computable Σ_α, with index in S_a^Σ, where $|a| = \alpha$. We have a c.e. set S of pairs giving rise to disjuncts of $\varphi(\overline{x})$. Then $\mathcal{A} \models \varphi(\overline{a})$ if and only if there exist a pair $(i, \overline{u}) \in S$ and a tuple $\overline{b}$ in $\mathcal{A}$ such that for some $b <_O a$, $i \in S_b^\Pi$ is the code for a formula ψ such that

$$\mathcal{A} \models \psi \left(\frac{\overline{b}}{\overline{u}} \right) \left(\frac{\overline{a}}{\overline{x}} \right).$$

Note that $\overline{u}$ may include some of $\overline{x}$—the notation above indicates that we first substitute $\overline{b}$ for $\overline{u}$ and then substitute elements of $\overline{a}$ for what remains of $\overline{x}$. By H.I., we can pass effectively from i to an index for $\psi^\mathcal{A}$ as a set co-c.e. relative to Δ_b^0, and from this we can pass effectively to an index for $\psi^\mathcal{A}$ as a set c.e. relative to Δ_a^0. From this, it is not difficult to see that $\varphi^\mathcal{A}$ is c.e. relative to Δ_a^0, so it is Σ_a^0. The statement for computable Π_α follows by considering complements.

Recall that the satisfaction predicate on $\mathcal{A}$ for a class of formulas is the relation which holds between between a formula in the class, or an index for the formula, and a tuple satisfying the formula. Let Sat_a^Σ, Sat_a^Π denote the satisfaction predicates for formulas with indices in S_a^Σ, S_a^Π, respectively. The uniformity in Theorem 7.5 is such that if $|a| = \alpha$, then for all structures $\mathcal{A}$, Sat_a^Σ is $\Sigma_\alpha^0(\mathcal{A})$ and Sat_a^Π is $\Pi_a^0(\mathcal{A})$, with an index which we can determine from a, independent of $\mathcal{A}$.

Remark: Theorem 7.5 is characteristic of the classes of computable Σ_α and Π_α formulas. As we shall see in Chapter 8, for any extension of the language—obtained by adding quantifiers, allowing more general disjunctions and conjunctions, etc.—if Theorem 7.5 (b) holds, then there are no new definable relations.

7.5 Further hierarchies of computable formulas

Recall that in a c.e. quotient structure, equality is c.e., while the other relations (corresponding to non-logical relation symbols) are computable. There is a variant of Theorem 7.5 for c.e. quotient structures (see [100]). We must change the classification of formulas, so that, in particular, atomic formulas involving equality are counted as Σ_1. Below, we give the definition only informally, without describing the indices.

7.5. FURTHER HIERARCHIES OF COMPUTABLE FORMULAS

The *computable* Σ_0^* and Π_0^* *formulas* are the finitary open formulas with no occurrences of $=$. The *computable* Σ_1^* *formulas* are the c.e. disjunctions of formulas $\exists \overline{u}\,(\psi_1\,\&\,\psi_2)$, where ψ_1 is a finitary open formula and ψ_2 is a finite conjunction of atomic formulas involving $=$. The *computable* Π_1^* *formulas* are the c.e. conjunctions of formulas $\forall \overline{u}\,(\psi_1 \vee \psi_2)$, where ψ_1 is finitary open and ψ_2 is a finite disjunction of negations of atomic formulas involving $=$. For $\alpha > 1$, the *computable* Σ_α^* *formulas* are the c.e. disjunctions of formulas $\exists \overline{u}\,\psi$, where ψ is computable Π_β^*, for some $\beta < \alpha$. The *computable* Π_α^* *formulas* are the c.e. conjunctions of formulas $\forall \overline{u}\,\psi$, where ψ is computable Σ_β^*, for some $\beta < \alpha$.

Here is the analogue of Theorem 7.5 (a) for c.e. quotient structures.)

Theorem 7.6 *If $\varphi(x_1,\ldots,x_n)$ is a computable Σ_α^* (or Π_α^*) formula, then for c.e. quotient structures $\mathcal{A}/\sim$, $\{(a_1,\ldots,a_n) : \mathcal{A}/\sim \models \varphi(a_1/\sim,\ldots,a_n/\sim)\}$ is Σ_α^0 (or Π_α^0. Moreover, this is true with the same uniformity as in Theorem 7.5 (a).*

In Chapter 3, we defined more general kinds of quotient structures. For an arbitrary structure $\mathcal{A}$ and a congruence relation $\sim$ on $\mathcal{A}$, we obtain a quotient structure structure $\mathcal{A}/\sim$. The quotient structure is said to be Σ_α^0, Π_α^0, or Δ_α^0 if $\mathcal{A}$ is computable and the congruence relation $\sim$ is Σ_α^0, Π_α^0, or Δ_α^0. In Chapter 9, we shall see that if $\mathcal{A}$ is a linear ordering or a Boolean algebra, then $\mathcal{A}$ has a Δ_2^0 copy if and only if it has a copy which is a c.e. quotient structure. In Chapter 18, we shall generalize this to Σ_α^0 quotients.

Exercise: Give the analogue of Theorem 7.5 (a) for Σ_α^0 quotient structures. Begin by describing the appropriate new classification of computable infinitary formulas. Do the same for Π_α^0 quotient structures.

We may consider generalized computable structures in which equality is computable and other relations have assigned complexities, Σ_β^0 or Π_β^0, for various computable ordinals β. Let L be a relational language. For each $R \in L$, let $\overline{R}$ be a new symbol, standing for $\neg R$. Let Γ be a computable function from the set of symbols $\{R : R \in L\} \cup \{\overline{R} : R \in L\}$ to computable ordinals. To be precise, the values of Γ are notations for ordinals along some fixed path through O, but we identify the ordinals with their notations. A Γ-*structure* is an L-structure $\mathcal{A}$ such that the universe, $|\mathcal{A}|$, is computable, and for all $R \in L$, if $\Gamma(R) = \beta$, then $R^\mathcal{A}$ is Σ_β^0, and if $\Gamma(\overline{R}) = \beta$, then $R^\mathcal{A}$ is Π_β^0, all uniformly (meaning that we can effectively compute the appropriate Σ_α^0 and Π_α^0 indices). Note that if $\Gamma(R) = \Gamma(\overline{R}) = \beta$, then $R^\mathcal{A}$ is Δ_β^0.

Example: Let $\mathcal{A}$ be a linear ordering with an added predicate S for the successor relation. Let $\Gamma(<) = 1$, $\Gamma(S) = 1$. The values of $\Gamma(\overline{<})$ and $\Gamma(\overline{S})$ are unimportant. If $\mathcal{A}$ is a Γ-structure, then the ordering and the successor relation are both c.e. It follows that both are actually computable.

There is an analogue of Theorem 7.5 (a) for Γ-structures. Again we must change the classification of formulas (see [10]). If φ is an atomic formula involving R, where $\Gamma(R) = \beta$ (for $\beta > 0$) then φ is *elementary computable* Σ_β^Γ and $\neg\varphi$ is *elementary computable* Π_β^Γ. The general definition, which we give only informally, proceeds by induction. The *computable* Σ_0^Γ and Π_0^Γ *formulas* are the finitary Boolean combinations of $\top$, $\bot$, and atomic formulas involving $=$ or symbols R such that $\Gamma(R) = 0$. For $\beta > 0$, a computable Σ_β^Γ formula is a c.e. disjunction of formulas

$$\exists \overline{u} \, (\, (\psi(\overline{x},\overline{u}) \, \& \, \rho(\overline{x},\overline{u})) \,) \, ,$$

where $\psi(\overline{x},\overline{u})$ is computable Π_γ^Γ for some $\gamma < \beta$, and $\rho(\overline{x},\overline{u})$ is a finite conjunction of elementary Σ_β^Γ formulas. A computable Π_β^Γ formula is a c.e. conjunction of formulas

$$\forall \overline{u} \, (\, \psi(\overline{x},\overline{u}) \lor \rho(\overline{x},\overline{u}) \,) \, ,$$

where $\psi(\overline{x},\overline{u})$ is computable Σ_γ^Γ for some $\gamma < \beta$ and $\rho(\overline{x},\overline{u})$ is a finite disjunction of elementary Π_β^Γ formulas.

Here is the analogue of Theorem 7.5 (a).

Theorem 7.7 *Suppose that $\mathcal{A}$ is a Γ-structure. If $\varphi(\overline{x})$ is a computable Σ_β^Γ formula, then $\varphi^{\mathcal{A}}$ is Σ_β^0. If $\varphi(\overline{x})$ is a computable Π_β^Γ formula, then $\varphi^{\mathcal{A}}$ is Π_β^0. Moreover, this is true with all possible uniformity.*

Exercise: Prove Theorem 7.7.

We may relativize the notion of Γ-structure. Suppose that Γ is defined on R and $\overline{R}$, for all $R \in L^{\mathcal{A}}$. For a structure $\mathcal{A}$ to be $\Gamma(X)$, we require that if $\Gamma(R) = \beta$, then $R^{\mathcal{A}}$ is $\Sigma_\beta(X)$, uniformly. We then have the natural analogue of Theorem 7.5 (b), saying that if $\mathcal{A}$ is $\Gamma(X)$ and the formula φ is computable Σ_β^Γ (or Π_β^Γ), then $\varphi^{\mathcal{A}}$ is $\Sigma_\beta^0(X)$ (or $\Pi_\beta^0(X)$). (See [11].)

7.6 Computable propositional formulas

Fix a computable propositional language P. We define the class of computable propositional formulas by induction on the complexity, where this is a computable ordinal. The informal definition is as follows. The *computable Σ_0 and Π_0 formulas* are just the finitary propositional formulas. For $\alpha > 0$, a *computable Σ_α formula* φ is the disjunction of a c.e. set of formulas, each computable Π_β for some $\beta < \alpha$. A computable Π_α formula is the conjunction of a c.e. set of formulas, each computable Σ_β for some $\beta < \alpha$.

To make this definition precise, we must assign indices, as we did in the case of predicate languages. Using the same notation as for predicate languages, we define new sets of indices S_a^Σ and S_a^Π, for $a \in O$, as follows. The set of Gödel numbers for finitary propositional formulas serves as both S_1^Σ and S_1^Π. For $a \in O$ such that $|a| > 0$, we let

$$S_a^\Sigma = \{\langle \Sigma, a, e \rangle : e \in \omega\} \, ,$$

7.6. COMPUTABLE PROPOSITIONAL FORMULAS

and we let
$$S_a^\Pi = \{\langle \Pi, a, e\rangle : e \in \omega\} \ .$$

These sets of indices are computable.

For $S_1^\Sigma = S_1^\Pi$, the correspondence between indices and the formulas they represent is the natural one. Suppose $|a| = \alpha > 0$. Let

$$S_{<_O a}^\Sigma = \cup_{b<_O a} S_b^\Sigma \ , \ S_{<_O a}^\Pi = \cup_{b<_O a} S_b^\Pi \ .$$

Then $\langle \Sigma, a, e\rangle$, an element of S_a^Σ, represents the computable Σ_α formula whose disjuncts have indices in $W_e \cap S_{<_O a}^\Pi$. Similarly, $\langle \Pi, a, e\rangle$, an element of S_a^Π, represents the computable Π_α formula whose conjuncts have indices in $W_e \cap S_{<_O a}^\Sigma$. We have the analogue of Theorem 7.1, saying that for a formula φ with an index in S_a^Σ, we can find an index in S_a^Π for a formula $neg(\varphi)$ that is logically equivalent to $\neg \varphi$, etc.

In what follows, we shall use infinitary propositional formulas frequently. Here we give some sample formulas for the language $P = \omega$. Let A be a structure for the language (i.e., $A \subseteq \omega$). We have

$$n \in A \Leftrightarrow A \models n \ .$$

Then for $\sigma \in 2^{<\omega}$, $\chi_A \supseteq \sigma$ if and only if

$$A \models \bigwedge_{\sigma(n)=1} n \ \& \ \bigwedge_{\sigma(n)=0} \neg n \ .$$

We can talk about the jump of A. For a fixed n, we have $n \in A'$ if and only if A satisfies the disjunction of the formulas saying $\chi_A \supseteq \sigma$, for σ in the set

$$\{\sigma \in 2^{<\omega} : \varphi_e^\sigma(n) \downarrow\} \ .$$

Continuing in this vein, we obtain the following result. Recall that if a is a notation for α, then Δ_a^0 denotes the complete Δ_α^0 oracle associated with a.

Proposition 7.8 *Let $a \in O$, where $|a| \geq 1$.*

(a) Given $n \in \omega$, we can find a computable infinitary formula $\delta(a,n)$, with index in $S_{<_O a}^\Sigma$, such that

$$A \models \delta(a,n) \ \Leftrightarrow \ n \in \Delta_a^0(A) \ .$$

(b) Given $\sigma \in 2^{<\omega}$, we can find computable infinitary formulas $\eta_0(a,\sigma)$, and $\eta_1(a,\sigma)$, the former with index in S_a^Σ and the latter with index in S_a^Π, both with the feature that

$$A \models \eta_i(a,s) \ \Leftrightarrow \ \chi_A \supseteq \sigma \ .$$

(c) Given e and n in ω, we can find a computable infinitary formula $\theta(a,n,e)$, with index in S_a^Σ, such that

$$A \models \theta(a,n,e) \ \Leftrightarrow \ n \in W_e^{\Delta_a^0(A)} \ .$$

(d) Given $e \in \omega$, we can find a computable infinitary formula $\tau(a,e)$, with index in $S_{2^a}^\Pi$, such that

$$A \models \tau(a,e) \Leftrightarrow \varphi_e^{\Delta_a^0(A)} \text{ is total}.$$

Proof: Formally, we define functions giving indices for the formulas in (a), (b), and (c) simultaneously, by computable transfinite induction on ordinal notation. We obtain (indices for) the formulas $\eta_i(a,\sigma)$ in (b) from (indices for) the formulas $\delta(a,n)$ in (a)—the formula

$$\bigwedge_{\sigma(n)=1} \delta(a,n) \, \& \, \bigwedge_{\sigma(n)=0} \neg\delta(a,n)$$

is easily transformed into the desired formulas $\eta_i(n,\sigma)$

We obtain (indices for) the formulas $\delta(a,n)$ in (a) from (indices for) the formulas $\theta(b,n,e)$ in (c), for $b <_O a$. There are three cases to consider.

Case 1: Suppose $|a| = 1$.

In this case $\Delta_a^0(A) = A$. We let $\delta(a,n) = n$.

Case 2: Suppose $a = 2^b$.

In this case, $\Delta_a^0 = (\Delta_b^0)'$. Taking e such that $(\Delta_b^0)' = W_e^{\Delta_b^0}$, we let $\delta(a,n)$ be $\theta(b,e,n)$.

Case 3: Suppose $a = 3 \cdot 5^e$.

In this case, $\Delta_a^0 = H(a) = \{\langle u,v \rangle : u <_O a \,\&\, v \in H(u)\}$, where

$$H(u) = \begin{cases} \Delta_u^0 & \text{if } |u| \geq \omega \\ \Delta_{2^u}^0 & \text{if } |u| < \omega \end{cases}.$$

Given n, we can find u,v such that $n = \langle u,v \rangle$. If $u \in O$, then we can determine whether $u \geq \omega$. We can find a c.e. index for the set

$$\begin{cases} \{\delta(u,v)\} & \text{if } u <_O a \,\&\, |u| \geq \omega \\ \{\delta(2^u,v)\} & \text{if } u <_O a \,\&\, |u| < \omega \\ \emptyset & \text{otherwise}. \end{cases}$$

The disjunction of this set of formulas is the desired $\delta(a,n)$.

We obtain (indices for) the formulas $\tau(a,e)$ in (d) from (indices for) the formulas $\theta(a,e,n)$ in (c), letting

$$\tau(a,e) = \bigwedge_n \theta(a,e,n).$$

7.7 The simplest language

We now consider formulas built up just out of $\bot$ and $\top$. The formulas do not say much—each is logically equivalent to $\bot$ or $\top$. They turn out to be surprisingly useful. These formulas are represented in all of our languages, both predicate and propositional—with different indices, of course. We give a uniform treatment of the basic result below, without specifying the kind of language. We then give applications to both predicate and propositional languages.

Theorem 7.9 *For any hyperarithmetical set S, we can find a sequence $(\theta_n)_{n \in \omega}$ of computable infinitary formulas, built up just out of $\bot$ and $\top$, such that θ_n is logically equivalent to*

$$\begin{cases} \top & \text{if } n \in S \\ \bot & \text{otherwise} \end{cases}.$$

To prove Theorem 7.9, we establish the following.

Claim: Let $a \in O$, where $|a| \geq 1$.
(a) For $n \in \omega$, we can find an index, in S_b^Σ for some $b <_O a$, for a formula $\delta(a, n)$ that is logically equivalent to $\top$ if $n \in \Delta_a^0$, and to $\bot$, otherwise.
(b) For $\sigma \in 2^{<\omega}$, we can find indices, one in S_a^Σ and one in S_a^Π, for a formula $\eta(a, \sigma)$ that is logically equivalent to $\top$ if $\chi_{\Delta_a^0} \supseteq \sigma$, and to $\bot$, otherwise.
(c) For n and e in ω, we can find an index, in S_a^Σ, for a sentence $\theta(a, e, n)$ that is logically equivalent to $\top$ if $n \in W_e^{\Delta_a^0}$, and to $\bot$, otherwise.

Before proving the claim, we note that if S is Σ_α^0, then it has the form $W_e^{\Delta_a^0}$, where $|a| = \alpha$, and the formulas $\theta(a, e, n)$ in (c) satisfy the conclusion of Theorem 7.9 If S is Π_α^0, we obtain the desired formulas by considering the complement of S and the negations of the formulas associated with it.

Proof of Claim: The proof is much like that for the previous result. We obtain (indices for) the sentences $\eta(a, \sigma)$ in (b) from (indices for) the sentences $\delta(a, n)$ in (a). Similarly, we obtain (indices for) the sentences $\theta(a, e, n)$ in (c) from (indices for) the sentences $\eta(a, \sigma)$ in (b). This is exactly as above. We obtain (indices for) the sentences $\delta(a, n)$ from (indices for) the sentences $\theta(b, e, n)$ for $b <_O a$. For $|a| = 1$, $\Delta_a^0 = \emptyset$, and we let $\delta(a, n) = \bot$. For $a = 2^b$, and for $a = 3 \cdot 5^e$, we obtain $\delta(a, n)$ as above.

Below is a consequence of Theorem 7.9 for a propositional language.

Corollary 7.10 *Consider the propositional language ω, with natural numbers n as propositional variables. For any hyperarithmetical set S, there is a computable propositional formula φ such that S is the unique model of φ.*

Proof: Let $(\theta_n)_{n\in\omega}$ be a computable sequence of propositional formulas as in Theorem 7.9, where θ_n is logically equivalent to $\top$ if $n \in S$, and to $\bot$ otherwise. The formula

$$\varphi = \bigwedge_n (\theta_n \leftrightarrow n)$$

has the desired meaning, and it is easily transformed into the desired computable infinitary propositional formula.

7.8 Hyperarithmetical formulas

In defining the class of computable infinitary formulas, for both predicate and propositional languages, we allowed disjunctions and conjunctions over c.e. sets. We could define the class of *hyperarithmetical* infinitary formulas, allowing disjunctions and conjunctions over hyperarithmetical sets. We might expect some gain in expressive power. However, it turns out that there is none—every hyperarithmetical infinitary formula is logically equivalent to a computable infinitary formula.

The next two results say that the disjunction, or conjunction of a Σ_α^0 set of computable Σ_α (or Π_α) formulas can be replaced by a c.e. disjunction, or conjunction, and, hence, by a computable Σ_α (or Π_α) formula. There is one result for propositional formulas and one for predicate formulas. In these results, the disjuncts or conjuncts have indices forming a Σ_α^0 subset of S_a^Σ, or S_a^Π, for some fixed notation a for α. Here is the result for propositional formulas.

Proposition 7.11 *Let $|a| = \alpha$.*

(a) Suppose φ is the disjunction of the computable propositional formulas with indices in a Σ_α^0 set $S \subseteq S_a^\Sigma$. Then φ is logically equivalent to a computable propositional formula with an index in S_a^Σ.

(b) Suppose φ is the conjunction of computable propositional formluas with indices in a Σ_α^0 set $S \supseteq S_a^\Pi$. Then φ is logically equivalent to a computable propositional formula with an index in S_a^Π.

Proof of (a): By Theorem 7.9, we have a computable sequence of computable Σ_α formulas θ_n, built up out of $\top$ and $\bot$, with indices in S_a^Σ, such that θ_n is logically equivalent to $\top$ if $n \in S$, and to $\bot$ otherwise. Let φ^* be the disjunction of the formulas $(\theta_i \& \psi)$, where $i \in S_a^\Sigma$ is an index for ψ—this is a c.e. disjunction that we can readily transform into a formula with an index in S_a^Σ. Clearly, φ^* is logically equivalent to φ.

(b) Given an index in $S_{<_o b}^\Sigma$ for ψ, we can find an index in $S_{<_o b}^\Pi$ for a formula that is logically equivalent to $\neg\psi$. Bearing this in mind, we can apply (a) to get a computable Σ_α formula logically equivalent to $\neg\varphi$.

Here is the analogous result for predicate formulas.

7.8. HYPERARITHMETICAL FORMULAS

Proposition 7.12 *Let $|a| = \alpha$.*

(a) Suppose $\varphi(\overline{x})$ is the disjunction of computable infinitary formulas, all having variables $\overline{x}$, and with indices in a Σ^0_α set $S \subseteq S^\Sigma_a$. Then $\varphi(\overline{x})$ is logically equivalent to a computable infinitary formula with index in S^Σ_a.

(b) Suppose $\varphi(\overline{x})$ is the conjunction of computable infinitary formulas, all having variables $\overline{x}$, and with indices in a Σ^0_α set $S \subseteq S^\Pi_a$. Then $\varphi(\overline{x})$ is logically equivalent to a computable infinitary formula with index in S^Π_a.

The proof is the same as for Proposition 7.11.

Most of the time, we work along a single path in O, where all of our notations are comparable, and it is easy to combine computable infinitary formulas. Occasionally, however, we are forced to consider incomparable notations. It turns out that a c.e. disjunction (or conjunction) of computable infinitary formulas can be effectively transformed into a computable infinitary formula even when the disjuncts (or conjuncts) come from many different sets S^Π_b (or S^Σ_b), for notations b which are not on the same path in O. In making these transformations, we may lose control of the complexity.

We give one result for propositional formulas and one for predicate formulas. Here is the one for propositional formulas.

Proposition 7.13 *Any c.e. disjunction or conjunction of computable propositional formulas is logically equivalent to a computable propositional formula.*

Proof: We prove the result for disjunctions. Let S be a c.e. set of indices for computable infinitary formulas. Let $(i_n)_{n \in \omega}$ be a computable list of the elements of S, say φ_n is the formula with index i_n. We may suppose that $i_n \in S^\Pi_{a_n}$ (replacing an index in S^Σ_b by one in S^Π_{2b}). We show the existence of a formula logically equivalent to $\bigwedge_n \varphi_n$, with an index in S^Σ_a for some a.

First, applying Proposition 4.16, we determine $a \in O$ and a partial computable function g such that
  (a) if $b \leq_O a_n$, then $g(n,b) <_O a$, and
  (b) if $d <_O b \leq_O a_n$, then $g(n,d) <_O g(n,b)$.
Next, we define a computable partial function f on the set

$$\{(n,i) : n \in \omega \;\&\; (\exists b \leq_O a_n)\, i \in S^\Sigma_b \cup S^\Pi_b\}$$

such that the formulas with indices i and $f(n,i)$ are logically equivalent, but $f(n,i) \in S^\Sigma_{<_O a} \cup S^\Pi_{<_O a}$.

We proceed by computable transfinite induction on $b \in O$.

(i) If $b = 1$ and i is in $S^\Sigma_1 = S^\Pi_1$ (so i is the Gödel number of a finitary propositional formula), then
$$f(n,i) = i\ .$$

(ii) Let $|b| > 0$, and suppose that we have a computable index for the function giving the values of $f(n,j)$ for all j in $S^\Sigma_{<_O b} \cup S^\Pi_{<_O b}$. If $i = \langle \Sigma, b, e \rangle$ is in S^Σ_b, then
$$f(n,i) = \langle \Sigma, g(n,b), e' \rangle\ ,$$

where e' is a c.e. index for the set $\{f(n,j) : j \in S^{\Pi}_{<_O b} \cap W_e\}$. If $i \in S^{\Pi}_b$, we define $f(n,i)$ similarly.

For each n, the formula φ_n with index i_n in $S^{\Pi}_{a_n}$ is logically equivalent to the one with index $f(n,i_n)$ in $S^{\Pi}_{g(n,a_n)}$, where $g(n,a_n) <_O a$. Let e^* be a c.e. index for the set
$$\{f(n,i_n) : n \in \omega\}\ .$$
Then $\langle \Sigma, a, e^* \rangle \in S^{\Sigma}_a$ is an index for a formula logically equivalent to $\bigwedge_n \varphi_n$. This completes the proof of Proposition 7.13.

Here is the result for predicate formulas.

Proposition 7.14 . *Let $\overline{x}$ be a finite sequence of variables and let S be a c.e. set of pairs, each in $P(\Pi, b, \overline{x})$ for some $b \in O$. Then there is a formula $\varphi(\overline{x})$, with an index in S^{Σ}_a, for some $a \in O$, such that $\varphi(\overline{x})$ is logically equivalent to the disjunction of the formulas $\exists \overline{u}\, \psi$ corresponding to the pairs in S. Similarly, if S is a c.e. set of pairs, each in $P(\Sigma, b, \overline{x})$ for some $b \in O$, then there is a formula $\varphi(\overline{x})$, with an index in S^{Π}_a, for some $a \in O$, such that $\varphi(\overline{x})$ is logically equivalent to the conjunction of the formulas $\forall \overline{u}\, \psi$ corresponding to pairs in S. Moreover, in either case, given $\overline{x}$ and a c.e. index for S, we can find an index for $\varphi(\overline{x})$.*

The proof is essentially the same as for Proposition 7.13.

For any computable ordinal α, we can transform any Σ^0_α disjunction or conjunction of computable propositional formulas into a computable propositional formula. First, we replace the Σ^0_α disjunction or conjunction by a c.e. one, as in the proof of Proposition 7.11. Then we apply Proposition 7.13. Similarly, we can transform any Σ^0_α disjunction or conjunction of computable predicate formulas, with a fixed tuple of variables, into a computable predicate formula. First, we replace the Σ^0_α disjunction or conjunction by a c.e. one, as in the proof of Proposition 7.12. Then we apply Proposition 7.14.

Thus, we have the following.

Theorem 7.15 *For any hyperarithmetical set Γ of computable infinitary propositional formulas, or computable infinitary predicate formulas with a fixed tuple of variables, there is a computable infinitary formula φ that is logically equivalent to the disjunction, and there is a computable infinitary formula ψ that is logically equivalent to the conjunction. Moreover, from an index for Γ, we can effectively determine $a \in O$ and indices for φ and ψ.*

7.9 X-computable formulas

For an arbitrary set X, we may consider infinitary formulas in which the disjunctions and conjunctions are required to be c.e. relative to X, as opposed to c.e. We may classify the X-computable formulas as X-computable Σ_α and X-computable Π_α, for various computable ordinals α. We have indices as before.

7.9. X-COMPUTABLE FORMULAS

Theorem 7.16 *For a structure $\mathcal{A}$ that is computable relative to X, satisfaction of X-computable Σ_α formulas is $\Sigma_\alpha^0(X)$. For arbitrary structures $\mathcal{A}$, satisfaction of X-computable Σ_α formulas is $\Sigma_\alpha^0(X,\mathcal{A})$. Similarly, satisfaction of X-computable Π_α formulas is $\Pi_\alpha^0(X)$, or $\Pi_\alpha^0(X,\mathcal{A})$.*

For some sets X, there are well orderings computable relative to X but not computable. The order types of these ordinals are the X-*computable ordinals*. It would be natural to extend the hierarchies (of sets and formulas) through the X-computable ordinals. We refrain from doing this since our main interest is in the hyperarithmetical hierarchy. For a discussion of the larger set of ordinals associated with a given set X, the associated family of subsets of ω, and the corresponding "fragment" of $L_{\omega_1\omega}$, see Barwise [22].

Chapter 8

The Barwise-Kreisel Compactness Theorem

The usual Compactness Theorem (given in Chapter 3), says that for a set of finitary sentences, if every finite subset is consistent, then the whole set is consistent, where consistency means having a non-empty model. The corresponding statement for sets of infinitary sentences is false. For example, take the set of sentences, in the language of arithmetic with an added constant e, consisting of sentences saying $e \neq S^{(n)}(0)$, for $n \in \omega$, and one further sentence saying $\forall x \bigvee_n x = S^{(n)}(0)$. Kreisel suggested that the Compactness Theorem might become true with a different notion of "finite" and some restrictions on the set of sentences. One result in this direction, for "ω-logic" rather than $L_{\omega_1\omega}$, is sketched in a footnote in [90]. Barwise [22] proved a very general result, for arbitrary "admissible" fragments of $L_{\omega_1\omega}$.

The Compactness Theorem in the present chapter says that for a Π_1^1 set of computable infinitary sentences, if every Δ_1^1 subset is consistent, then the whole set is consistent. This theorem is useful for priority constructions with a Π_1^1 set of requirements, where the requirements are given by computable infinitary sentences. If we can satisfy every Δ_1^1 set of requirements, then, by Compactness, we can satisfy the whole set.

8.1 Model existence and paths through trees

In Chapter 3, we gave a consistency criterion for sets of finitary sentences. We showed that given a c.e. set Γ of finitary sentences, we can determine a computable tree P such that Γ is consistent if and only if P has a path. There was more than one version of the tree. One, infinitely branching, had paths representing different (complete diagrams of) models of Γ. Another, binary branching, had paths representing different completions of Γ.

In Chapter 6, we gave a consistency criterion for sets of $L_{\omega_1\omega}$ sentences. We now aim for a result saying that if Γ is a "nice" set of computable infinitary sentences, then we can find a "nice" tree P such that Γ is consistent if and only if P has a path. We consider first the case where Γ consists of a single computable infinitary sentence.

Theorem 8.1 *Given a computable infinitary sentence φ, we can determine a computable tree P (infinitely branching) such that φ is consistent if and only if P has a path. In fact, there is a partial computable function g such that if i is an index for a computable infinitary sentence, then $g(i)$ is a computable index for an appropriate tree.*

Proof: We may suppose that the language of φ, say L, is relational. (For any operation symbol, we treat it as a relation symbol, and we add to φ a conjunct saying that the relation denoted by the symbol is a total function.) Let B be an infinite computable set of new constants. Given a finite set of finitary open $(L \cup B)$-sentences, we can effectively determine consistency—it is enough to check a finite collection of finite structures. (For a language with operation symbols, this might not be possible.)

Let $(\alpha_n)_{n \in \omega}$ be a computable list of the finitary open $(L \cup B)$-sentences. Let $F(\{\varphi\}, B)$ be as in Chapter 6, the set of sentences which result from substituting constants from B for the free variables in finitary open L-formulas and subformulas of φ. We identify the sentences from $F(\{\varphi\}, B)$ with their indices, and we use the standard codes for finite sets and sequences. Let $\mathcal{C}$ be the set of finite subsets of $F(\{\varphi\}, B)$ in which the set of finitary open sentences is consistent. The tree P will consist of finite sequences $S_0, S_1, S_2, \ldots$ of elements of $\mathcal{C}$, where $S_0 = \{\varphi\}$, and for each n, S_n and S_{n+1} are related as follows:

(1) if $\sigma \in S_n$, where σ is the conjunction of sentences $\forall \overline{u}\, \psi$ for pairs $(\overline{u}, \psi)$ in a c.e. set U (with index determined from σ, or the index for σ), then for all pairs $(\overline{u}, \psi) \in U_{n+1}$ (the stage $n+1$ approximation to U) and all tuples $\overline{d}$ (corresponding to $\overline{u}$) chosen from among the first $n+1$ constants in B, we have $\psi(\overline{d}) \in S_{n+1}$,

(2) if $\sigma \in S_n$, where σ is the disjunction of formulas $\exists \overline{u}\, \psi$, for $(\overline{u}, \psi)$ in the c.e. set U (with index determined from σ), then for some $(\overline{u}, \psi) \in U$ and some $\overline{d}$ corresponding to $\overline{u}$, we have $\psi(\overline{d}) \in S_{n+1}$,

(3) one of α_n, $\neg \alpha_n$ is in S_{n+1},

The tree P is clearly computable. Using Theorem 6.4, we can see that for any path through P, the union yields a non-empty model of φ. Now, suppose that φ has a non-empty model $\mathcal{A}$, and let f be a function from B onto $\mathcal{A}$. Then f makes each sentence of $F(\{\varphi\}, B)$ true or false in $\mathcal{A}$. We can determine a path $S_0, S_1, S_2, \ldots$ through P such that f makes all sentences of $\cup_n S_n$ true.

We can extend Theorem 8.1 to hyperarithmetical sets of sentences.

Theorem 8.2 *For any hyperarithmetical set Γ of computable infinitary sentences (all in the same computable language), there is a computable tree P (infinitely branching) such that Γ is consistent if and only if P has a path. Moreover, given $a \in O$ and an index for Γ as a set computable relative to $H(a)$, we can find a computable index for P.*

8.2 The Compactness Theorem

Proof: By Theorem 7.15, we can find a single computable infinitary sentence that is logically equivalent to the conjunction of Γ. Then we apply Theorem 8.1 to obtain the desired tree.

8.2 The Compactness Theorem

The usual version of the Barwise Compactness Theorem (see [22]) involves "admissible" sets in both the statement and the proof, where these sets are the standard models of a system of set theoretic axioms due to Kripke and Platek. Our version involves no special set theory in the statement, and the proof that we give uses only basic notions and facts from computability, all of which were reviewed in Chapter 5. This proof must be very similar to the one that Kreisel [90] had in mind.

Neither Kreisel nor Barwise considered computable infinitary sentences—this class of formulas was introduced by Ash. However, the discussion of hyperarithmetical formulas in Chapter 7 should convince anyone who is familiar with admissible sets that the computable infinitary sentences are essentially the same as the sentences in the least admissible fragment of $L_{\omega_1\omega}$. It follows that the result here is a special case of Barwise's.

Theorem 8.3 (Compactness) *Let Γ be a Π^1_1 set of computable infinitary sentences (in a fixed computable language). If every Δ^1_1 subset of Γ is consistent, then Γ is consistent.*

Proof: By Corollary 5.15, $\Gamma \leq_m O$ via some computable function h. For each computable ordinal α, let

$$\Gamma_\alpha = \{x \in \Gamma : |h(x)| < \alpha\} \ .$$

We have

$$\Gamma_\alpha \leq_m \{b \in O : |b| < \alpha\} \ ,$$

so by Theorem 5.18, Γ_α is Δ^1_1. Moreover, for any Δ^1_1 set $\Lambda \subseteq \Gamma$, there is some α such that $\Lambda \subseteq \Gamma_\alpha$. To see why this is so, note that by Corollary 5.9, $ran(h|\Lambda)$ is Δ^1_1. Then by Lemma 5.16, $ran(h|\Lambda)^*$ is also Δ^1_1, and this must have the form Γ_α.

As in the proof of Theorem 8.1, we may suppose that the language of Γ, say L, is relational. Let B be an infinite computable set of new constants (not in L), and let $F(\Gamma, B)$ be as in Chapter 6. Let $\mathcal{C}$ consist of the finite sets $S \subseteq F(\Gamma, B)$ such that for all computable ordinals α, $S \cup \Gamma_\alpha$ is consistent (so it has a model with elements named by constants from B). By Theorem 6.4, to show that Γ is consistent, it is enough to show that $\mathcal{C}$ is a consistency property for Γ.

The only condition that is not immediately clear is the one involving a disjunction of existential formulas. Let $S \in \mathcal{C}$, and suppose that $\varphi \in S$, where

φ is a disjunction of formulas $\exists \overline{u}\,\psi(\overline{u})$, for $\psi(\overline{u})$ in some c.e. set U. We show that for some $\psi(\overline{u}) \in U$ and some tuple $\overline{d}$ in B,

$$S \cup \{\psi(\overline{d})\} \in \mathcal{C}\ .$$

Let $(\theta_i)_{i \in \omega}$ be a computable list of the sentences $\psi(\overline{d})$, for $\psi(\overline{u})$ in U and $\overline{d}$ in B (possible witnesses for φ). Let

$$\Gamma_{\alpha,i} = \Gamma_\alpha \cup S \cup \{\theta_i\}\ .$$

By our hypothesis, for each α, $\Gamma_\alpha \cup S$ has a model, so for some i, $\Gamma_{\alpha,i}$ has a model. We must show that a *single* i works for all α.

By Lemma 5.2, if $|a| = \alpha$, then the set $\{b \in O : |b| < \alpha\}$ is computable relative to $H(2^a)$. Given $a \in O$ and $i \in \omega$, we can find an index for $\Gamma_{|a|,i}$ as a set computable in $H(2^a)$. Then by Theorem 8.2, we can find an index for a computable tree $P(a,i)$ such that $P(a,i)$ has a path if and only if $\Gamma_{|a|,i}$ has a model. Let g be a partial computable function such that for all $a \in O$ and $i \in \omega$, $g(a,i)$ is an index for $P(a,i)$.

Let Q be the relation such that

$$Q(i,a) \Leftrightarrow (a \in O\ \&\ P(a,i) \text{ has no path})\ .$$

Then we have

$$Q(i,a) \Leftrightarrow (a \in O\ \&\ \forall f\, \exists t\, \varphi_{g(a,i)}(f|t) = 0)\ .$$

The relation $\varphi_{g(a,i)}(f|t) = 0$ is c.e. Applying Theorem 5.8 and the remarks before Theorem 5.6, Q is Π^1_1. By Theorem 5.19 (Single-Valuedness), there is a Π^1_1 function F such that for all i, if there exists a such that $Q(i,a)$, then $Q(i,F(i))$.

Suppose that for all i, $S \cup \{\theta_i\} \notin \mathcal{C}$ (we expect a contradiction). Then $\forall i\,\exists a\, Q(i,a)$, so F is total. By Corollary 5.9, $ran(F)$ is Δ^1_1, so it is contained in the set $\{b \in O : |b| < \alpha\}$, for some computable ordinal α. For some i, $\Gamma_{\alpha,i}$ has a model. However, if $F(i) = b$, then $\Gamma_{|b|,i}$, which is a subset of $\Gamma_{\alpha,i}$, has no model. This is a contradiction. We have finished the proof of the theorem.

The corollary below gives conditions for the existence of a *computable* model. The proof uses what has been called "Keisler's method of expansions" (because of Keisler's many proofs that involve expanding some known structure in a clever way).

Corollary 8.4 *Let Γ be a Π^1_1 set of computable infinitary sentences. If every Δ^1_1 subset of Γ has a computable model, then Γ has a computable model.*

Proof: Let L be the language of Γ, and let B be an infinite computable set of new constants. Below, we shall describe a sentence φ, in the language $L \cup B$,

8.2. THE COMPACTNESS THEOREM

whose models are exactly the computable structures $\mathcal{B}$, with elements named by constants in B. First, let

$$\psi = \forall x \, (\bigvee_{b \in B} x = b) \,.$$

Let At be the set of atomic sentences in the language $L \cup B$. For each $\alpha \in At$ and each $e \in \omega$, we can find a pair of computable Σ_1 sentences $\eta(e, \alpha, 1)$, $\eta(e, \alpha, 0)$, built up out of $\top$ and $\bot$, such that

$$\eta(e, \alpha, 1) \text{ is logically equivalent to } \begin{cases} \top & \text{if } \varphi_e(\alpha) = 1 \\ \bot & \text{otherwise} \end{cases},$$

$$\eta(e, \alpha, 0) \text{ is logically equivalent to } \begin{cases} \top & \text{if } \varphi_e(\alpha) = 0 \\ \bot & \text{otherwise} \end{cases}.$$

Let

$$\psi_e = \bigwedge_{\alpha \in At} ([\alpha \to \eta(e, \alpha, 1)] \,\&\, [\neg \alpha \to \eta(e, \alpha, 0)]) \,.$$

Finally, let

$$\varphi = (\psi \,\&\, \bigvee_e \psi_e) \,.$$

Claim: For any set Λ of L-sentences, Λ has a (non-empty) computable model if and only if $\Lambda \cup \{\varphi\}$ has a model.

Proof of Claim: First, suppose that Λ has a (non-empty) computable model $\mathcal{A}$. There is a computable function f from B onto $\mathcal{A}$. Then f interprets the constants in such a way that we can decide which sentences of At are true. We now forget the original names of the elements (in $|\mathcal{A}|$), and keep only the new names (in B). We have converted $\mathcal{A}$ into a model $\mathcal{B}$, still computable, with elements named by constants in B. Taking e to be a computable index for the set of true sentences of At, we see that $\mathcal{B}$ satisfies ψ_e, so it satisfies φ.

Now, suppose that $\Lambda \cup \{\varphi\}$ has a model $\mathcal{A}$. Each element of $\mathcal{A}$ is named by some constant from B, and we can decide the truth of atomic sentences involving these constants. Therefore, the model $\mathcal{A}$ is computable. We have proved the claim.

We return to the proof of the corollary. Since every Δ_1^1 subset of Γ has a computable model, the claim says that every Δ_1^1 subset of $\Gamma \cup \{\varphi\}$ has a model. By Theorem 8.3, $\Gamma \cup \{\varphi\}$ has a model. Therefore, by the claim, Γ has a computable model. This completes the proof.

It is clear that Π_1^1 sets of sentences are important. It may be difficult at first to see how we might establish that there is a Π_1^1 set of sentences with some desired meaning. One common way is as follows

Proposition 8.5 *Suppose g is a partial computable function that assigns to each $a \in O$ a computable infinitary sentence (or an index for such). Then $ran(g|O)$ is Π_1^1.*

This is immediate from Theorem 5.8.

The next result says that computable structures are expandable.

Theorem 8.6 *Let $\mathcal{A}$ be a computable structure. Let Γ be a Π_1^1 set of sentences involving one or more new relation symbols (not in the language of $\mathcal{A}$).*

(a) If for every Δ_1^1 set $\Lambda \subseteq \Gamma$, $\mathcal{A}$ can be expanded to a model of Λ, then $\mathcal{A}$ can be expanded to a model of Γ.

(b) If for every Δ_1^1 set $\Lambda \subseteq \Gamma$, $\mathcal{A}$ can be expanded to a computable model of Λ, then $\mathcal{A}$ can be expanded to a computable model of Γ.

Proof: First, let φ be the conjunction of $D(\mathcal{A})$ (the atomic diagram of $\mathcal{A}$) and the sentence

$$\forall x \bigvee_{a \in \mathcal{A}} x = a \ .$$

For any structure $\mathcal{B}$ for the language of $\mathcal{A}$, $\mathcal{B} \cong \mathcal{A}$ if and only if $\mathcal{B}$ can be expanded to a model $\mathcal{B}'$ of φ. If $\mathcal{B}'$ is computable, then we have a computable isomorphism taking each $a \in \mathcal{A}$ to its interpretation in $\mathcal{B}'$. Let $\Gamma^* = \Gamma \cup \{\varphi\}$. For (a), we apply Theorem 8.3 to get a model of Γ^*. Since some copy of $\mathcal{A}$ can be expanded to a model of Γ, $\mathcal{A}$ itself can be. For (b), we apply Corollary 8.4 to get a computable model $\mathcal{B}'$ of Γ^*. There is a computable isomorphism from $\mathcal{A}$ onto the appropriate reduct $\mathcal{B}$ of $\mathcal{B}'$. From this, it follows that $\mathcal{A}$ can be expanded to a computable model of Γ.

Remarks: Theorem 8.6 extends directly to hyperarithmetical structures $\mathcal{A}$, and to a still more general class of structures, which we describe in the next section. We can vary Theorem 8.6 further to produce a model of a Π_1^1 set of sentences (possibly an expansion of a given hyperarithmetical structure) in which a certain (new) relation is computable, or Σ_α^0, for some computable ordinal α.

The next result says that computable, or hyperarithmetical, structures have a certain "homogeneity" property. This result provides some evidence of the expressive power of computable infinitary formulas.

Theorem 8.7 *If $\mathcal{A}$ is a computable structure, then for any pair of tuples satisfying the same computable infinitary formulas in $\mathcal{A}$, there is an automorphism taking one to the other. The same is true if $\mathcal{A}$ is a hyperarithmetical structure.*

Proof: Let $\mathcal{F}$ be the set of finite functions between tuples of $\mathcal{A}$ that satisfy the same computable infinitary formulas. It is enough to show that $\mathcal{F}$ has the back-and-forth property. Say $p \in \mathcal{F}$ maps $\bar{a}$ to $\bar{b}$, and consider $c \in \mathcal{A}$. Let e be a new constant. Let $\bar{u}$ be a tuple of variables of the same length as $\bar{a}$ and $\bar{b}$.

8.3. HYPERARITHMETICAL SATURATION

For each computable infinitary formula $\varphi(\overline{u}, x)$, we have a computable infinitary sentence saying

$$(\varphi(\overline{a}, c) \leftrightarrow \varphi(\overline{b}, e)) .$$

Let Γ be the Π_1^1 set of these sentences.

We must show that $\mathcal{A}$ can be expanded to a model of Γ. By Theorem 8.6, it is enough to show that for every Δ_1^1 set $\Lambda \subseteq \Gamma$, $\mathcal{A}$ has an expansion satisfying Λ. We have a Δ_1^1 set Σ of formulas such that for each of the formulas φ giving rise to a sentence in Λ, Σ includes both φ and $neg(\varphi)$ (where $neg(\varphi)$ is logically equivalent to $\neg\varphi$). Let $\Theta = \{\varphi(\overline{u}, x) \in \Sigma : \mathcal{A} \models \varphi(\overline{a}, c)\}$.

Claim: Θ is Δ_1^1.

Proof of Claim: By Theorem 7.5, given an index for φ in S_b^Σ or S_b^Π, we can find an index for $\varphi^{\mathcal{A}}$, as a set c.e. or co-c.e. relative to Δ_b^0, and we can replace this by a Σ_1^1 index, $g(\varphi)$, or a Π_1^1 index $h(\varphi)$. Then

$$\varphi \in \Theta \ \Leftrightarrow \ (\varphi \in \Lambda \ \& \ E_\Sigma^1(g(\varphi), \overline{a}, c))$$
$$\Leftrightarrow \ (\varphi \in \Lambda \ \& \ E_\Pi^1(h(\varphi), \overline{a}, c)) .$$

Therefore, by Theorem 5.8, Θ is Δ_1^1.

By the comments after Theorem 7.5, there is a computable infinitary formula $\psi(\overline{u}, x)$ that is logically equivalent to the conjunction of Θ. Since

$$\mathcal{A} \models \exists x\, \psi(\overline{a}, x) ,$$

we have

$$\mathcal{A} \models \exists x\, \psi(\overline{b}, x) .$$

Letting e be an element of $\mathcal{A}$ that satisfies $\psi(\overline{b}, x)$, we have an expansion of $\mathcal{A}$ satisfying Γ'.

For a hyperarithmetical structure $\mathcal{A}$, in the proof of the claim, we use Proposition 5.21 to pass from an index for $\varphi^{\mathcal{A}}(\overline{u}, x)$ as a set c.e. or co-c.e. relative to $\Delta_b^0(\mathcal{A})$ (for some $b \in O$) to an index as a set c.e. or co-c.e. relative to Δ_c^0, where c (also in O) can be computed from b.

In the next section, we say how the conclusions of Theorem 8.6 and Theorem 8.7 both follow from a certain "saturation" property, which is trivially satisfied by all hyperarithmetical structures.

8.3 Hyperarithmetical saturation

In Chapter 3, we defined a notion of saturation for elementary first order logic. We are about to define "hyperarithmetical" saturation, for infinitary logic. This is a special case of Ressayre's notion of Σ-*saturated* [133].

A structure $\mathcal{A}$ is *hyperarithmetically saturated* if it satisfies the following conditions:

(a) for all tuples $\bar{a}$ in $\mathcal{A}$ and all Π^1_1 sets $\Gamma(\bar{a},x)$ of computable infinitary formulas with parameters $\bar{a}$, if every Δ^1_1 set $\Gamma'(\bar{a},x) \subseteq \Gamma(\bar{a},x)$ is satisfied in $\mathcal{A}$, then $\Gamma(\bar{a},x)$ is satisfied,

(b) for all tuples $\bar{a}$ in $\mathcal{A}$ and all Π^1_1 sets Λ of pairs $(i, \gamma(\bar{a}))$, where $i \in \omega$ and $\gamma(\bar{a})$ is a computable infinitary sentence with parameters $\bar{a}$, if for every Δ^1_1 set $\Lambda' \subseteq \Lambda$,

$$\mathcal{A} \models \bigvee_{i} \bigwedge_{(i,\gamma) \in \Lambda'} \gamma(\bar{a})$$

then

$$\mathcal{A} \models \bigvee_{i} \bigwedge_{(i,\gamma) \in \Lambda} \gamma(\bar{a}) \ .$$

Finite structures trivially satisfy the definition of saturation from Chapter 3. Similarly, hyperarithmetical structures trivially satisfy the definition of hyperarithmetical saturation above.

Remark: If a structure $\mathcal{A}$ is hyperarithmetically saturated, then so is $(\mathcal{A}, \bar{c})$, for any finite tuple $\bar{c}$.

Theorems 8.6 and 8.7 extend to arbitrary hyperarithmetically saturated structures. We are almost ready to state the generalizations, but we need one more definition. If L' is an extension of the language L, and Γ is a set of L'-sentences, then the *consequences* of Γ, in the language L, are the L-sentences true in all models of Γ. Note that this definition does not involve any system of proof.

Theorem 8.8 *Suppose that $\mathcal{A}$ is a hyperarithmetically saturated structure, and Γ is a Π^1_1 set of sentences involving symbols from the language $L^{\mathcal{A}}$, plus finitely many new symbols. If the consequences of Γ in the language of $\mathcal{A}$ are all true in $\mathcal{A}$, then $\mathcal{A}$ can be expanded to a model $\mathcal{A}'$ of Γ. Moreover, $\mathcal{A}'$ can be taken to be hyperarithmetically saturated.*

Exercise: Prove Theorem 8.8. In the construction of $\mathcal{A}'$, it is useful to maintain the condition that what has been determined at each stage is a Δ^1_1 set of sentences $\Sigma(\bar{a})$, in the language L' with finitely many added parameters $\bar{a}$, such that the consequences of $\Gamma \cup \Sigma(\bar{a})$, in the language L with the added parameters $\bar{a}$, are all true in $\mathcal{A}$.

Applying Theorem 8.8 to the trivial structure, the set ω with no relations, we get the fact that if a Π^1_1 set of computable infinitary sentences has a model, then it has a hyperarithmetically saturated model.

Theorem 8.9 *If the structure $\mathcal{A}$ is hyperarithmetically saturated, and $\bar{a}$ and $\bar{b}$ are tuples in $\mathcal{A}$ satisfying the same computable infinitary formulas, then there is an automorphism of $\mathcal{A}$ taking $\bar{a}$ to $\bar{b}$.*

8.4 Orderings and trees

The following somewhat surprising result is due to Spector [147].

Theorem 8.10 (Spector) *For any well ordering $\mathcal{A}$, the following are equivalent:*

(a) $\mathcal{A}$ has a computable copy,
(b) $\mathcal{A}$ has a hyperarithmetical copy.

Proof: Clearly, (a) $\Rightarrow$ (b). To show that (b) $\Rightarrow$ (a), we suppose not, expecting a contradiction. Say $\mathcal{A} = (A, <_A)$ is a hyperarithmetical well ordering of order type $\alpha \geq \omega_1^{CK}$. Take B to be an infinite computable set disjoint from A (altering A, if necessary). Let

$$\mathcal{A}^* = (A \cup B, A, <_A, B) \ .$$

We define a Π_1^1 set Γ of computable infinitary sentences involving new binary relation symbols $<_B$ and F. First, Γ includes a finite set Γ_0 of sentences saying that $<_B$ is a computable linear ordering on B and F is an isomorphism from an initial segment of $(A, <_A)$ onto $(B, <_B)$. In addition, Γ includes, for each $a \in O$, a sentence ψ_a saying that $(B, <_B)$ has an initial segment of order type $|a|$ (these sentences were described in Chapter 7).

Each Δ_1^1 subset of Γ is contained in

$$\Gamma_\beta = \Gamma_0 \cup \{\psi_c : c \in O \ \& \ |c| < \beta\} \ ,$$

for some computable ordinal β (see Proposition 5.20). There is an expansion of $\mathcal{A}^*$ in which (letting $<_B$ and F be the names for their interpretations) F is the unique isomorphism between the initial segment of $\mathcal{A}$ of type β and $(B, <_B)$. Thus, each Δ_1^1 subset of Γ is satisfied in some expansion of $\mathcal{A}^*$. Then by Theorem 8.6, $\mathcal{A}^*$ has an expansion satisfying all of Γ. In this expansion, $(B, <_B)$ must be a well ordering, since it is isomorphic under F to an initial segment of $\mathcal{A}$. The order type of $(B, <_B)$ is greater than any computable ordinal. Since $(B, <_B)$ is a computable ordering, this is a contradiction.

While the ordering ω_1^{CK} has no computable (or hyperarithmetical) copy, it can be embedded as an initial segment of a computable ordering (see [138]).

Theorem 8.11 (Harrison) *There is a computable linear ordering of type*

$$\omega_1^{CK} \cdot (1 + \eta) \ .$$

Proof: For each $a \in O$, we can find a computable infinitary sentence σ_a saying, in linear orderings, that for each x, there exists y such that the half-open interval $[x, y)$ has order type $|a|$. For each $a \in O$ and $e \in \omega$, we can find a computable infinitary sentence $\rho_{a,e}$ saying that $\varphi_e^{\Delta_1^0}$ is not an infinite decreasing sequence. Let Γ consist of the usual axioms for linear orderings and the sentences σ_a and $\rho_{a,e}$. The set Γ is Π_1^1, and every Δ_1^1 subset has a computable model. Therefore, by Corollary 8.4, Γ has a computable model $\mathcal{A}$.

We must show that $\mathcal{A}$ has the desired order type.

Claim: Each $a \in \mathcal{A}$ belongs to a maximal well-ordered interval of type ω_1^{CK}. (The interval is not unbounded, and it has no least upper bound in $\mathcal{A}$.)

Proof of Claim: It is clear from the axioms that each element a belongs to an interval I of type ω_1^{CK}, where I has first point a. Clearly, the interval I is not unbounded and has no least upper bound in $\mathcal{A}$. It may not be maximal well-ordered. There is no well-ordered interval of order type greater than ω_1^{CK}, so we cannot extend I by adding elements on the right. However, there may exist $b \leq a$ such that the interval $[b, a]$ is well ordered, and then the union of I and $[a, b]$ is well-ordered. If there is no maximal well ordered interval containing a, then for any $b \leq a$ such that the interval $[b, a]$ is well ordered, there exists c such that $c < b$ and $[c, a]$ is also well ordered. Let

$$Q = \{(b, c) : c < b \leq a \,\&\, [c, a] \text{ has order type } \alpha \text{ for some } \alpha < \omega_1^{CK}\} \ .$$

It is easy to see that Q is Π_1^1. By Theorem 5.19 (Single-Valuedness), there is a partial Π_1^1 function $F \subseteq Q$ such that for all b, if there exists c such that $Q(b, c)$, then $Q(b, F(b))$. Let f be the function on ω such that $f(0) = a$, and for all n, $f(n+1) = F(f(n))$. For each n, the interval $[f(n), a]$ is well ordered, so $F(f(n))$ is defined. There is a computable relation $C(\sigma, x, y)$ such that

$$f(x) = y \Leftrightarrow (\exists \sigma \subseteq F)\, C(\sigma, x, y) \ .$$

Then by Theorem 5.8, f is Π_1^1. Since $dom(f) = \omega$, f is Δ_1^1, by Corollary 5.9. This is a contradiction.

Since each element of $\mathcal{A}$ lies on a maximal well ordered interval of order type ω_1^{CK}, $\mathcal{A}$ has order type $\omega_1^{CK} \cdot \rho$, for some ρ. Now, $\mathcal{A}$ has a first element, for otherwise, we could determine an infinite computable sequence that is strictly decreasing. It follows that ρ has a first element. There is no last element of ρ, for if there were, then $\mathcal{A}$ would have an element a such that the interval to the right of a has order type ω_1^{CK}. Similarly, there is no pair of successors in ρ, for if there were, then $\mathcal{A}$ would have elements a and b such that the interval $[a, b)$ has order type ω_1^{CK}. Therefore, ρ has order type $1 + \eta$. This completes the proof of the theorem.

The construction in Theorem 8.11 yields the following.

Corollary 8.12 *There is a computable linear ordering $\mathcal{A}$ such that $\mathcal{A}$ is not well ordered, and $\mathcal{A}$ has no infinite hyperarithmetical decreasing sequence.*

8.5. BOOLEAN ALGEBRAS

Here is another simple corollary of Theorem 8.11.

Corollary 8.13 *There is a computable linear ordering $\mathcal{A}$ such that $\mathcal{A}$ has $2^{\aleph_0}$ automorphisms but there is no non-trivial hyperarithmetical automorphism.*

Proof: Let $\mathcal{A}$ be the linear ordering from Theorem 8.11. Clearly, $\mathcal{A}$ has $2^{\aleph_0}$ automorphisms—induced by the automorphisms of η. Suppose f is a hyperarithmetical automorphism of $\mathcal{A}$ such that $f(a) \neq a$, for some $a \in \mathcal{A}$. If $f(a) < a$, then since f preserves the ordering,

$$f(f(a)) < f(a), f(f(f(a))) < f(f(a)) \text{, etc.}$$

Therefore, there is an infinite hyperarithmetical decreasing sequence

$$a > f(a) > f(f(a)) > \ldots.$$

If $a < f(a)$, then again there is an infinite hyperarithmetical decreasing sequence, defined in the same way except that f^{-1} replaces f. This completes the proof of the corollary.

Suppose that P is a hyperarithmetical subtree of $\omega^{<\omega}$ with no path. Then the Kleene-Brouwer ordering on P is a hyperarithmetical well ordering, so by Theorem 8.10, the order type is a computable ordinal.

Proposition 8.14 (Kleene) *There is a computable tree P, infinitely branching, such that P has no hyperarithmetical path.*

Proof: Take the set of finite decreasing sequences in the ordering from Theorem 8.11. This is a tree P with the required features.

8.5 Boolean algebras

There is a close connection between linear orderings and Boolean algebras. Given a linear ordering $\mathcal{A}$, we pass effectively to the *interval algebra* $I(\mathcal{A})$ (see the appendix). The next result says that we can also go in the other direction, although not uniquely.

Proposition 8.15 *For any Boolean algebra $\mathcal{A}$, there is a linear ordering $\mathcal{B}$ such that $I(\mathcal{B}) \cong \mathcal{A}$ and $\mathcal{B} \leq_T \mathcal{A}$.*

Proof: If $\mathcal{A}$ is the one-element Boolean algebra ($1^{\mathcal{A}} = 0^{\mathcal{A}}$), then we give $\mathcal{B}$ order type 0, so that $\emptyset$ is the only element of $I(\mathcal{B})$. If $\mathcal{A}$ is the two-element Boolean algebra, then $\mathcal{B}$ will have order type 1. For more general Boolean algebras $\mathcal{A}$, we obtain an ordering $\mathcal{B}$ with the desired property in stages, as the union of an increasing chain of finite orderings $\mathcal{B}_s$. The elements come from an infinite computable set of constants B.

We choose $\mathcal{B}_s$ such that for the algebra $\mathcal{A}_s$ generated by the first s elements of $\mathcal{A}$, $I(\mathcal{B}_s) \cong \mathcal{A}_s$. The universe consists of the first few constants from $\mathcal{B}$. Each atom of $\mathcal{A}_s$ corresponds to an interval in $\mathcal{B}_s$ of the form $[b, b')$, where b' is the successor of b (or $[b, \infty)$ if b is the last point). If the atom is split in $\mathcal{A}_{s+1}$, then we put into $\mathcal{B}_{s+1}$ a new element b'' such that $b < b'' < b'$ (or $b < b''$, if there is no b').

A countable superatomic Boolean algebra has isomorphism type $I(\alpha)$, for some countable ordinal α. When α is expressed in Cantor normal form, the leading term $\omega^\beta \cdot n$ determines the Boolean algebra, since for $\gamma < \beta$, any γ-atoms can be absorbed by the β-atoms. Thus, any countable superatomic Boolean algebra has isomorphism type $I(\omega^\beta \cdot n)$, for some countable ordinal β and some $n \in \omega$.

Proposition 8.16 *For a superatomic Boolean algebra $\mathcal{A}$, the following are equivalent:*
  (a) $\mathcal{A}$ has a computable copy,
  (b) $\mathcal{A}$ has a hyperarithmetical copy.

Proof sketch: Clearly, (a) $\Rightarrow$ (b). Suppose that $\mathcal{A}$ has a hyperarithmetical copy. Since $\mathcal{A}$ is superatomic, $\mathcal{A} \cong I(\alpha)$, for some countable ordinal α. If $\alpha \geq \omega_1^{CK}$, then $I(\omega_1^{CK})$ has a hyperarithmetical copy $\mathcal{B}$. Note that for each element a of $I(\omega_1^{CK})$, either a or its complement is a join of finitely many β-atoms, for some computable ordinal β. Let e be a new constant. We have a Π_1^1 set Γ made up of computable infinitary sentences $\varphi_a(e)$, for $a \in O$, saying that neither e nor its complement is the join of finitely many $|a|$-atoms. Since each Δ_1^1 subset of Γ is satisfied in an expansion of $\mathcal{B}$, the whole set is, by the remark after Theorem 8.6. This is a contradiction. Therefore, α is a computable ordinal, and $\mathcal{A}$ has a computable copy.

8.6 Groups

The mathematical structure of countable Abelian p-groups is well understood. A *reduced* Abelian p-group G (an Abelian p-group in which every element except the identity has a height) is characterized, up to isomorphism, by its *Ulm sequence* $u_\alpha(G)_{\alpha < \lambda(G)}$, where $\lambda(G)$ is the *length* of G. An *arbitrary* (not necessarily reduced) Abelian p-group G can be expressed as the direct sum of a divisible part, consisting of elements that do not have a height, and a reduced part. The divisible part is unique. The reduced part is not unique (as a set), but its isomorphism type is unique. A countable Abelian p-group is characterized up to isomorphism by the "dimension" of the divisible part and the Ulm sequence of the reduced part. (See the appendix for a more complete description of these mathematical invariants.)

The next result says what lengths are possible for computable Abelian p-groups, both reduced and arbitrary.

Theorem 8.17 *Let G be a computable Abelian p-group. Then*
  (a) $\lambda(G) \leq \omega_1^{CK}$,
  (b) if G is reduced, then $\lambda(G) < \omega_1^{CK}$.

Proof sketch: (a) Suppose that $\lambda(G) > \omega_1^{CK}$. There exists $a \in G$ such that $height(a) = \omega_1^{CK}$. There is a Π_1^1 set $\Gamma(a,x)$ of computable infinitary formulas saying that $px = a$, and for each computable ordinal α, $height(x) \geq \alpha$. Each Δ_1^1 subset of $\Gamma(a,x)$ is satisfied in G. Therefore, by Corollary 8.6, some $b \in G$ satisfies all of $\Gamma(a,x)$. Then $height(a) > height(b) \geq \omega_1^{CK}$, a contradiction.

(b) Suppose that $\lambda(G) = \omega_1^{CK}$. There is a Π_1^1 set $\Lambda(x)$ of computable infinitary formulas saying that $px = 0$, and for each computable ordinal α, $height(x) \geq \alpha$. Each Δ_1^1 subset of $\Lambda(x)$ is satisfied. Therefore, by Theorem 8.6, some $a \in G$ satisfies all of $\Lambda(x)$. Then $height(a) \geq \omega_1^{CK}$, a contradiction.

8.7 Priority constructions

We shall give many priority constructions in succeeding chapters. In Chapters 13 and 14, we attempt a formal description of a fairly general class of these constructions. Informally, the features typical of a priority construction are as follows. A c.e. set is being enumerated while a list of requirements is satisfied. The requirements use some high-level information, and there is a systematic scheme for generating computable approximations for the high-level information. There are attempts to satisfy the requirements, based on approximate information. Finally, there is a system of priorities that determines which requirement receives attention, and which are "injured"—previous work on them being scrapped—as the approximations change.

Chapter 14 has a general "metatheorem", with conditions guaranteeing the success of a priority construction in which the information for satisfying the requirements is Δ_α^0, for an arbitrary computable ordinal α. We see the Barwise-Kreisel Compactness Theorem as a similar result. It gives conditions guaranteeing the success of a priority construction in which the list of requirements, with relevant information, is Π_1^1. To apply the Compactness Theorem, we must formulate our requirements as computable infinitary sentences, making up a Π_1^1 set, and we must show that every Δ_1^1 subset can be satisfied.

The metatheorem in Chapter 14 will provide one possible method for satisfying Δ_1^1 sets of requirements. In principal at least, when we apply the methods of Chapter 14, we can see everything that happens on the computable level—the computable sequence of approximations, the actions, and the injuries. When we apply Compactness, the proof picks out some cohesive set of partial constructions for us. The theorem does not show us the computable sequence of steps, with approximations of information, actions, and injuries.

This is all rather vague. We hope that the following three examples will illustrate the kind of construction that we have in mind. The first example concerns the image of a relation in computable copies of a structure. In Chapter 16, we give conditions, for a computable ordinal α, a computable structure $\mathcal{A}$, and a relation R on $\mathcal{A}$, guaranteeing that $\mathcal{A}$ has a computable copy in which the image of R is not Σ_α^0. If $\mathcal{A}$ and R satisfy these conditions for all computable ordinals α, then the result below gives a computable copy of $\mathcal{A}$ in which the image of R is not hyperarithmetical.

Theorem 8.18 *Let $\mathcal{A}$ be a computable structure, and let R be a further relation on $\mathcal{A}$. Suppose that for each computable ordinal α, there is a computable copy of $\mathcal{A}$ in which the image of R is not Σ_α^0. Then there is a computable copy of $\mathcal{A}$ in which the image of R is not hyperarithmetical.*

Proof: We suppose that R is hyperarithmetical. Otherwise, there is nothing to prove. Let L be the language of $\mathcal{A}$. Let A be the universe of $\mathcal{A}$, considered as a set of constants, and let B be an infinite computable set of constants, disjoint from $\mathcal{A}$. We describe a set Γ of sentences in the language $L \cup A \cup B$ such that any model will yield a computable structure $\mathcal{B}$ with an isomorphism F from $\mathcal{A}$ to $\mathcal{B}$ such that $F(R)$ is not hyperarithmetical.

First, there is a computable infinitary sentence φ, in the language $L \cup B$, such that the models of φ are the computable structures with elements named by constants from B. Second, there is a computable infinitary sentence ψ, in the language $L \cup A$, such that for all L-structures $\mathcal{B}$,

$$\mathcal{B} \cong \mathcal{A} \Leftrightarrow \mathcal{B} \text{ can be expanded to a model of } \psi \ .$$

The sentence ψ is the conjunction of the atomic diagram of $\mathcal{A}$ and one further sentence saying

$$\forall x \bigvee_{a \in A} x = a \ .$$

Now, $\mathcal{B}$ is a computable copy of $\mathcal{A}$ if and only if $\mathcal{B}$ can be expanded to a model of $\varphi \ \& \ \psi$. We could have added a symbol for the isomorphism F, but we did not. For $c \in \mathcal{A}$ and $d \in \mathcal{B}$, the sentence $c = d$ has the meaning $F(c) = d$. Then the sentence

$$\bigvee_{c \in R} c = d$$

has the meaning $d \in F(R)$. For all $a \in O$, we can find a computable infinitary sentence θ_a saying that $F(R)$ is not c.e. relative to $H(a)$.

We let Γ be the Π_1^1 set of sentences consisting of φ, ψ, and the sentences θ_a, for $a \in O$. By our hypothesis, every Δ_1^1 subset of Γ has a model. Therefore, by Theorem 8.3, Γ has a model, and this yields the desired computable copy $\mathcal{B}$ of $\mathcal{A}$, with the isomorphism F as described.

The next result concerns isomorphisms between computable copies of a structure. In Chapter 17, we give conditions, for a computable ordinal α, and a computable structure $\mathcal{A}$, guaranteeing that $\mathcal{A}$ has a computable copy with some isomorphism that is not Δ_α^0. If $\mathcal{A}$ satisfies the conditions for *all* computable ordinals α, then the result below gives a computable copy of $\mathcal{A}$ with an isomorphism that is not hyperarithmetical.

Theorem 8.19 *Let $\mathcal{A}$ be a computable structure. Suppose that for each computable ordinal α, there is an isomorphism F from $\mathcal{A}$ onto a computable copy $\mathcal{B}$ such that F is not Δ_α^0. Then there is an isomorphism F from $\mathcal{A}$ onto a computable copy $\mathcal{B}$ such that F is not hyperarithmetical.*

8.7. PRIORITY CONSTRUCTIONS

Proof: We proceed as in the proof of Theorem 8.18, modifying the set Γ. Here Γ consists of the sentences φ, ψ (as above) and, for each $a \in O$ and $e \in \omega$, a new sentence $\theta_{a,e}$ saying that

$$\varphi_e^{\Delta_a^0} \neq F,$$

where F is the isomorphism described above, such that for $c \in \mathcal{A}$ and $d \in \mathcal{B}$, $F(c) = d$ if and only if the sentence $c = d$ holds in the expansion of $\mathcal{B}$ that we are building.

The third application of Compactness to a priority construction also concerns isomorphisms between computable copies of a structure. Theorem 8.19 gave conditions guaranteeing the existence of a copy with *some* isomorphism that is not hyperarithmetical. Of course, there may be further isomorphisms, and some of these may be hyperarithmetical. The next result gives conditions guaranteeing the existence of a copy with *no* hyperarithmetical isomorphism.

Theorem 8.20 *Let $\mathcal{A}$ be a computable structure. Suppose that for each computable ordinal α, there is a computable $\mathcal{B} \cong \mathcal{A}$ with no Δ_α^0 isomorphism from $\mathcal{A}$ onto $\mathcal{B}$. Then there is a computable $\mathcal{B} \cong \mathcal{A}$ with no hyperarithmetical isomorphism from $\mathcal{A}$ onto $\mathcal{B}$.*

Proof: We proceed as in the proofs of Theorems 8.18 and 8.19, again modifying the set Γ. Here Γ consists of φ and ψ (as above), and, for each $a \in O$ and $e \in \omega$, a new sentence $\theta_{a,e}$ saying that $\varphi_e^{\Delta_a^0}$ is not an isomorphism from $\mathcal{A}$ onto $\mathcal{B}$.

In Chapter 3, we mentioned the lattice $\mathcal{E}$ of c.e. sets, which can be obtained as a Π_2^0 quotient structure in which the elements of the underlying computable structure are natural numbers—indices of c.e. sets, and the congruence relation is

$$a \sim b \Leftrightarrow W_a = W_b \ .$$

An automorphism F of $\mathcal{E}$ is considered to be Δ_α^0 if it is induced by a Δ_α^0 function on the natural numbers—i.e., there is a Δ_α^0 function $f : \omega \to \omega$ such that for all a, $F(W_a) = W_{f(a)}$. Cholak and Harrington have announced the following result [29].

Theorem 8.21 *For each computable ordinal α, there exist a and b such that some automorphism of $\mathcal{E}$ takes W_a to W_b, but no Δ_α^0 automorphism does this.*

Combining Theorem 8.20 with Compactness (Theorem 8.6), we obtain the following.

Corollary 8.22 *There exist c and c' such that some automorphism of $\mathcal{E}$ takes W_c to $W_{c'}$, but no hyperarithmetical automorphism does this.*

Proof: We let Γ be a set of sentences, saying in expansions of the standard model of arithmetic $\mathcal{N}$, with added unary function F and constants c and c', that F induces an automorphism of $\mathcal{E}$, $F(c) = c'$, and for each $a \in O$ and each $e \in \omega$, if $\varphi_e^{\Delta_a^0}$ induces an automorphism of $\mathcal{E}$, then $\varphi_e^{\Delta_a^0}(c) \not\sim c'$. By Theorem 8.21, every Δ_1^1 subset of Γ is satisfied (in an expansion of $\mathcal{N}$). Then by Theorem 8.6, the whole set Γ is satisfied. Therefore, we have the desired elements of $\mathcal{E}$.

8.8 Ranks

Recall the notions of rank defined in Chapter 6. We can set upper bounds on the ranks for computable, or hyperarithmetical, structures.

Corollary 8.23 *Suppose $\mathcal{A}$ is a computable structure. Then $r(\mathcal{A}) \leq \omega_1^{CK}$, $R(\mathcal{A}) \leq \omega_1^{CK} + 1$, and $SR(\mathcal{A}) \leq \omega_1^{CK} + 1$. The same is true if $\mathcal{A}$ is hyperarithmetical, or if it is hyperarithmetically saturated.*

Proof: By Theorem 8.7 (or Theorem 8.9), if all computable infinitary formulas true of $\bar{a}$ are true of $\bar{b}$, then all infinitary formulas true of $\bar{a}$ are true of $\bar{b}$. From this, it follows that $r(\mathcal{A}) \leq \omega_1^{CK}$. Then by Proposition 6.8, $R(\mathcal{A}) \leq \omega_1^{CK} + 1$. Since $SR(\mathcal{A}) \leq R(\mathcal{A})$, we have $SR(\mathcal{A}) \leq \omega_1^{CK} + 1$.

In Chapter 15, we shall see that the maximal possible ranks are attained by the ordering of type $\omega_1^{CK}(1 + \eta)$; i.e.,

$$r(\omega_1^{CK}(1 + \eta)) = \omega_1^{CK}, \text{ and}$$
$$R(\omega_1^{CK}(1 + \eta)) = SR(\omega_1^{CK}(1 + \eta)) = \omega_1^{CK} + 1 \ .$$

Other familiar examples of hyperarithmetical structures $\mathcal{A}$ with $r(\mathcal{A}) = \omega_1^{CK}$ behave in the same way; that is,

$$R(\mathcal{A}) = SR(\mathcal{A}) = \omega_1^{CK} + 1 \ .$$

However, Makkai [111] showed that there is a hyperarithmetical structure $\mathcal{A}$ such that $SR(\mathcal{A}) = \omega_1^{CK}$. This example is actually arithmetical. There are as yet no computable examples.

Chapter 9

Existence of computable structures

For a class of structures that is well understood mathematically, it is natural to ask which members have computable copies. (We consider only countable structures.) For algebraically closed fields of a given characteristic, or vector spaces over a given computable field, the answer is trivial—they all have computable copies. For models of PA, the answer is again trivial. We saw in Chapter 3 that only the standard model has a computable copy. For equivalence structures, we can give a complete answer in terms of the natural mathematical invariants of these structures. For reduced Abelian p-groups, we can give at least a partial answer, in terms of the natural mathematical invariants.

We give several other results on structures with and without computable copies. We close with a result on homogeneous structures with *decidable* copies.

9.1 Equivalence structures

An *equivalence structure* has the form $\mathcal{A} = (A, \sim)$, where $\sim$ is an equivalence relation on the set A (see the Appendix). Clearly, each equivalence structure $\mathcal{A}$ is characterized up to isomorphism by the number of equivalence classes of various sizes. Let $c_n(\mathcal{A})$ be the number of classes of size n, and let $c_\infty(\mathcal{A})$ be the number of infinite classes. These invariants characterize an equivalence structure up to isomorphism. Let

$$R(\mathcal{A}) = \{(n, k) : c_n(\mathcal{A}) \geq k\}.$$

The result below characterizes the equivalence structures that have computable copies, in terms of the invariants.

Theorem 9.1 *Let $\mathcal{A}$ be an equivalence structure. Then $\mathcal{A}$ has a computable copy if and only if:*
  (a) $R(\mathcal{A})$ is Σ_2^0, and

(b) one of the following holds:
  (i) $\{n : c_n(\mathcal{A}) \neq 0\}$ is finite,
  (ii) $c_\infty(\mathcal{A}) = \infty$,
  (iii) there is a computable function f such that for each n, $f(n,s)$ is non-decreasing in s, with limit $f^*(n) \geq n$, where $c_{f^*(n)}(\mathcal{A}) \neq 0$.

Proof: Supposing that $\mathcal{A}$ has a computable copy, we show that (a) and (b) hold. For simplicity, we take $\mathcal{A}$ itself to be computable. From the definition, it is clear that $R(\mathcal{A})$ is Σ_2^0. Suppose $\mathcal{A}$ has only finitely many infinite classes. For simplicity, say there are none. Suppose that there are finite classes of infinitely many different sizes. Then we obtain a function f, as described above, in the following way. For each n, we enumerate $D(\mathcal{A})$ until we locate an element a_n whose equivalence class has size at least n, and we compute $f(n,s)$ by consulting the stage s approximation to the diagram and seeing how many of the elements already mentioned are equivalent to a_n.

Now, supposing that (a) and (b) hold, we show that $\mathcal{A}$ has a computable copy. We consider three cases.

Case 1: Suppose $\{n : c_n(\mathcal{A}) \neq 0\}$ is finite.

In this case, the full sequence of invariants is computable (trivially), so $\mathcal{A}$ has a computable copy.

Case 2: Suppose $\{n : c_n(\mathcal{A}) \neq 0\}$ is infinite and $c_\infty(\mathcal{A})$ is finite.

Again, for simplicity, we suppose that $c_\infty(\mathcal{A}) = 0$. Let B be an infinite computable set of constants, which we may identify with ω. We shall refer to the usual ordering on this set. Let f be as in (b), and let $f^*(n)$ be the limit of $f(n,s)$. Let R be as in (a), and let ρ be a Δ_2^0 function enumerating the pairs in $R(\mathcal{A})$, without repetition, such that if $\rho(i) = (n, k+1)$, then there is some $j < i$ such that $\rho(j) = (n, k)$. Guessing initial segments of ρ ($\rho|k$, for various k), we determine a computable $\mathcal{B} \cong \mathcal{A}$ with universe B.

At stage s, we have a guess ρ_s at a finite initial segment of ρ. Based on this, we determine an equivalence structure $\mathcal{B}_s$, with universe a finite initial segment of B. The number of classes of various sizes agrees with ρ_s, except that there may be additional classes "designated" for size $f^*(m)$, for distinct m. We do not know the value of $f^*(m)$, but we can make sure that a class has this size in the end by adding elements to the class as $f(m,s)$ grows. If, at stage s, a class is newly designated for size $f^*(m)$, then m is chosen to be greater than any size from ρ_s. We give the class size $f(m,s)$. At a later stage $t+1$, we may increase the size of a class designated for size $f^*(m)$ if $f(m, t+1) \neq f(m,t)$, so that the class has size $f(m, t+1)$.

If ρ_s has length k, then ρ_{s+1} is either an extension of ρ_s of length $k+1$ or the result of dropping back and changing the value on some $k' < k$. If there are too *few* classes of some size, then we add some. If there are too *many* classes of some size, say n, then we choose new, distinct numbers m, larger than any of the current classes, and we designate the extra classes of size n for size $f^*(m)$, for these m. We may give a designation and later be forced to replace it by another. When deciding which equivalence classes of size n to leave undisturbed, we give priority to those which were created "earlier"—i.e., the ones whose first element is smaller.

It is not difficult to see that the size of each equivalence class eventually stabilizes. Some classes have designations $f^*(m)$, and some do not. There is a $1-1$ correspondence between pairs (n,k) in R, and equivalence classes, where the pair (n,k) corresponds to the kth class of eventual size n, in the order created. From this, it should be clear that $\mathcal{B} = \cup_s \mathcal{B}_s$ is the desired computable copy of $\mathcal{A}$.

Case 3: Suppose $\{n : c_n(\mathcal{A}) \neq 0\}$ is infinite and $c_\infty(\mathcal{A})$ is infinite.

Then we can determine a computable structure $\mathcal{B} \cong \mathcal{A}$ as in Case 2, except that we make the extra classes infinite instead of designating them for size $f^*(m)$, for various finite m.

9.2 Abelian p-groups

In Chapter 8, we mentioned the invariants for countable Abelian p-groups (for more detailed information, see the Appendix). Khisamiev [76], [77] gave a partial characterization of the reduced Abelian p-groups with computable copies. The result is for groups of of length less than ω^2. In the case of groups G of length ω, the conditions resemble those for equivalence structures. They involve a Σ_2^0 relation
$$R_0 = \{(n,k) : u_n(G) \geq k\}$$
and a computable function $f_0(n,s)$, non-decreasing in s, with limit $f_0^*(n) \geq n$, where $u_{f_0^*(n)} \neq 0$.

Theorem 9.2 (Khisamiev) *Let G be a reduced Abelian p-group such that $\lambda(G) < \omega^2$. Then G has a computable copy if and only if for each i such that $\omega \cdot (i+1) \leq \lambda(G)$, the following hold:*

(a) $R_i = \{(n,k) : u_{\omega \cdot i + n}(G) \geq k\}$ is Σ_{2i+2}^0.

(b) there is a Δ_{2i+1}^0 function f_i such that for all n, $f_i(n,s)$ is non-decreasing in s, with limit $f_i^*(n) \geq n$, where $u_{\omega \cdot i + f_i^*(n)}(G) \neq 0$.

Proof: Suppose that $\mathcal{A}$ is computable. For each i, n, and k, the natural sentence saying that $u_{\omega \cdot i + n}(G) \geq k$ is computable Σ_{2i+2}. It follows that R_i is Σ_{2i+2}^0. We determine functions f_i according to cases. We write pa as an abbreviation for $\overbrace{a + \ldots + a}^{p}$—a term in the language of Abelian groups.

Case 1: Suppose $\omega \cdot (i+1) = \lambda(G)$.

First, using Δ^0_{2i+1}, we find b such that $pb = 0$ and $height(b) \geq \omega \cdot i + n$. Let $f_i(n, 0) = n$. Given $f_i(n, s)$, we search $s + 1$ steps for c such that $pc = b$ and $height(c) \geq \omega \cdot i + f_i(n, s)$. Then

$$f_i(n, s+1) = \begin{cases} f_i(n, s) + 1 & \text{if we find such a } c, \\ f_i(n, s+1) = f_i(n, s) & \text{otherwise}. \end{cases}$$

Case 2: Suppose $\omega \cdot (i + 1) < \lambda(G)$.

First, we (non-effectively) choose an element a of height $\omega \cdot (i + 1)$. Next, using Δ^0_{2i+1}, we find b such that $pb = a$ and $height(b) \geq \omega \cdot i + n$. After this, we proceed as in Case 1.

Now, we show that if the group G is equipped with relations R_i as in (a) and functions f_i as in (b), then G has a computable copy. We obtain our group using a tree. We consider trees that are isomorphic to subtrees of $\omega^{<\omega}$, but we make no use of the fact that the tree elements are finite sequences. We are really thinking of a tree as a structure with a named top element (corresponding to $\emptyset$) and a successor relation. For such a tree P, let $G(P)$ be the Abelian p-group generated by the elements of P under the relations
  (i) $e = 0$, and
  (ii) $pb = a$, where b is a successor of a.

Note: The group $G(P)$ is reduced just in case the tree P has no path.

There is a notion of *height* for elements of P, matching the notion of height in $G(P)$. If a is a *terminal* node—one with no successor, then $height(a) = 0$. If a has successors, then $height(a)$ is the least α such that $\alpha > height(b)$ for all successors b of a. The Ulm sequence for $G(P)$ is determined from P in the following way (see [135], [136]).

Lemma 9.3 (Rogers, Oates) *Let P be a tree (isomorphic to a subtree of $\omega^{<\omega}$). If $G = G(P)$, then $u_\beta(G)$ is the number of nodes a in P such that $height(a) = b$ and one of the following holds:*
  (a) a is at level 1,
  (b) a is at a level greater than 1, and a is a successor of an element b such that $height(b) > \beta + 1$,
  (c) a is at a level greater than 1, and a is a successor of an element b such that $height(b) = \beta + 1$, but a is not the first successor of b—that one is thought of as witnessing the height of b.

We omit the proof of Lemma 9.3 except to note that if a and b are related to each other as in either (b) or (c), then the element $a - b$ has order p, with the same height as a.

9.2. ABELIAN P-GROUPS

Clearly, if P is a computable tree of the kind we are considering, then there is a computable Abelian p-group $G \cong G(P)$. To prove Theorem 9.2, it is enough to show that there is a computable tree P such that $G(P)$ has the given Ulm sequence. We proceed by induction.

Open problem: If G is a computable reduced Abelian p-group, must there be a computable tree P (as above) such that $G \cong G(P)$?

From the proof of Theorem 9.2, we shall see that the answer is positive for groups of length less than ω^2.

The following lemma is the special case of Theorem 9.2 for groups of length ω.

Lemma 9.4 Suppose G is a reduced Abelian p-group such that $\lambda(G) = \omega$,

$$R = \{(n,k) : u_n(G) \geq k\}$$

is Σ_2^0, and f is a computable function such that for each n, $f(n,s)$ is non-decreasing in s, with limit $f^*(n)$, where $u_{f^*(n)}(G) \neq 0$. Then there is a computable tree P such that $G(P) \cong G$.

Sketch of proof: The proof is similar to that for Theorem 9.1. Guessing at a function enumerating the pairs in R, and using the function f to take care of wrong guesses, we produce a tree with nodes at level 1 corresponding to pairs in R, such that if a corresponds to the pair (n,k), then there is a chain of length n below a.

The next lemma is the inductive part of the proof of Theorem 9.2

Lemma 9.5 Suppose G is a reduced Abelian p-group such that $G_\omega \cong G(P)$, where P is a Δ_3^0 tree, the relation

$$R = \{(n,k) : u_n(G) \geq k\}$$

is Σ_2^0, and there is a computable function f such that for each n, $f(n,s)$ is non-decreasing in s, with limit $f^*(n) \geq n$, where $u_{f^*(n)} \neq 0$. Then there is a computable tree P^* such that $G(P^*) \cong G$.

Sketch of proof: If P is a Δ_2^0 subtree of $\omega^{<\omega}$, then there is another subtree Q of $\omega^{<\omega}$, with Π_1^0 universe, such that $Q \cong P$. Using computable guesses at Q, we produce an embedding g of Q into a computable tree P^* in which the elements of infinite height are just those in $ran(g)$. While doing this, we make sure that there is a $1-1$ correspondence between pairs (n,k) in R and elements c of P^* that count towards the Ulm sequence, as in Lemma 9.3, such that if $c \in P^*$ corresponds to $(n,k) \in R$, then $height(c) = n$. Note that if $height(b)$ is infinite, then all successors of finite height count. If $height(b)$ is finite, then all but one successor counts—where that one has maximum possible length.

If we wish b to have height ω, then we give it successors c with arbitrarily long finite chains below. We make each such c correspond (eventually) to some pair $(n, k) \in R$, where the chain below c has length n. We use the function f as we did in Theorem 9.1 and Lemma 9.4, to take care of wrong guesses at the pairs in R. Suppose our guess at P changes, so that we no longer wish to give b height ω. We vow not to continue indefinitely adding successors with arbitrarily long chains. However, if the predecessor of b has height ω, then b itself will count toward the Ulm sequence, and we add one last successor c, with a chain which is allowed to grow longer than the others. We make b correspond (eventually) to a pair $(n, k) \in R$, where the chain below c has length $n - 1$. If the predecessor of b has finite height, we must trace things back further to assign an appropriate length to the final chain.

Now, let us see how Lemmas 9.4 and 9.5, relativized, combine to complete the proof of Theorem 9.2. Let G be a reduced Abelian p-group, with

$$\omega \cdot n < \lambda(G) \leq \omega \cdot (n + 1) .$$

If $\lambda(G) = \omega \cdot n + k$, where k is finite, then $\lambda(G_{\omega \cdot n}) = k$, and we have a computable tree P_n such that

$$G(P_n) \cong G_{\omega \cdot n} .$$

If $\lambda(G) = \omega \cdot (n + 1)$, then $\lambda(G_{\omega \cdot n}) = \omega$. We apply Lemma 9.4, in relativized form, to the group $G_{\omega \cdot n}$, the Δ^0_{2n+2} relation R_n, and the Δ^0_{2n+1} function f_n, to obtain a Δ^0_{2n+1} tree P_n such that

$$G(P_n) \cong G_{\omega \cdot n} .$$

So, in either case, we have a top tree P_n.

For $0 < i \leq n$, having determined a Δ^0_{2i+1} tree P_i such that

$$G(P_i) \cong G_{\omega \cdot i} ,$$

we apply Lemma 9.5, in relativized form, to the group $G_{\omega \cdot (i-1)}$, the Δ^0_{2i} relation R_{i-1}, and the Δ^0_{2i-1} function f_{i-1}, to obtain a Δ^0_{2i-1} tree P_{i-1} such that

$$G(P_{i-1}) \cong G_{\omega \cdot (i-1)} .$$

We arrive at a Δ^0_1 tree P_0 such that

$$G(P_0) \cong G_0 = G .$$

This completes the proof of Theorem 9.2.

9.3 Linear orderings

The well orderings with computable copies are those whose order type is a computable ordinal. In Chapter 4, we saw that the computable ordinals are

9.3. LINEAR ORDERINGS

exactly the ordinals with notations in O. In Chapter 8, we saw that if a well ordering is hyperarithmetical, then it has a computable copy. So, we have some kind of mathematical understanding of the computable (or hyperarithmetical) well orderings. We do not have simple invariants, however—nothing simpler than the orderings themselves.

We described the equivalence structures with computable copies, in terms of mathematical invariants. It seems hopeless to try to describe the linear orderings with computable copies. For linear orderings as a whole, there are no obvious invariants. Moreover, a model theorist would know not to look for a system of invariants, on the grounds that there are *too many* linear orderings—2^κ in power κ, up to isomorphism, for each infinite cardinal κ. It would be desirable to have a concrete result saying that it is impossible to give a mathematical description of the linear orderings with computable copies.

We can give some "transfer" theorems, saying that a linear ordering has a copy which is Δ_2^0 or Δ_3^0 just in case a certain *related* ordering has a computable copy. Finally, we show that for any non-computable set S, there is a linear ordering $\mathcal{A}$, computable in S, such that $\mathcal{A}$ has *no* computable copy.

For a structure $\mathcal{A}$, if the universe is c.e. then there is a computable partial 1-1 function from the universe of $\mathcal{A}$ onto ω, or some initial segment, so there is a copy of $\mathcal{A}$ with computable universe. If $\mathcal{A}$ is a substructure of a computable structure $\mathcal{B}$, and the universe of $\mathcal{A}$ is c.e., then $\mathcal{A}$ has a computable copy. This reasoning yields the following observation of Downey.

Proposition 9.6 *Let $\mathcal{U}$ be a computable linear ordering of type η. Then for any linear ordering $\mathcal{A}$, there is a computable copy if and only if there is a copy whose universe is a c.e. subset of $\mathcal{U}$.*

A c.e. quotient structure always has a Δ_2^0 copy. Various people, including Feiner [50], Selivanov [141], Love [100], and Downey and Jockusch [38] observed that the converse holds for Boolean algebras. It also holds for linear orderings, provided that we use the binary operation symbol *min* instead of the usual binary relation symbol $<$. This is the first of several "transfer theorems" that we shall give.

Theorem 9.7 *If $\mathcal{A}$ is a Δ_2^0 linear ordering, then there is a c.e. quotient ordering $\mathcal{B}/\sim \,\cong \mathcal{A}$.*

Proof: Let B be an infinite computable set of constants, for the universe of $\mathcal{B}$. At stage s in the construction, we have an approximation $\mathcal{A}_s$ to a finite substructure of $\mathcal{A}$, and we have determined a finite substructure $\mathcal{B}_s$ of $\mathcal{B}$, together with an equivalence relation on $\mathcal{B}_s$, such that there is a one-one correspondence between equivalence classes in $\mathcal{B}_s$, which are intervals, and elements of $\mathcal{A}_s$. At stage $s + 1$, if our guess at the position of some element of $\mathcal{A}$ changes, then we extend $\mathcal{B}_s$ to $\mathcal{B}_{s+1}$, with a new element b starting a new equivalence class which corresponds to a. We destroy the equivalence class that formerly corresponded to a by adding the elements of the class to an adjacent one.

In Chapter 18, we shall extend Theorem 9.7 through the hyperarithmetical hierarchy. Below is a variant of Theorem 9.7.

Theorem 9.8 *If $\mathcal{A}$ is a Δ_2^0 linear ordering, then there is a computable ordering of type $\omega \cdot \mathcal{A}$ in which the relation $x \sim_1 y$ (saying that the interval between x and y is finite) is c.e.*

Exercise: Prove Theorem 9.8.

The next transfer theorem is due to Downey [39].

Theorem 9.9 *For any linear ordering $\mathcal{A}$, the following are equivalent:*
(1) $(\eta + 2 + \eta) \cdot \mathcal{A}$ has a computable copy,
(2) $\mathcal{A}$ has a Δ_2^0 copy.

Proof: (1) $\Rightarrow$ (2) In a computable copy of $(\eta + 2 + \eta) \cdot \mathcal{A}$, the set of elements with immediate successors is Σ_2^0. Using the remarks before 9.6 in relativized form, we see that $\mathcal{A}$ has a Δ_2^0 copy.

(2) $\Rightarrow$ (1) Suppose that $\mathcal{A}$ is Δ_2^0. Guessing at $\mathcal{A}$, we start building an interval of type $\eta + 2 + \eta$ corresponding to each element. At each stage, we have put only finitely many elements into the ordering. To create an interval corresponding to an element a of $\mathcal{A}$, we begin with the successor pair, and we add elements to the dense intervals at each stage. If our guess changes, then we destroy each unwanted interval, incorporating the elements into an adjacent one.

The following transfer theorem, due to Watnik [153], has been rediscovered by several people. The proof requires an infinite-injury priority construction.

Theorem 9.10 *For any linear ordering $\mathcal{A}$, the following are equivalent:*
(1) there is a copy of $\mathcal{A}$ that is Δ_3^0,
(2) there is a copy of $(\omega^* + \omega) \cdot \mathcal{A}$ that is computable.

Proof: (1) $\Rightarrow$ (2) Let $x \sim y$ if the interval between x and y is finite. This relation is Σ_2^0 in any computable ordering. It follows that a computable copy of $(\omega^* + \omega) \cdot \mathcal{A}$ yields a Δ_3^0 copy of $\mathcal{A}$.

(2) $\Rightarrow$ (1) Suppose that $\mathcal{A}$ is Δ_3^0, with computable universe $A = \{a_0, a_1, \ldots\}$. There is a Δ_2^0 sequence of finite linear orderings $(\alpha_k)_{k \in \omega}$, where α_k has universe $A_{<n} = \{a_0, \ldots, a_{n-1}\}$, for some n, and
 (a) if α_k has universe $A_{<n}$, then either
  (i) $\alpha_{k+1} \supseteq \alpha_k$, where α_{k+1} has universe $A_{<n+1}$, or else
  (ii) for some $m < n$, $\alpha_{k+1} \supseteq \alpha_k | A_{<m}$, where α_{k+1} is an ordering of $A_{<m+1}$,
 (b) for each n, there is some $k(n)$ such that $\alpha_{k(n)}$ is an ordering of $A_{<n}$, and for all $k > k(n)$, $\alpha_k \supseteq \alpha_{k(n)}$.

We enumerate a chain of orderings $\mathcal{B}_s$ while guessing the sequence $(\alpha_k)_{k \in \omega}$. Suppose at stage $s + 1$, k is first such that the guess at α_{k+1} is new or different.

9.3. LINEAR ORDERINGS

If $\mathcal{B}_s$ has intervals corresponding to the elements of α_k, then we extend $\mathcal{B}_s$ to $\mathcal{B}_{s+1}$ as follows. If α_{k+1} appears to be an extension of α_k with one new element, then we start a new interval corresponding to this element. If α_{k+1} appears to be the result of dropping back and putting some old element in a new place, then we start a new interval in the appropriate place. In either case, we incorporate the elements of unwanted intervals into adjacent intervals so that each interval corresponds to some element of α_{k+1}.

Now that the intervals are properly assigned, we add an element to each end of each one. Note that if $\alpha_{k+1} \supseteq \alpha_k|A_{<n}$, then the intervals in $\mathcal{B}_{s+1}$ corresponding to elements of $A_{<n}$ are enlarged in $\mathcal{B}_{s+1}$ in such a way that all of the new elements come at the beginning or end.

We have described $\mathcal{B}_s$, for all s. Let $\mathcal{B} = \cup_s \mathcal{B}_s$. For each k, there is some $s(k)$ such that for all $s \geq s(k)$, our guess at $\alpha_0, \ldots, \alpha_k$ is unchanged. We have $\mathcal{B} = \cup_k \mathcal{B}_{s(k)}$, where the sequence $(\mathcal{B}_{s(k)})_{k\in\omega}$ preserves the relation "b and b' lie on the same designated interval", and it preserves the successor relation on designated intervals. We also have $\mathcal{B} = \cup_n \mathcal{B}_{s(k(n))}$, where the sequence $(B_{s(k(n))})_{n\in\omega}$ preserves the relation "b lies on the interval corresponding to a_i". From this, it is clear that each element of $\mathcal{B}$ lies on a designated interval. If $i < n$, then there is an interval in $\mathcal{B}_{s(k(n))}$ corresponding to a_i, and this interval grows into a copy of $\omega^* + \omega$ in $\mathcal{B}$.

In Theorem 9.10, other orderings may be substituted for $\omega^* + \omega$ (see [5]).

Theorem 9.11 *For any linear ordering $\mathcal{A}$, the following are equivalent:*
(1) there is a copy of $\mathcal{A}$ that is Δ^0_3,
(2) there is a copy of ω) $\cdot$ $\mathcal{A}$ that is computable.

Exercise: Prove Theorem 9.11.

Theorems 9.9, 9.10, and 9.11 relativize. They can also be iterated.

Exercise: Show that for all positive integers n, all sets X, and all linear orderings $\mathcal{A}$, the following are equivalent:
(1) there is a copy of $\mathcal{A}$ that is $\Delta^0_{2n+1}(X)$,
(2) there is a copy of $\omega^{(n)} \cdot \mathcal{A}$ that is computable relative to X.

Exercise: Show that for all positive integers n, all sets X, and all linear orderings $\mathcal{A}$, the following are equivalent:
(1) there is a copy of $\mathcal{A}$ that is $\Delta^0_{n+1}(X)$,
(2) there is a copy of $(\eta + 2 + \eta)^{(n)} \cdot \mathcal{A}$ that is computable relative to X.

There are many schemes for coding a set in a linear ordering. We follow [6] and use a *shuffle sum*, with densely many copies of each ordering in some given family (see the Appendix). Lerman [99] gives a different coding scheme.

Theorem 9.12 *For any set S, there is a linear ordering $\mathcal{A}(S)$ such that for all sets X, the following are equivalent:*
(a) there exists $\mathcal{B} \cong \mathcal{A}(S)$ such that $\mathcal{B} \leq_T X$,
(b) S is $\Sigma_3^0(X)$.

Proof sketch: For an arbitrary set S, let $\mathcal{A}(S) = \sigma(\mathcal{F})$, where

$$\mathcal{F} = \{\omega\} \cup \{n+1 : n \in S\} \ .$$

For simplicity, we suppose that X is computable and show that $\mathcal{A}(S)$ has a computable copy if and only if S is Σ_3^0. There are no further ideas needed in the proof of the full result.

(a) $\Rightarrow$ (b)

For each n, the natural sentence saying "there exists a maximal discrete set of size $n+1$" is finitary Σ_3. From this, it follows that if $\mathcal{B} \cong \mathcal{A}(S)$ is computable, then S is Σ_3^0.

(b) $\Rightarrow$ (a) Let $R(x, y, z, n)$ be a computable relation such that

$$n \in S \Leftrightarrow \exists x \, \forall y \, \exists z \, R(x, y, z, n) \ .$$

We have the following requirements.

$R_{n,x}$: Give $\mathcal{B}$ densely many maximal discrete intervals of type $n+1$ if and only if $\exists y \, \forall z \, R(n, x, y, z)$.

At each stage, we have a finite ordering in which each element belongs to an interval associated with some pair (n, x). At stage s, we add new intervals between all existing intervals, for all pairs (n, x) such that $n, x \leq s$. An interval associated with (n, x) initially has order type $n+1$, but it may eventually have type ω. Consider the intervals (associated with (n, x)) that have been created before stage s. We adjust these intervals as follows.

Case 1: Suppose that for some first $y < x$, $(\forall z < s) \neg R(n, x, y, z)$, and $R(n, x, y, s)$.

If the intervals associated with (n, x) currently have $n+1$ elements, then we do nothing. If the intervals currently have more than $n+1$ elements, then leaving the initial $n+1$ elements alone, we incorporate the rest into new intervals being created.

Case 2: Suppose $(\forall y < s)(\exists z < s) R(n, x, y, z)$, and $(\exists z \leq s) R(n, x, s, z)$.

In this case, we proceed as in Case 1.

Case 3: Suppose we are not in either Case 1 or Case 2.

In this case, we simply add an extra element at the end of the interval.

9.3. LINEAR ORDERINGS

We have described the construction. An interval associated with (n, x) ends up having type n if $\forall y\, \exists z\, R(n, x, y, z)$, since we have infinitely many chances to remove extra elements. The interval has type ω if $\exists y\, \forall z\, \neg R(n, x, y, z)$, since we add one element to the end of the interval at stage s, for all sufficiently large s. This completes the proof of the theorem.

If we can code an arbitrary set in a linear ordering, then we can code a set and its complement.

Corollary 9.13 *For any set S, there is a linear ordering $\mathcal{A}(S)$ such that*

$$\{X : \exists \mathcal{B} \leq_T X : \mathcal{B} \cong \mathcal{A}(S)\} = \{X : S \text{ is } \Delta_3^0(X)\}\,.$$

Proof: We apply Theorem 9.12, replacing S by $S \oplus \neg S$.

Jockusch and Soare [71] showed that for any c.e. non-computable set S, there is a linear ordering $\mathcal{A}$, computable in S, such that $\mathcal{A}$ has no computable copy. Seetapun (unpublished) and Downey [36] extended this result to sets S that are Δ_2^0 and not computable.

Theorem 9.14 (Jockusch-Soare, Seetapun, Downey) *If S is Δ_2^0 and not computable, then there is a linear ordering $\mathcal{A} \leq_T S$ such that $\mathcal{A}$ has no computable copy.*

Proof sketch: For any computable linear ordering, there is a copy with elements named by constants in ω. Then there is a computable function which on input s gives the ordering on $\{0, \ldots, s-1\}$. It is convenient, just in this proof, to consider e to be the index of a computable linear ordering if φ_e is such a function. The ordering $\mathcal{A}$ will have the form

$$\sum_e (2 + \eta + (e+2) + \eta + 2 + \mathcal{A}_e)\,,$$

where $\mathcal{A}_e$ has no dense subinterval.

It is enough to satisfy the following requirements:

R_e: If e is the index of a linear ordering with a unique subinterval of form

$$2 + \eta + (e+2) + \eta + 2 + \mathcal{B}_e + 2 + \eta + 2\,,$$

then

$$\mathcal{B}_e \not\cong \mathcal{A}_e\,.$$

In the ordering $\mathcal{A}$ that we are building, we suppose that the interval of type 2 to the left of $\mathcal{A}_e$ and the one of type $e + 3 = (e+1) + 2$ to the right of $\mathcal{A}_e + 2 + \eta$ have been designated before we begin work on requirement R_e. The part of $\mathcal{A}$ that we focus on for requirement R_e will be

$$\mathcal{A}_e + 2 + \eta\,.$$

Within this interval, the elements that we designate for "2", and "η" may vary during the construction, but in the end, we will have an ordering $\mathcal{A}_e$, followed by an interval of type 2, and then a dense interval, such that one of the following holds:

(a) $\mathcal{A}_e$ has order type $k + \omega^*$, for some finite k, while $\mathcal{B}_e$ has an initial segment of type $k + 1$, or

(b) $\mathcal{A}_e$ has order type ω, while $\mathcal{B}_e$ has an element with infinitely many predecessors.

One difficulty is that we do not know the "markers" for $\mathcal{B}_e$—the intervals of type 2, $e+2$, and 2 separated by η's on the left, and the intervals of type 2, $e+3$, and 2, separated by η's on the right. The tuple of markers satisfies a finitary Π_2 formula. By Proposition 2.13, there is a computable guessing function that gives the true markers infinitely often, and gives any other particular tuple only finitely often.

We designate a pair for the "2", and start building $\mathcal{A}_e + 2 + \eta$, based on our first guess at the markers. When we have a new guess at the markers, we start building a new $\mathcal{A}_e + 2 + \eta$, using the elements already enumerated into the ordering as an initial segment of $\mathcal{A}_e$, and we designate a new pair for the "2", to the right of this initial segment. We may return to an old guess at the markers. If the corresponding old version of $\mathcal{A}_e + 2 + \eta$ forms an initial segment of the current one, then we take up building the old version where we left off, using the old "2", and we incorporate the rest of the current version of $\mathcal{A}_e + 2 + \eta$ into the dense interval to the right of this pair, adding new elements for density.

We may return to an old guess at the markers, where the corresponding old version of $\mathcal{A}_e + 2 + \eta$ is *not* an initial segment of the current one, but has been incorporated into the dense part at the end of the current one. In this case, we proceed as if we had a totally new guess at the markers. In particular, we designate a new "2" at the end. The result of this strategy is a version of $\mathcal{A}_e + 2 + \eta$ in which a finite initial segment of $\mathcal{A}_e$ may result from early wrong guesses at the markers, but the rest is the result of correct guesses. Later wrong guesses result only in additions to the dense interval at the end.

The universe of $\mathcal{A}$ will be an infinite computable set of constants. At stage s, we work on $\mathcal{A}_e + 2 + \eta$, for $e < s$. We may suppose that the rest of $\mathcal{A}$ (outside the intervals $\mathcal{A}_e + 2 + \eta$) is constructed in advance. We determine a tentative ordering on a set constants (including the first s). By Proposition 2.12, there is a computable function $g(x, s)$ such that

$$lim_{s \to \infty} g(x, s) = \chi_S(x) .$$

To make $\mathcal{A}$ computable relative to S, we let the ordering on the first k constants be the one determined at the first stage s at which $\chi_S|k$ is approximated correctly. Thus, if $g(x, s)$ agrees with $g(x, t)$ for all $x < k$, then the orderings at stages s and t must agree on the first k constants.

For simplicity, we focus on a single e, and a single guess at the markers, which we assume to be correct. Thus, we can watch elements being enumerated into the ordering $\mathcal{B}_e$. We suppose that in $\mathcal{A}_e + 2 + \eta$, the "2" has been determined,

9.3. LINEAR ORDERINGS

along with an initial segment, say of type k_0, in $\mathcal{A}_e$. We describe the procedure for putting further elements into $\mathcal{A}_e$, and ignore the rest of the construction.

We put into $\mathcal{A}_e$ the first new constant, say a_0. Next, we wait for some b_0 to appear in $\mathcal{B}_e$, with k_0 predecessors. If this never happens, then $\mathcal{B}_e \not\cong \mathcal{A}_e$. We then start building an interval of type ω^* just to the left of a_0, adding one element at each stage. We continue until we find a stage s_0 at which, for some $k_1 > k_0$, b_0 has k_1 predecessors and $\chi_S|k_1$ has not yet been guessed correctly. Suppose there is no such stage. If b_0 has infinitely many predecessors, then we would have a computable function σ, where $\sigma(k)$ is a stage by which $\chi_S|k$ has been guessed correctly, and by Proposition 2.18, S is computable, a contradiction. Therefore, b_0 has only finitely many predecessors but at least k_0, and no element of $\mathcal{A}$ has exactly the same number of predecessors.

Suppose that at some stage s_1, b_0 has k_1 predecessors, and $\chi_S|k_1$ has not yet been guessed correctly. Then we remove all but the first $k_1 - 1$ constants to the left of a_0, and we put the first unused constant, say a_1, to the right of a_0. Now, a_1 has k_1 predecessors. We start building an interval of type ω^* just to the left of a_1, adding one element at each stage, as above. We continue until we find a stage s_1 at which, for some $k_2 > k_1$, b_0 has k_1 predecessors and $\chi_S|k_2$ has not yet been guessed correctly. If there is no such stage, then, as above, b_0 has only finitely many predecessors, at least k_1, and no element of $\mathcal{A}_e$ has the same number.

Suppose at some stage s_2, b_0 has k_2 predecessors, and $\chi_S|k_2$ has not yet been guessed correctly. Then we remove all but the first $k_2 - 1$ constants to the left of a_1, we put the first unused constant, say a_2 to the right of a_1, and we start building an interval of type ω^* just to the left of a_2, as above. We continue the construction in this way.

One of two things happens.

Case 1: $\mathcal{A}_e$ has order type $k_n + \omega^*$, for some n.

In this case, b_0 has only finitely many predecessors, and at least k_n, and no element of $\mathcal{A}_e$ has exactly the same number of predecessors.

Case 2: $\mathcal{A}_e$ has order type ω.

In this case, b_0 has infinitely many predecessors. Again, no element of $\mathcal{A}$ has the same number.

This is all we shall say about the proof of Theorem 9.14.

Remark: Theorem 9.14 relativizes.

Combining Theorem 9.9, Corollary 9.13, and Theorem 9.14, we obtain the following.

Theorem 9.15 *For any non-computable set S, there is a linear ordering $\mathcal{A}$ such that $\mathcal{A} \leq_T S$ and $\mathcal{A}$ has no computable copy.*

Proof: There are three cases.

Case 1: Suppose that S is low.

In this case, the statement follows immediately from Theorem 9.14.

Case 2: Suppose that S is not low_2.

In this case, we form $\mathcal{A}(S'')$ as in Corollary 9.13. There exists $\mathcal{B} \cong \mathcal{A}(S'')$ such that $\mathcal{B} \leq_T S$. Moreover, if there were a computable $\mathcal{B} \cong \mathcal{A}(S'')$, then we would have $S'' \leq_T \mathcal{B}'' \leq \emptyset''$, contradicting our assumption about S.

Case 3: Suppose that S is low_2 but not low.

In this case, we apply Theorem 9.14 relativized to $\emptyset'$. Since S' is $\Delta_2^0(\emptyset')$ but not computable in $\emptyset'$, there is a a linear ordering $\mathcal{B} \leq_T S'$, with no copy which is computable in $\emptyset'$. Then by Theorem 9.9, $(\eta + 2 + \eta) \cdot \mathcal{B}$ has a copy computable in S, but no computable copy.

Miller [121] has shown that there is a linear ordering $\mathcal{A}$ with the feature that for all non-computable Δ_2^0 sets X, there is a copy of $\mathcal{A}$ computable in X, but there is no computable copy.

9.4 Boolean algebras

Boolean algebras are closely related to linear orderings. Goncharov pointed out that the analogue of Proposition 9.6 holds for Boolean algebras.

Proposition 9.16 *Let $\mathcal{U}$ be a computable atomless Boolean algebra. Then for any Boolean algebra $\mathcal{A}$, there is a computable copy if and only if there is a copy whose universe is a c.e. subset of $\mathcal{U}$.*

The following is the analogue of Theorem 9.9. As we mentioned earlier, it was observed by many people, including Feiner [50], Selivanov [141], Love [100], and Downey and Jockusch [38].

Theorem 9.17 *Any Δ_2^0 Boolean algebra is isomorphic to a c.e. quotient algebra.*

Proof sketch: By Proposition 8.15, we can pass from a Δ_2^0 Boolean algebra $\mathcal{A}$ to a Δ_2^0 linear ordering $\mathcal{C}$ such that the interval algebra $I(\mathcal{C})$ is isomorphic to $\mathcal{A}$. Applying Theorem 9.7, we obtain a computable linear ordering $\mathcal{B}$ and a computably enumerable equivalence relation $\sim$ on $\mathcal{B}$ such that $\mathcal{B}/_\sim \cong \mathcal{C}$. To complete the proof, we note that $\sim$ induces a c.e. congruence relation $\sim'$ on $I(\mathcal{B})$ such that $I(\mathcal{B})/_{\sim'} \cong I(\mathcal{B}/_\sim)$. Suppose $b \leq b'$, and consider $[b, b')$ as an element of $I(\mathcal{B})$. If $b \sim b'$, then $[b, b') \sim' 0$. More generally, for x and y in $I(\mathcal{B})$.

9.4. BOOLEAN ALGEBRAS

we have $x \sim' y$ if and only if x and y differ by a finite join of intervals, each $\sim'$-equivalent to 0.

Exercise: Give a direct proof of Theorem 9.17, along the lines of the proof of Theorem 9.7.

Not *all* results on linear orderings carry over to Boolean algebras. We have seen that there are low linear orderings with no computable copy. By contrast, Downey and Jockusch [38] showed that every low Boolean algebra has a computable copy. They conjectured that for all n, every low_n Boolean algebra has a computable copy. Thurber [151] proved this for $n = 2$. He conjectured that when the statement was proved for $n = 3$, the general pattern would become clear. In [88], it is shown that every Boolean algebra which is low_3 or even low_4 has a computable copy, but the pattern is not yet clear.

The idea behind the known results, for $n = 1, 2, 3$, and 4, is to isolate certain important predicates and show that if a Boolean algebra $\mathcal{A}$ is Δ^0_{n+1} with these added predicates, then it has a computable copy. The predicates are all definable by computable Π_n or Σ_n formulas, so they will necessarily be Δ^0_{n+1} in a low_n Boolean algebra.

In the simplest case, where $n = 1$, the important predicate is $atom(x)$ (x is an atom). This is defined by a (finitary) Π_1 formula. If a Boolean algebra $\mathcal{A}$ is low, then it is Δ^0_2 with the added predicate $atom(x)$. Then the result below implies that that $\mathcal{A}$ has a computable copy.

Theorem 9.18 *If a Boolean algebra $\mathcal{A}$ is Δ^0_2 with the predicate $atom(x)$, then it has a computable copy.*

The proof depends on two lemmas. The first is an embedding theorem, very similar to Theorem 9.17.

Lemma 9.19 (Embedding Theorem) *Let $\mathcal{A}$ be a Boolean algebra which is Δ^0_2 with $atom(x)$. Then there is an embedding f of $\mathcal{A}$ into a computable Boolean algebra $\mathcal{B}$, where $\mathcal{B}$ is generated by elements of $ran(f)$ and additional atoms, each lying below $f(a)$ for some element a that is an atom a of $\mathcal{A}$.*

Proof sketch: Guessing $\mathcal{A}$, and attempting to build a computable copy, we succeed except for the fact that some atoms are split into finitely many parts.

The second lemma is an isomorphism theorem, due to Vaught and Remmel [132], [152].

Lemma 9.20 (Isomorphism Theorem) *Suppose $\mathcal{A}$ and $\mathcal{B}$ are Boolean algebras such that $\mathcal{A}$ has infinitely many atoms, $\mathcal{A} \subseteq \mathcal{B}$, $\mathcal{B}$ is generated by elements of $\mathcal{A}$ and atoms, and each atom of $\mathcal{B}$ lies below some atom of $\mathcal{A}$. Then $\mathcal{A} \cong \mathcal{B}$.*

Proof sketch: Let $\mathcal{F}$ be the set of functions p mapping a finite subalgebra of $\mathcal{A}$ isomorphically onto a finite subalgebra of $\mathcal{B}$ such that for all $a \in dom(p)$, the number of atoms below a (in $\mathcal{A}$) matches the number of atoms below $p(a)$ (in $\mathcal{B}$) and $p(a)\Delta a$ is a finite join of atoms in $\mathcal{B}$. We can show that $\mathcal{F}$ has the back-and-forth property. Therefore, there is an isomorphism f such that for all $a \in \mathcal{A}$, $a\Delta f(a)$ is a finite join of atoms in $\mathcal{B}$.

Using the two lemmas, we can complete the proof of Theorem 9.18. If $\mathcal{A}$ has only finitely many atoms, then clearly $\mathcal{A}$ has a computable copy. Suppose $\mathcal{A}$ has infinitely many atoms. Let f embed $\mathcal{A}$ into a computable Boolean algebra $\mathcal{B}$ as in Lemma 9.19. Then $ran(f)$ and $\mathcal{B}$ satisfy the conditions of Lemma 9.19. Since $ran(f) \cong \mathcal{B}$, we have $\mathcal{A} \cong \mathcal{B}$.

In the proof that every low_2 Boolean algebra has a computable copy, the important predicates turn out to be $atom(x)$, $atomless(x)$ (x is atomless), and $inf(x)$ (x is not a finite join of atoms). If $\mathcal{A}$ is low_2, then it is Δ^0_3 with these predicates. We can show that $\mathcal{A}$ has a copy which is Δ^0_2 with $atom(x)$. Hence, by Theorem 9.18, it has a computable copy.

Jockusch and Soare [72] proved the following very intriguing result.

Theorem 9.21 *For any Boolean algebra $\mathcal{A}$, and any n, there is a computable Boolean algebra $\mathcal{B}$ such that all Π_n sentences true of $\mathcal{A}$ are true of $\mathcal{B}$.*

The proof gives predicates such that if $\mathcal{A}$ and $\mathcal{B}$ agree agree on the numbers of elements satisfying these predicates, then they satisfy the same Π_n sentences. The predicates are definable by computable infinitary formulas, but these have complexity greater than n.

9.5 Results of Wehner

Lempp asked whether a structure $\mathcal{A}$ that has copies in all non-zero degrees must have a computable copy. Below, we state some related questions in pure computability theory. We begin with a definition that will be important here and later. Let $\mathcal{S}$ be a family of subsets of ω. An *enumeration* of $\mathcal{S}$ is a binary relation R such that $\mathcal{S} = \{R_n : n \in \omega\}$, where $R_n = \{x : (n,x) \in R\}$.

Question 1: If for each non-computable set X, $\mathcal{S}$ has an enumeration that is c.e. in X, must $\mathcal{S}$ have a c.e. enumeration?

Question 2: If for each non-computable set X, $\mathcal{S}$ has an enumeration that is computable in X, must $\mathcal{S}$ have a computable enumeration?

The result below shows how a negative answer to either Question 1 or Question 2 easily yields a negative answer to Lempp's question. Wehner [154] gave negative answers to Questions 1 and 2. Independently of Wehner, and slightly later, Slaman [143] gave a different construction answering Lempp's question.

For a countable family $\mathcal{S}$ of subsets of ω, we let $En(\mathcal{S})$ be the set of all enumerations of $\mathcal{S}$.

9.5. RESULTS OF WEHNER

Proposition 9.22 *Let S be a countable family of sets. Then*
(a) there is a stucture $\mathcal{A}(S)$ such that

$$\{deg(\mathcal{B}) : \mathcal{B} \cong \mathcal{A}(S)\} = \{deg(X) : (\exists R \in En(S)) \ R \text{ is c.e. relative to } X\} \ ,$$

(b) there is a structure $\mathcal{A}(S)$ such that

$$\{deg(\mathcal{B}) : \mathcal{B} \cong \mathcal{A}\} = \{deg(R) : R \in En(S)\} \ .$$

Proof: (a) We let $\mathcal{A}(S)$ be a directed graph with infinitely many identical connected components for each $X \in S$, where each connected component associated with X has the following form—we use $\to$ for the directed edge relation. First, there is a special point a such that $a \to a$. In addition, for each $n \in X$, there are additional points $a_1, \ldots, a_n$, forming an $(n+1)$-cycle

$$a \to a_1 \to \ldots \to a_n \to a \ .$$

(b) Let S^* consist of the sets $X \oplus \neg X$, for $X \in S$. Take the directed graph $\mathcal{A}(S^*)$, constructed as in (a).

Here is Wehner's result on Question 1.

Theorem 9.23 *There is a family S, consisting of finite sets, such that for each non-computable set X, there is an enumeration of S that is c.e. relative to X, but there is no c.e. enumeration.*

It may be difficult to imagine how we can do *anything* below an arbitrary non-computable set. Wehner showed that we can do the following. Here we let Fin denote the set of all finite sets.

Lemma 9.24 *There is a uniform effective procedure which, for any i and any non-computable set X, yields an enumeration R of $Fin - \{W_i\}$ such that R is c.e. relative to X.*

Sketch of proof: Recall that D_n is the nth finite set in a canonical list, and R_k denotes the set of all x such that $(k, x) \in R$. We have the following requirements.

P_n: $D_n \neq W_i \Rightarrow \exists k \ D_n = R_k$,

N_k: $R_k \neq W_i$

The construction of R proceeds in stages. At stage s, we have enumerated only finitely many pairs into R. We write $R_{k,s}$ for the set of elements that have entered R_k by stage s. To satisfy P_n, if $D_n \neq W_{i,s}$, then we choose a new k and let $R_{k,s} = D_n$. We refrain from adding more elements to R_k until/unless we come to a stage t at which $W_{i,t} = D_n$, and then we add a new element to $R_{k,t}$. Now, we leave R_k unchanged until/unless we come to a stage $t' > t$ such that $W_{i,t'} = R_{k,t'-1}$. At this stage, requirement N_k needs attention again, and requirement P_n also needs attention. We can satisfy P_n by choosing a new k' and letting $R_{k',t'} = D_n$. For N_k, we add another new element to $R_{k',t'}$.

To make sure that N_k is satisfied in the end, we need a strategy for adding elements so that the requirement will not need attention infinitely often. We note that the set
$$D_n \cup \{\langle x, \chi_X(x)\rangle : x \in \omega\}$$
is not c.e., and is therefore different from W_i. At each stage t when we act on N_k, because $W_{i,t}$ is equal to $R_{k,t-1}$, the element that we add to $R_{k,t-1}$, to form $R_{k,t}$, is $\langle x, \chi_X(x)\rangle$, for the first x such that this is not already in $R_{k,t-1}$.

Making use of some flexibility in the proof of Lemma 9.24, we obtain the following.

Lemma 9.25 *For any partial computable function g, there is a uniform effective procedure which, for all i and all non-computable sets X, yields an enumeration, c.e. relative to X, of the family*
$$\mathcal{F}_i = \begin{cases} Fin - \{W_{g(i)}\} & \text{if } g(i) \downarrow \\ Fin & \text{otherwise} \end{cases}.$$

Proof sketch: We proceed as above, considering only the requirements P_n until/unless $g(i) \downarrow$. Then we consider the requirements N_k, letting $W_{g(i)}$ play the role of W_i. This is all we shall say about the proof of Lemma 9.25.

We return to the proof of Theorem 9.23. We must determine $\mathcal{S}$ such that for all non-computable sets X, $\mathcal{S}$ has an enumeration c.e. in X. We must also satisfy the following requirements.

Q_e: W_e is not an enumeration of $\mathcal{S}$

We may think of W_e as a set of pairs (i, a)—identifying (i, a) with $\langle i, a\rangle$. Then W_e is an enumeration of a family of sets, which we denote by $\mathcal{S}_e$. We refer to a set of the form
$$\{\langle e, x\rangle : x \in X\}$$
as an *e-set*. To keep different requirements Q_e from interfering with each other, we try to make $\mathcal{S}$ differ from $\mathcal{S}_e$ on some e-set.

Each element of $\mathcal{S}$ will be a finite e-set, for some e, and for each e, $\mathcal{S}$ will contain either all finite e-sets, or all but one. We enumerate W_e, searching for an element of form $(i, \langle e, x\rangle)$. If the search does not halt, then there is no non-empty e-set in $\mathcal{S}_e$, and Q_e is satisfied trivially. If the search halts, then we take the number i from the first element $(i, \langle e, x\rangle)$ that we find in W_e. We satisfy Q_e by leaving out of $\mathcal{S}$ the e-set $\{\langle e, x\rangle : (i, \langle e, x\rangle) \in W_e\}$.

There is a computable partial function g such that
$$W_{g(e)} = \begin{cases} \{x : (i, \langle e, x\rangle) \in W_e\} & \text{if the search halts and we have } i \text{ as above} \\ \emptyset & \text{otherwise} \end{cases}.$$

9.6. DECIDABLE HOMOGENEOUS STRUCTURES

By Lemma 9.25, there is a uniform effective procedure which, for each e and each non-computable set X, yields an enumeration R_e of $Fin - \{W_{g(e)}\}$, such that R_e is c.e. relative to X. We can turn the family of enumerations R_e into an enumeration R of $\mathcal{S}$, letting

$$(\langle i, e \rangle, \langle e, x \rangle) \in R \Leftrightarrow (i, x) \in R_e .$$

This completes the proof of Theorem 9.23.

Having answered Question 1, we have already obtained the answer to Lempp's question. Wehner proved the following variant of Theorem 9.23, which simultaneously answered Questions 1 and 2.

Theorem 9.26 *There is a family $\mathcal{S}$, consisting of finite sets, such that for all non-computable sets X, $\mathcal{S}$ has an enumeration computable in X, but no c.e. enumeration.*

Exercise: Prove Theorem 9.26. Hints: The outline is the same as for Theorem 9.23, but the lemmas must be changed so that the enumerations are computable relative to X. In the new version of Lemma 9.24, when acting on requirement N_k at stage t, instead of adding $\langle x, \chi_X(x) \rangle$ to R_k, you will add $\langle \langle x, \chi_X(x) \rangle, t \rangle$.

9.6 Decidable homogeneous structures

Recall that $\mathcal{A}$ is decidable if the complete diagram $D^c(\mathcal{A})$ is computable. There is quite a body of work, by Morley [122], Millar [119], [120], and others, on the existence of decidable copies of a given structure $\mathcal{A}$, where $\mathcal{A}$ is saturated, or prime, or prime over some finite tuple, etc. Below, we give a result of Goncharov [55] and Peretyat'kin [131], which seems to be the ultimate result of this kind. Recall that a (countable) structure $\mathcal{A}$ is *homogeneous* if for any tuples $\bar{a}$ and $\bar{b}$ realizing the same complete type, for any further element c, there exists d such that $\bar{a}, c$ and $\bar{b}, d$ realize the same type. If $\mathcal{A}$ is either prime or saturated, then it is homogeneous.

We shall consider computable enumerations R of the set all of complete types realized in a structure $\mathcal{A}$. We need to be able to determine, given an R-index i for a type, the set of free variables in the type. We can effectively tell whether a given x is one of the free variables of the type R_i, by checking whether the formula $x = x$ is an element of R_i. Determining the *set* of free variables is a problem. We could require that the tuple of variables in a type is always *initial*—consisting of the first k variables, for some k. Or, we could require that a complete type in a particular tuple of free variables consist of formulas in which all of these variables appear free. Or, we replace the given enumeration R by another enumeration with the desired feature. We shall come back to this point.

CHAPTER 9. EXISTENCE OF COMPUTABLE STRUCTURES

Theorem 9.27 (Goncharov, Peretyat'kin) *Let $\mathcal{A}$ be a homogeneous structure. Then $\mathcal{A}$ has a decidable copy if and only if there exist a computable enumeration R of the set of all complete types realized in $\mathcal{A}$, and a computable function $f(i,\varphi)$ such that if R_i is a type in variables $\bar{u}$ and $\exists x\, \varphi(\bar{u},x) \in R_i$, then $f(i,\varphi)$ is an R-index j for a type, in variables $\bar{u}, x$, such that $R_j \supseteq R_i \cup \{\varphi(\bar{u},x)\}$.*

Proof: First, suppose that $\mathcal{A}$ has a decidable copy $\mathcal{B}$. Let $(\bar{b}_i)_{i\in\omega}$ be a computable list of the tuples from $\mathcal{B}$, and let R consist of the pairs (i,φ) such that $\mathcal{B} \models \varphi(\bar{b}_i)$. Then R_i is the type of $\bar{b}_i$. If $\exists x\, \varphi(\bar{u},x) \in R_i$, then we can search for c such that $\mathcal{B} \models \varphi(\bar{b},c)$. We can then search for j such that $\bar{b}_j = \bar{b}_i, c$. We let $f(i,\varphi) = j$.

Now, suppose that R is a computable enumeration of the complete types realized in $\mathcal{A}$, and f is a computable function locating R-indices of extensions, as in the statement of the theorem. For now, we assume that R has the feature that given i, we can determine the free variables of the type R_i. (We shall see later that this assumption can be dropped.) We describe the construction of the required $\mathcal{B}$ in some detail, partly because the construction is delicate, and partly because some of the features will reappear in other, more complicated constructions.

Let B be an infinite computable set of constants. Let L be the set of pairs $(i, \bar{b})$, where $\bar{b}$ is a tuple of constants from B and R_i is a complete type in a sequence of variables $\bar{u}$ corresponding to $\bar{b}$. For $\ell \in L$, where $\ell = (i, \bar{b})$, let $E(\ell)$ be the set of sentences $\varphi(\bar{b})$ such that $\varphi(\bar{u}) \in R_i$ and $\varphi(\bar{u})$ has Gödel number less than $length(\bar{b})$. We also fix a computable list of the pairs $(j, \bar{d})$ such that R_j is a type in variables $\bar{v}, x$ and $\bar{d}$ is a tuple from B appropriate to substitute for $\bar{v}$ (leaving x free). We arrange that if $(j, \bar{d})$ is the nth pair on the list, then $\bar{d}$ is included among the first n constants.

Lemma 9.28 *Suppose $\ell_0 \ell_1 \ell_2 \ldots$ is an infinite sequence of elements of L such that, letting $\ell_n = (i_n, \bar{b}_n)$, we have the following:*
(a) $\bar{b}_n$ includes the first n constants from B,
(b) $R_{i_n}(\bar{b}_n) \subseteq R_{i_{n+1}}(\bar{b}_{n+1})$,
(c) for the nth pair $(j, \bar{d})$ on the list, if $R_j(\bar{d},x) \cup R_{i_n}(\bar{b}_n)$ is consistent, then $R_j(\bar{d},b) \subseteq R_{i_{n+1}}(\bar{b}_{n+1})$, for some b.
Then $\cup_n E(\ell_n)$ is the complete diagram of some $\mathcal{B} \cong \mathcal{A}$.

Idea of proof: It should be clear that the given sequence yields a homogeneous structure $\mathcal{B}$ realizing the same complete types as $\mathcal{A}$. By Theorem 3.12, two homogeneous structures realizing the same complete types are isomorphic. Therefore, $\mathcal{B} \cong \mathcal{A}$.

Now, we continue with the proof of Theorem 9.27. We can easily produce a Δ_2^0 sequence as in the lemma. Let $\ell_0 = (i_0, \emptyset)$, where i_0 is an index for $Th(\mathcal{A})$—this is just the type of $\emptyset$. Given $\ell_n = (i_n, \bar{b}_n)$, if $(j, \bar{d})$ is the nth pair on our list, we check whether $R_j(\bar{d},x) \cup R_{i_n}(\bar{b}_n)$ is consistent. If so, then we let $\ell_{n+1} = (i, b_{n+1})$, where $R_i(\bar{b}_{n+1}) \supseteq R_j(\bar{d},b) \cup R_{i_n}(\bar{b}_n)$, for some constant b.

9.6. DECIDABLE HOMOGENEOUS STRUCTURES

We have an effective procedure for guessing, eventually correctly, whether $R_j(\overline{d}, x) \cup R_{i_n}(\overline{b}_n)$ is consistent. At stage s, we consider the first s formulas in $R_j(\overline{d}, x)$ and for the conjunction, say $\psi(\overline{d}, x)$, we check whether

$$\exists x\, \psi(\overline{d}, x) \in R_{i_n}(\overline{b}_n) \,.$$

We also have an effective procedure for guessing, eventually correctly, whether $R_i(\overline{b}_{n+1}) \supseteq R_j(\overline{d}, b) \cup R_{i_n}(\overline{b}_n)$. At stage s, we check whether the first n sentences of $R_j(\overline{d}, b) \cup R_{i_n}(\overline{b}_n)$ are in $R_i(\overline{b}_{n+1})$.

Our goal is to produce an infinite sequence

$$\pi = \ell_0 \ell_1 \ell_2 \ldots$$

as in Lemma 9.28, such that $E(\pi) = \cup_n E(\ell_n)$ is c.e. At stage s, we have tentatively determined an initial segment of the desired sequence, say $\ell_0 \ell_1 \ldots \ell_r$, and we have permanently enumerated $E(\ell_r)$. Say $\ell_n = (i_n, \overline{b}_n)$, where for all $n < r$, $E(\ell_r) \cup R_{i_n}(\overline{b}_n)$ is consistent. At stage $s + 1$, we check the terms of the finite sequence $\ell_0 \ell_1 \ldots \ell_r$, using stage $s + 1$ guesses, and we cross off everything from the first change onwards. We are left with

$$\ell_0 \ell_1 \ldots \ell_n \,,$$

for some $n \leq r$ (in addition to $E(\ell_r)$). The stage $s + 1$ sequence will have the form

$$\ell_0 \ell_1 \ldots \ell_n \ell'_{n+1} \,.$$

The new term, ℓ'_{n+1}, is determined as follows. Suppose $(j, \overline{d})$ is the nth pair, and $R_j(\overline{d}, x) \cup R_{i_n}(\overline{b}_n)$ appears to be consistent. We find the first i such that $R_i(\overline{b}_n, x)$ seems to be an extension of $R_j(\overline{d}, x) \cup R_{i_n}(\overline{b}_n)$. We make sure that $R_i(\overline{b}_n, x) \cup E(\ell_r)$ is consistent as follows. Let b be the first constant which is not in $\overline{b}_n$ and is not mentioned in $E(\ell_r)$. Say the conjunction of $E(\ell_r)$ is $\psi(\overline{b}_n, d, d')$, where d is the first constant not in b_n. We make sure that

$$\exists y\, \exists \overline{u}\, \psi(\overline{b}_n, y, \overline{u}) \in R_i(\overline{b}_n, b) \,.$$

Using f, we can find some i' such that $R'_i(\overline{b}_n, b, y) \supseteq R_i(\overline{b}_n, b) \cup \{\exists \overline{u}\, \psi(\overline{b}_n, y, \overline{u})\}$. Then $\ell'_{n+1} = (i', b_n, b, d)$.

Above, we assumed that our computable enumeration of complete types had the feature that given an index for a type, we could effectively determine the set of free variables in the type. The proposition below says that we can replace the given enumeration by one with this feature.

Proposition 9.29 *Let R be a computable enumeration of the set of complete types realized in the structure $\mathcal{A}$. Suppose f is a computable function on pairs (i, φ) such that if $\exists x\, \varphi(\overline{u}, x) \in R_i$, then $f(i, \varphi)$ is an R-index for a completion of $R_i \cup \{\varphi(\overline{u}, x)\}$. There is another computable enumeration, R^*, of the same set of types, such that, given i, we can effectively determine the free variables of R_i^*, and there is another computable function, f^*, such that for all i, if $\exists x\, \varphi(\overline{u}, x) \in R_i^*$ (where x is not a variable of R_i^*), then $f^*(i, \varphi)$ is an R^*-index for a completion of $R_i^* \cup \{\varphi(\overline{u}, x)\}$.*

Sketch of proof: The idea is to replace an R-index i by a family of R^*-indices representing pairs $(i,\overline{u})$, where $\overline{u}$ is an arbitrary tuple of variables. Let $(\overline{u}_j)_{j\in\omega}$ be a computable list of all tuples of variables. Let $R^*_{\langle i,j\rangle}$ be the set of formulas of R_i with free variables among $\overline{u}$. Given $\langle i,j\rangle$, we can determine the tuple of variables $\overline{u}$ that actually appear free in the type. Given an index $\langle i,j\rangle$ and φ, if the free variables of φ are among $\overline{u}, x$ and $\exists x\, \varphi(\overline{u},x) \in R^*_{\langle i,j\rangle}$, then $f(i,\varphi)$ is an index i' for a complete extension of $R_i \cup \{\varphi(\overline{u},x)\}$. Say $\overline{u}_{j'} = \overline{u}, x$. Then we let $f^*(i,\varphi) = \langle i', j'\rangle$.

Chapter 10

Completeness and forcing

This chapter is devoted to a kind of "Completeness Theorem". The results do not involve rules of proof, nor do they say the set of sentences true in all structures in some class is c.e. Rather, they assert the *expressive completeness* of various classes of computable infinitary formulas. We have already seen one result in this direction. Theorem 8.7 said that the computable infinitary formulas true of a tuple in a hyperarithmetical structure determine the orbit of the tuple under automorphisms. The basic results that we shall present are taken from [16] or [27]. The main technique of proof is forcing.

10.1 Images of a relation

Problem 1 *Find syntactical conditions, on a computable structure $\mathcal{A}$ with an added relation R, guaranteeing that in all copies $\mathcal{B}$ of $\mathcal{A}$, the image of R is $\Sigma^0_\alpha(\mathcal{B})$.*

We are most interested in computable structures, and it may seem more natural to ask about the image of R in *computable* copies of $\mathcal{A}$, rather than in *arbitrary* copies. In fact, the syntactical conditions were isolated in results on computable copies, by Ash and Nerode [19] for the case where $\alpha = 1$, and by Barker [20] for arbitrary computable ordinals α. The result on arbitrary copies (Problem 1) came later, but the statement is cleaner, and the proof is simpler. Logically, it should have come first.

We give some simple examples, which suggest the general result.

Example 1: In all copies $\mathcal{B}$ of $(\omega, +)$, the image of the successor relation is computable relative to $\mathcal{B}$.

Proof: The successor relation is defined in $(\omega, +)$ by the atomic formula

$$y = x + 1 ,$$

with parameter 1.

Example 2: In all orderings $\mathcal{B}$ of type ω, the complement of the successor relation is c.e. relative to $\mathcal{B}$—the successor relation itself is co-c.e. relative to $\mathcal{B}$).

Proof: The complement of the successor relation is defined by the finitary Σ_1 formula
$$\varphi(x,y) = \exists z\, (x < z \,\&\, z < y)\,.$$
By Theorem 3.19, $\varphi^{\mathcal{B}}(x,y)$ is co-c.e. relative to $\mathcal{B}$, so the successor relation is co-c.e. relative to $\mathcal{B}$.

Example 3: For an infinite-dimensional vector space V over Q, the linear span of a finite tuple $\bar{b}$ is c.e. relative to V.

Proof: The linear span of $\bar{b}$ is defined by a computable Σ_1 formula $\varphi(\bar{b}, x)$—the disjunction of the atomic formulas $x = \tau(\bar{b})$ saying that x is equal to a particular linear combination. By Theorem 7.5, $\varphi^V(\bar{b}, x)$ is c.e. relative to V.

Example 4: For an ordering $\mathcal{B}$ of type $\omega + \omega^*$, the initial segment of type ω is $\Delta_2^0(\mathcal{B})$.

Proof: The initial segment of type ω is defined by a computable Σ_2 formula $\varphi(x)$—the disjunction over n of the natural finitary Σ_2 formulas saying that there are exactly n elements to the left of x. The terminal segment of type ω^* is defined by a similar computable Σ_2 formula $\psi(x)$. By Theorem 7.5, $\varphi^{\mathcal{B}}(x)$ and $\psi^{\mathcal{B}}(x)$ are both $\Sigma_2^0(\mathcal{B})$.

Example 5: Let $\mathcal{A}$ be the ordered field of real algebraic numbers, and let S be an arbitrary computable relation on $\mathcal{A}$. Then in any copy $\mathcal{B}$ of $\mathcal{A}$, the image of S is computable relative to $\mathcal{B}$.

Proof: There is a c.e. defining family for $\mathcal{A}$, consisting of finitary existential formulas (see Chapter 6). This is a family of formulas defining the *elements* of $\mathcal{A}$. By taking conjunctions, we obtain a family of formulas defining the *n-tuples*. Then S is defined by the disjunction of the defining formulas satisfied by tuples in S, a computable Σ_1 formula. The complementary relation $\neg S$ has a similar definition. By Theorem 7.5, if $\mathcal{B}$ is a copy of $\mathcal{A}$, then the images of S and $\neg S$ are both $\Sigma_1^0(\mathcal{B})$.

Here is the general result on Problem 1. The result was obtained first by Manasse and Slaman, then by the authors and, independently, by Chisholm [16], [27].

Theorem 10.1 *For a computable structure $\mathcal{A}$ with a further relation R, the following are equivalent:*
(1) R is definable in $\mathcal{A}$ by a computable Σ_α formula,
(2) in all copies $\mathcal{B}$ of $\mathcal{A}$, the image of R is $\Sigma_\alpha^0(\mathcal{B})$.

10.1. IMAGES OF A RELATION

Comments: The fact that (1) $\Rightarrow$ (2) follows from Theorem 7.5. The statement (2) $\Rightarrow$ (1) is the basic Completeness Theorem that we were aiming for. Note that if (2) holds, then the image of R under an automorphism of $\mathcal{A}$ must be Σ_α^0. It follows that there are fewer than $2^{\aleph_0}$ possible images. Therefore, by Theorem 6.14, R is definable by an $L_{\omega_1\omega}$ formula. To show that it is definable by a *computable* Σ_α formula, we use forcing.

The outline of the proof is as follows. Assuming (2), we build a "generic" copy $\mathcal{B}$ of $\mathcal{A}$. By hypothesis, the image of R in $\mathcal{B}$ is $\Sigma_\alpha^0(\mathcal{B})$. Then the statement asserting this must be "forced". From this fact, we will extract a definition of R, as in (1).

We begin a detailed description of the forcing machinery. Let L be the language of $\mathcal{A}$. Let B be an infinite computable set of new constants—for the universe of $\mathcal{B}$. Let $\mathcal{F}$ be the set of finite $1-1$ functions from B to $\mathcal{A}$. This is our set of *forcing conditions*. There is a natural partial ordering on $\mathcal{F}$, namely $\subseteq$. In what follows, p, q, etc. will always denote elements of $\mathcal{F}$. We shall choose a special sequence of forcing conditions. The union of this sequence will be a $1-1$ function F from $\mathcal{A}$ onto B. We take $\mathcal{B}$ to be the structure with universe B such that $\mathcal{A} \cong_F \mathcal{B}$.

We need an appropriate *forcing language*, in which we can talk about the generic copy $\mathcal{B}$ and the expansion $(\mathcal{B}, F(R))$, which is under construction. We must be able to say such things as "$F(R) = W_e^{\Delta_\alpha^0(\mathcal{B})}$". There are different possibilities for the forcing language. In [16], it was a predicate language, the result of adding to L the relation symbol R, an operation symbol F, for the isomorphism, and names for the elements of $\mathcal{A}$. Here the forcing language will be propositional. We take the language P in which the propositional symbols are just the atomic sentences of the predicate language $L \cup \{R\} \cup$ B. We also consider the sub-language P', in which the propositional symbols are the atomic sentences of the language $L \cup$ B—omitting R.

Let S be the set of computable infinitary formulas in the language P, and let S' be the set of computable infinitary formulas in the language P'. We indicate how propositional formulas from S, or S', describe the structures $\mathcal{B}$ and $(\mathcal{B}, F(R))$. Considering $(\mathcal{B}, F(R))$ as a structure for the language $L \cup \{R\}$, and taking the positive sentences from the atomic diagram, we obtain a structure $\mathcal{B}^*$ for the propositional language P. The positive sentences from the diagram of $\mathcal{B}$ form a structure for the sub-language P'. For $\varphi \in P$, we have

$$\mathcal{B}^* \models \varphi \Leftrightarrow (\mathcal{B}, F(R)) \models \varphi.$$

On the left, we think of φ as a propositional formula, while on the right, it is the result of substituting a tuple of elements (constants from B) for the free variables in a predicate formula.

We describe some formulas of special importance. For each n and e, we can find a computable Σ_1 formula φ, in S', such that

$$\mathcal{B}^* \models \varphi \Leftrightarrow n \in W_e^\mathcal{B}.$$

In fact, for each n and e, and each computable ordinal α (identified with its notation on some path in O), we can find a computable Σ_α formula, in S', with the meaning

$$n \in W_e^{\Delta_\alpha^0(\mathcal{B})} \ .$$

We write $n \in W_e^{D_\alpha}$ for this formula. In addition, we have a computable $\Pi_{\alpha+1}$ formula with the meaning

$$W_e^{\Delta_\alpha^0(\mathcal{B})} = F(R) \ .$$

We write $W_e^{D_\alpha} = F(R)$ for this formula.

We are ready to define forcing. That is, we define the relation p *forces* φ, denoted by $p \Vdash \varphi$, for $\varphi \in S$ and $p \in \mathcal{F}$, by induction on φ. While we are thinking of φ as a propositional formula, we remember that the propositional variables are actually atomic sentences involving constants from B, and we may speak of the "constants" of φ.

(1) If φ is finitary, then $p \Vdash \varphi$ if and only if the constants of φ are all in $dom(p)$ and p makes φ true in $(\mathcal{A}, R)$—for negations, this does not match the definition in other standard treatments of forcing,

(2) $p \Vdash \bigvee_i \varphi_i$ if and only if $p \Vdash \varphi_i$, for some i,

(3) $p \Vdash \bigwedge_i \varphi_i$ if and only if for each i and each $q \supseteq p$, there exists $r \supseteq q$ such that $r \Vdash \varphi_i$.

Below, we give slightly modified versions of the standard lemmas on forcing. For a computable Σ_α (or computable Π_α) formula φ, $neg(\varphi)$ is the naturally associated computable Π_α (or computable Σ_α) formula that is logically equivalent to $\neg \varphi$. We say that p *decides* φ if $p \Vdash \varphi$ or $p \Vdash neg(\varphi)$.

Lemma 10.2 *For any p (in $\mathcal{F}$), and any $\varphi \in S$, there exists $q \supseteq p$ such that q decides φ.*

Proof: The proof is by induction on the φ.

(1) If φ is finitary, then only finitely many propositional symbols appear in φ, and each of these, thought of as a predicate sentence, mentions only finitely many constants from B. If $q \supseteq p$, where $dom(q)$ includes all of these constants, then q decides φ.

(2) Let φ be computable Σ_α, say

$$\varphi = \bigvee_i \varphi_i \ ,$$

where for each i, φ_i is computable Π_β, for some $\beta < \alpha$, and the statement holds for all φ_i. Note that $neg(\varphi) = \bigwedge_i neg(\varphi_i)$. If no extension of p forces φ, then, by the definition of forcing, for all i and all $q \supseteq p$, $q \nVdash \varphi_i$. By H. I., there exists $r \supseteq q$ such that $r \Vdash neg(\varphi_i)$. Then by the definition of forcing, $p \Vdash neg(\varphi)$.

(3) Finally, let φ be computable Π_α, say

$$\varphi = \bigwedge_i \varphi_i \ ,$$

10.1. IMAGES OF A RELATION

where for each i, φ_i is computable Σ_β for some $\beta < \alpha$, and the statement holds for all φ_i. Then $neg(\varphi) = \bigvee_i neg(\varphi_i)$. If p does not force φ, then by the definition of forcing, for some i and some $q \supseteq p$, there is no $r \supseteq q$ such that $r \Vdash \varphi_i$. By H.I., for this i and q, there exists $r \supseteq q$ such that $r \Vdash neg(\varphi_i)$. Then by the definition of forcing, $r \Vdash neg(\varphi)$.

Lemma 10.3 *If $q \Vdash \varphi$ and $p \supseteq q$, then $p \Vdash \varphi$.*

Proof: The proof is an easy induction on φ.

Lemma 10.4 *For all $\varphi \in S$ and all p, it is not the case that*

$$p \Vdash \varphi \ \& \ p \Vdash neg(\varphi) \ .$$

Proof: Again the proof is by induction on φ.
(1) If φ is finitary, then the statement is clear.
(2) Suppose that

$$\varphi = \bigvee_i \varphi_i \ ,$$

where the statement holds for all φ_i. If $p \Vdash \varphi$ and $p \Vdash neg(\varphi)$, then by the definitions, together with Lemma 10.3, we obtain i and $q \supseteq p$ such that $q \Vdash \varphi_i$ and $q \Vdash neg(\varphi_i)$, contradicting H.I.
(3) Finally, suppose that

$$\varphi = \bigwedge_i \varphi_i \ ,$$

where the statement holds for all φ_i. Again we obtain i and $q \supseteq p$ such that $q \Vdash \varphi_i$ and $q \Vdash neg(\varphi_i)$, contradicting H.I.

A set $D \subseteq \mathcal{F}$ is said to be *dense* if for all $p \in \mathcal{F}$, there exists $q \supseteq p$ such that $q \in D$. The following sets are dense:

$$\begin{aligned} D_\varphi &= \{p : p \text{ decides } \varphi\} \ , \text{ for } \varphi \in S \ , \\ D'_a &= \{p : a \in ran(p)\} \ , \text{ for } a \in \mathcal{A} \ . \end{aligned}$$

Let $\mathcal{D}$ be the family consisting of these dense sets. A *complete forcing sequence*, or c.f.s., is a sequence of forcing conditions $(p_n)_{n \in \omega}$ such that for all n, $p_n \supseteq p_{n+1}$, and for each $D \in \mathcal{D}$, there exists n such that $p_n \in D$. Since $\mathcal{D}$ is countable, it is easy to see that such sequences exist.

We fix a c.f.s. $(p_n)_{n \in \omega}$, and describe the generic copy of $\mathcal{A}$ that it yields.

Claim: $\cup_n p_n$ is a $1-1$ function from B onto $\mathcal{A}$.

Proof of Claim: Since each p is a $1-1$ function, $\cup_n p_n$ is clearly a $1-1$ function from a subset of B to $\mathcal{A}$. The fact that $\mathcal{D}$ includes the sets D'_a implies that the range includes all of $\mathcal{A}$. The fact that $\mathcal{D}$ includes the sets $D_{b=b}$ implies that the domain includes all of B.

Let F be the inverse of the function above, mapping $\mathcal{A}$ 1 − 1 onto B. As planned, we form the structure $\mathcal{B}$ such that $\mathcal{A} \cong_F \mathcal{B}$. This is our generic copy of $\mathcal{A}$. We also form the expansion $(\mathcal{B}, F(R))$. From the latter, we obtain the propositional structure $\mathcal{B}^*$ consisting of the atomic sentences true in $(\mathcal{B}, F(R))$; i.e., those $\varphi \in P$ such that $(\mathcal{B}, F(R)) \models \varphi$. Recall that a finitary formula φ in S is also a finitary open sentence in the predicate language $L \cup \{R\} \cup$ B. Thus, for a finitary open formula, φ, we have

$$\mathcal{B}^* \models \varphi \Leftrightarrow (\mathcal{B}, F(R)) \models \varphi \ .$$

Lemma 10.5 (Truth and Forcing) *For all $\varphi \in S$,*

$$\mathcal{B}^* \models \varphi \Leftrightarrow \exists n \, p_n \, \Vert\!\!- \varphi \ .$$

Proof: We proceed by induction on φ.

(1) First, suppose φ is finitary. Then $\mathcal{B}^* \models \varphi$ if and only if for all n such that $dom(p_n)$ includes the constants appearing in φ, $p_n \models \varphi$, so the statement holds.

(2) Next, suppose that $\varphi = \bigvee_i \varphi_i$, where the statement holds for all φ_i. We have

$$\begin{aligned}\mathcal{B}^* \models \varphi &\Leftrightarrow \exists i \ \mathcal{B}^* \models \varphi_i \\ &\Leftrightarrow \exists i \ \exists n \ p_n \, \Vert\!\!- \varphi_i \\ &\Leftrightarrow \exists n \ p_n \, \Vert\!\!- \varphi \ .\end{aligned}$$

(3) Finally, suppose that $\varphi = \bigwedge_i \varphi_i$, where the statement holds for all φ_i. First, suppose that $\mathcal{B}^* \models \varphi$. Then for all i, $\mathcal{B}^* \models \varphi_i$. Take n such that p_n decides φ. If $p_n \, \Vert\!\!- neg(\varphi)$, then for some i, $p_n \, \Vert\!\!- neg(\varphi_i)$. By H.I., there exists m such that $p_m \, \Vert\!\!- \varphi_i$, and by Lemma 10.3, we may suppose that $m = n$ (taking the larger). By Lemma 10.4, this is a contradiction. Now, suppose that $p_n \, \Vert\!\!- \varphi$. Then for each i, there exists $m \geq n$ such that p_m decides φ_i. Since $p_n \, \Vert\!\!- \varphi$, there exists $q \supseteq p_m$ such that $q \, \Vert\!\!- \varphi_i$, so in fact $p_m \, \Vert\!\!- \varphi_i$. Then by H.I., $\mathcal{B}^* \models \varphi_i$. Since this is true for all i, $\mathcal{B}^* \models \varphi$.

So far, we have made no use of S'—the subset of S in which the propositional symbols (elements of P') are atomic sentences not involving R. The next result says that for the elements of S', forcing is definable in $\mathcal{A}$.

Lemma 10.6 (Definability of Forcing) *For each $\varphi \in S'$ and each tuple $\overline{b}$ in $\mathcal{A}$, we can find a computable infinitary predicate formula $Force_{\overline{b},\varphi}(\overline{x})$ in the language of $\mathcal{A}$, such that $\mathcal{A} \models Force_{\overline{b},\varphi}(\overline{a})$ if and only if for the forcing condition p that maps $\overline{b}$ to $\overline{a}$, $p \, \Vert\!\!- \varphi$. Moreover, if φ is computable Σ_α (or Π_α), then so is $Force_{\overline{b},\varphi}(\overline{x})$.*

Proof: The proof is by induction on φ.

(1) First, suppose that φ is finitary. If $\overline{b}$ includes the constants from B that appear in φ, then $Force_{\overline{b},\varphi}(\overline{x})$ is a finite conjunction of basic L-formulas (i.e.,

10.1. IMAGES OF A RELATION

atomic formulas and negations of atomic formulas) saying that the assignment of $\bar{b}$ to $\bar{x}$ makes φ true. Otherwise, $Force_{\bar{b},\varphi}(\bar{x}) = \bot$.

(2) Next, let φ is computable Σ_α. Suppose $\varphi = \bigvee_i \varphi_i$, where each φ_i is computable Π_β, for some $\beta < \alpha$, and we have determined $Force_{\bar{b},\varphi_i}(\bar{x})$ for all i. Then

$$Force_{\bar{b},\varphi}(\bar{x}) = \bigvee_i Force_{\bar{b},\varphi_i}(\bar{x}) .$$

(3) Finally, let φ be computable Π_α. Suppose $\varphi = \bigwedge_i \varphi_i$, where each φ_i is computable Σ_β for some $\beta < \alpha$, and we have determined $Force_{\bar{b},\bar{d},\varphi_i}(\bar{x})$, for all i and all $\bar{d}$. Then $Force_{\bar{b},\varphi}(\bar{x})$ is the conjunction over i and $\bar{b}_1$ of the formulas

$$\forall \bar{u}_1 \bigvee_{\bar{b}_2} \exists \bar{u}_2 \, Force_{\bar{b},\bar{b}_1,\bar{b}_2,\varphi_i}(\bar{x},\bar{u}_1,\bar{u}_2) .$$

Here the lengths of the tuples of variables $\bar{u}_1$, $\bar{u}_2$ match the lengths of the corresponding tuples of constants $\bar{b}_1$, $\bar{b}_2$.

We return to the proof of Theorem 10.1. Let $(p_n)_{n\in\omega}$ be our fixed c.f.s., and let F and $\mathcal{B}$ be the resulting isomorphism and copy of $\mathcal{A}$, as described above. Supposing that $F(R)$ is $\Sigma_\alpha^0(\mathcal{B})$, we must produce a computable Σ_α formula that defines R. For some e, we have $F(R) = W_e^{\Delta_\alpha^0(\mathcal{B})}$. By Lemma 10.5 (Truth and Forcing), there exists n such that

$$p_n \Vdash W_e^{D_\alpha} = F(R) ,$$

say p_n maps $\bar{d}$ to $\bar{c}$. From the definition of forcing, it follows that $a \in R$ if and only if there exist $q \supseteq p_n$ and $b \in B$ such that $q(b) = a$ and $q \Vdash b \in W_e^{D_\alpha}$.

By Lemma 10.6 (Definability of Forcing), for all b and $\bar{b}_1$ in B, we can find a computable Σ_α formula $Force_{\bar{d},b,\bar{b}_1,b\in W_e^{D_\alpha}}(\bar{c},x,\bar{u})$, in the predicate language L, such that

$$\mathcal{A} \models Force_{\bar{d},b,\bar{b}_1,b\in W_e^{D_\alpha}}(\bar{c},a,\bar{a}_1)$$

if and only if for q taking $\bar{d},b,\bar{b}_1$ to $\bar{c},a,\bar{a}_1$, we have $q \Vdash b \in W_e^{D_\alpha}$. Let $\varphi(\bar{c},x)$ be the disjunction over all $b,\bar{b}_1$ in B, of the formulas

$$\exists \bar{u} \, Force_{\bar{d},b,\bar{b}_1,b\in W_e^{D_\alpha}}(\bar{c},x,\bar{u}) .$$

Then $\varphi(\bar{c},x)$ defines R in $\mathcal{A}$, and $\varphi(\bar{c},x)$ is easily converted into the required computable Σ_α formula. This completes the proof of Theorem 10.1.

We mention some terminology associated with Problem 1. For a computable structure $\mathcal{A}$ and a further relation R, we say that R is *relatively intrinsically* Σ_α^0 on $\mathcal{A}$ if for all isomorphisms F from $\mathcal{A}$ onto a copy $\mathcal{B}$, $F(R)$ is $\Sigma_\alpha^0(\mathcal{B})$. If R is definable in $\mathcal{A}$ by a computable Σ_α formula, then we say that R is *formally* Σ_α^0 on $\mathcal{A}$. Thus, Theorem 10.1 says that R is relatively intrinsically Σ_α^0 on $\mathcal{A}$ if and only if it is formally Σ_α^0 on $\mathcal{A}$. Similarly, we say that R is *relatively intrinsically* Δ_α^0 on $\mathcal{A}$ if for all isomorphisms F from $\mathcal{A}$ onto a copy $\mathcal{B}$, $F(R)$ is $\Delta_\alpha^0(\mathcal{B})$. The next result gives conditions under which R is relatively intrinsically Δ_α^0.

Corollary 10.7 *For a computable structure $\mathcal{A}$ with a further relation R, R is relatively intrinsically Δ^0_α on $\mathcal{A}$ if and only if R and $\neg R$ are both formally Σ^0_α on $\mathcal{A}$.*

In the proof of Theorem 10.1, when we extracted the computable Σ^0_α definition of R, we did not look at *arbitrary* copies of $\mathcal{A}$. We considered a single generic copy. Chisholm [27] observed that there is a $\Delta^0_{\alpha+1}$ copy that is sufficiently generic for the argument to go through.

Theorem 10.8 (Chisholm) *For a computable structure $\mathcal{A}$ with an added relation R, the following are equivalent:*
  (1) R is definable in $\mathcal{A}$ by a computable Σ_α formula,
  (2) in all $\Delta^0_{\alpha+1}$ copies $\mathcal{B}$ of $\mathcal{A}$, the image of R is $\Sigma^0_\alpha(\mathcal{B})$.

Sketch of proof: If $\mathcal{A}$ is computable, then forcing of computable Σ_α formulas is Σ^0_α, and forcing of computable Π_α formulas is Π^0_α, with all possible uniformity. As above, we have a forcing condition p such that

$$p \Vdash W_e^{D_n} = F(R) \ .$$

Now, we restrict S to the computable infinitary sentences of the language P that are either Σ_β or Π_β, for $\beta \leq \alpha$. Similarly, we restrict S' to the computable infinitary sentences of the language P' that are either Σ_β or Π_β, for $\beta \leq \alpha$. The lemmas (Truth and Forcing, etc.) still hold. We can choose a $\Delta^0_{\alpha+1}$ c.f.s. $(p_n)_{n\in\omega}$, with $p_0 = p$. This yields a $\Delta^0_{\alpha+1}$ isomorphism F from $\mathcal{A}$ onto a structure $\mathcal{B}$, which must also be $\Delta^0_{\alpha+1}$. As above, we obtain a computable Σ_α formula defining R.

Problem 2 *Find syntactical conditions, on a computable structure $\mathcal{A}$ and a further relation R, guaranteeing that in all copies $\mathcal{B}$ of $\mathcal{A}$, if the image of R is $\Sigma^0_\alpha(\mathcal{B})$, then it is actually $\Delta^0_\alpha(\mathcal{B})$.*

The syntactical conditions were isolated by Harizanov [61], for the case where $\alpha = 1$, in the context of computable structures.

Example: Let $\mathcal{A}$ be an infinite-dimensional vector space over an infinite computable field, and let R be a subspace of finite co-dimension. Then in any copy $\mathcal{B}$ of $\mathcal{A}$ in which the image of $\neg R$ is c.e. relative to $\mathcal{B}$, it is computable relative to $\mathcal{B}$.

To see why, say that $\overline{a}$ is a basis for $\mathcal{A}$ over R; i.e., $\overline{a}$ is linearly independent over R, and the elements of R and $\overline{a}$ span $\mathcal{A}$. Let $(\tau_n(\overline{a}))_{n\in\omega}$ be a computable list of terms representing "non-trivial" linear combinations of elements of $\overline{a}$ (those in which the coefficient of some a_i is not zero). Then $\neg R$ has a definition of the form

$$\varphi(x) = \exists u \, \exists v \, (\, x = u + v \, \& \, Ru \, \& \, \bigvee_n v = \tau_n(\overline{a}) \,) \ ,$$

Then in any copy $\mathcal{B}$, the image of $\neg R$ is enumeration reducible to $\mathcal{B}$ and the image of R.

10.2. ERSHOV'S HIERARCHY

Theorem 10.9 *For a computable structure $\mathcal{A}$ with a further relation R, the following are equivalent:*
(1) in all copies $\mathcal{B}$ of $\mathcal{A}$ in which the image of R is $\Sigma^0_\alpha(\mathcal{B})$, it is $\Delta^0_\alpha(\mathcal{B})$,
(2) R is definable in $\mathcal{A}$ by a computable Σ_α formula

$$\varphi(\overline{c}, \overline{x}) = \bigvee_i \exists \overline{u}_i \left(\psi_i(\overline{c}, \overline{x}, \overline{u}_i) \ \& \ \rho_i(\overline{c}, \overline{x}, \overline{u}_i) \right),$$

where for each i, ψ_i is computable Π_β for some $\beta < \alpha$, in the language L, and ρ_i is a finite conjunction of positive atomic formulas involving R.

Exercise: Prove Theorem 10.9.

10.2 Ershov's hierarchy

Recall Ershov's hierarchy of Δ^0_2 sets, described in Chapter 5.

Problem 3 *Find syntactical conditions on a structure $\mathcal{A}$ and a further relation R, guaranteeing that in all copies $\mathcal{B}$ of $\mathcal{A}$, the image of R is d-c.e., or α-c.e. relative to $\mathcal{B}$.*

It turns out that we can extend Theorem 10.1 through Ershov's hierarchy. The syntactical conditions are fairly obvious. They were first isolated in [13], in the context of computable structures. The result below, proved by McCoy [105] and, independently, by McNicholl [108], characterizes the relations that are *intrinsically d-c.e.!relatively* on $\mathcal{A}$ (where this has the obious meaning). McNicholl proved the natural extension for relations that are *intrinsically n-c.e.!relatively* on $\mathcal{A}$.

Theorem 10.10 (McCoy, McNicholl) *For a computable structure $\mathcal{A}$ and a further relation R on $\mathcal{A}$, the following are equivalent:*
(1) in all copies $\mathcal{B}$ of $\mathcal{A}$, the image of R is d-c.e. relative to $\mathcal{B}$,
(2) R is definable in $\mathcal{A}$ by a formula of the form

$$\varphi(\overline{c}, \overline{x}) \ \& \ \neg \psi(\overline{c}, \overline{x}),$$

where φ and ψ are computable Σ_1.

Sketch of proof: The fact that (2) implies (1) is clear from Theorem 7.5. For simplicity, suppose that R is unary. To show that (1) implies (2), we consider an isomorphism F from $\mathcal{A}$ onto a generic copy $\mathcal{B}$, as in the proof of Theorem 10.1. We must have

$$F(R) = W^\mathcal{B}_{e_1} - W^\mathcal{B}_{e_2},$$

for some e_1 and e_2, . Then for some forcing condition p, we have

$$p \Vdash F(R) = W^D_{e_1} - W^D_{e_2}.$$

Note that $a \in R$ if and only if

$$(\exists q \supseteq p)\, \exists b\, (\, q(b) = a\ \&\ q \Vdash \varphi^D_{e_1}(b) \downarrow\,)$$

and

$$(\forall q \supseteq p)\, \forall b\, \neg(\,(q(b) = a\ \&\ q \Vdash \varphi^D_{e_2}(b) \downarrow)\,).$$

Say p takes $\overline{d}$ to $\overline{c}$. Then we have computable Σ_1 formulas $\varphi_i(\overline{c}, x)$ saying

$$\bigvee_b (\exists q \supseteq p)\,(\, q(b) = a\ \&\ q \Vdash \varphi^D_{e_i}(b) \downarrow\,).$$

for $i = 1, 2$. Then R is defined by the formula $\varphi_1(\overline{c}, x)\ \&\ \neg \varphi_2(\overline{c}, x)$.

Example: Let $\mathcal{A}$ be a reduced Abelian p-group, and let R be the set of elements of order p and height 1. Then R is intrinsically relatively d-c.e. on $\mathcal{A}$.

To see why, note that there is a finitary existential formula $\varphi(x)$, saying that x has order p and is divisible by p, and there is another finitary existential formula $\psi(x)$, saying that x is divisible by p^2. Then R is defined by the formula $\varphi\ \&\ \neg \psi$.

The next result characterizes the relations that are relatively intrinsically α-c.e. on $\mathcal{A}$. The result is notation-dependent. Nonetheless, we suppress the ordinal notation, identifying ordinals $\beta \leq \alpha$ with their notations on a fixed path in O.

Theorem 10.11 *For a computable structure $\mathcal{A}$ and a further relation R on $\mathcal{A}$, the following are equivalent:*

(1) in all copies $\mathcal{B}$ of $\mathcal{A}$, the image of R is α-c.e. relative to $\mathcal{B}$,

(2) there are computable sequences of computable Σ_1 formulas

$$(\varphi_\beta(\overline{c}, x))_{\beta < \alpha},\quad (\psi_\beta(\overline{c}, x))_{\beta < \alpha},$$

with a fixed finite tuple of parameters $\overline{c}$, such that

 (a) for any $a \in \mathcal{A}$, and any $\beta < \alpha$, if a satisfies $(\varphi_\beta(\overline{c}, x)\ \&\ \psi_\beta(\overline{c}, x))$, then it also satisfies $(\varphi_\gamma(\overline{c}, x) \vee \psi_\gamma(\overline{c}, x))$, for some $\gamma < \beta$ (this means that a cannot satisfy $\varphi_0(\overline{c}, x)\ \&\ \psi_0(\overline{c}, x)$), and

 (b) R is defined by the formula

$$\bigvee_{\beta < \alpha} (\varphi_\beta(\overline{c}, x)\ \&\ \neg \bigvee_{\gamma < \beta} \psi_\gamma(\overline{c}, x)).$$

Sketch of proof: The fact that $(2) \Rightarrow (1)$ is immediate from Proposition 5.23. We show that $(1) \Rightarrow (2)$. As above, we consider an isomorphism F from $\mathcal{A}$ onto a generic copy $\mathcal{B}$. We include in the forcing language computable infinitary sentences $\theta_{e,e'}$ saying that the families of sets

$$A_\beta = W^{\mathcal{B}}_{\varphi_e(\beta)},\ B_\beta = W^{\mathcal{B}}_{\varphi_{e'}(\beta)}$$

10.3. IMAGES OF A PAIR OF RELATIONS

satisfy the following conditions:
  (a) if $b \in A_\beta \cap B_\beta$, then for some $\gamma < \beta$, $b \in A_\gamma \cup B_\gamma$,
  (b) $F(R) = \cup_{\beta < \alpha} (A_\beta - \cup_{\gamma < \beta} B_\gamma)$.

For some e and e', the sentence $\theta_{e,e'}$ is true, so it is forced by some p. Then $a \in R$ if and only if the following two conditions hold:

(i) there is some $\beta < \alpha$ such that

$$(\exists q \supseteq p) \bigvee_b (q(b) = a \ \& \ q \Vdash b \in W^D_{\varphi_e(\beta)})$$

(ii) there is no $\gamma < \beta$ such that

$$(\exists q \supseteq p) \bigvee_b (q(b) = a \ \& \ q \Vdash b \in W^D_{\varphi_{e'}(\gamma)}) \ .$$

If p takes $\bar{d}$ to $\bar{c}$, then we have computable Σ_1 formulas $\varphi_\beta(\bar{c}, x)$ and $\psi_\beta(\bar{c}, x)$, saying

$$(\exists q \supseteq p) \bigvee_b (q(b) = x \ \& \ q \Vdash b \in W^D_{\varphi_e(\beta)}) \ ,$$

$$(\exists q \supseteq p) \bigvee_b (q(b) = x \ \& \ q \Vdash b \in W^D_{\varphi_{e'}(\beta)})$$

(the difference between the two is in the index e). Then R is defined by the formula

$$\bigvee_{\beta < \alpha} (\varphi_\beta(\bar{c}, x) \ \& \ \neg \bigvee_{\gamma < \beta} \psi_\gamma(\bar{c}, x)) \ .$$

There are further results with syntactical conditions guaranteeing that in all copies of a given computable structure, the complexity of the image of a relation is tied in some specific way to that of the copy. In addition to the result cited above, McNicholl has results on some further reducibilities, which we have not defined [108].

10.3 Images of a pair of relations

We change the setting slightly, considering a computable structure $\mathcal{A}$, with two further relations, R and S, of the same arity. Recall that a *separator* for R and S is a relation Q that contains all $\bar{a} \in R$ and no $\bar{a} \in S$.

Problem 4 *Find syntactical conditions on a computable structure $\mathcal{A}$ with further relations R and S, guaranteeing that for any isomorphism F from $\mathcal{A}$ onto a copy $\mathcal{B}$, there is a separator for $F(R)$ and $F(S)$ that is $\Delta^0_\alpha(\mathcal{B})$.*

The syntactical conditions were isolated by Davey [33], in the context of computable structures. The result below was proved by McCoy [105].

Theorem 10.12 *Let $\mathcal{A}$ be a computable structure, with further relations R and S, of the same arity. Then the following are equivalent:*

(1) for any isomorphism F from $\mathcal{A}$ onto a copy $\mathcal{B}$, $F(R)$ and $F(S)$ have a separator that is $\Delta^0_\alpha(\mathcal{B})$.

(2) there exist computable Σ_α formulas $\varphi(\overline{c}, \overline{x})$ and $\psi(\overline{c}, \overline{x})$ such that

  (a) every tuple $\overline{a}$ (of the appropriate length) satisfies $(\varphi(\overline{c}, \overline{x}) \vee \psi(\overline{c}, \overline{x}))$,
  (b) $\mathcal{A} \models \varphi(\overline{c}, \overline{a})$ for all $\overline{a}$ in R and no $\overline{a}$ in S, and
  (c) $\mathcal{A} \models \psi(\overline{c}, \overline{a})$ for all $\overline{a}$ in S and no $\overline{a}$ in R.

Remark: Theorem 10.12 implies Theorem 10.1—take S to be $\neg R$.

Proof that $(2) \Rightarrow (1)$ Suppose that we have formulas as in (2), with parameters $\overline{c}$. If $F(\overline{c}) = \overline{d}$, then $\varphi^{\mathcal{B}}(\overline{d}, \overline{x})$ contains all of $F(R)$ and none of $F(S)$, and $\psi^{\mathcal{B}}(\overline{d}, \overline{x})$ contains all of $F(S)$ and none of $F(R)$. Both $\varphi^{\mathcal{B}}(\overline{d}, \overline{x})$ and $\psi^{\mathcal{B}}(\overline{d}, \overline{x})$ are $\Sigma^0_\alpha(\mathcal{B})$. By Theorem 1.8 (relativized), together with the fact that being $\Sigma^0_\alpha(\mathcal{B})$ is the same as being c.e. relative to $\Delta^0_\alpha(\mathcal{B})$, there exist disjoint relations Q_1 and Q_2, both $\Sigma^0_\alpha(\mathcal{B})$, such that

$$Q_1 \subseteq \varphi^{\mathcal{B}}(\overline{d}, \overline{x}),$$
$$Q_2 \subseteq \psi^{\mathcal{B}}(\overline{d}, \overline{x}),$$
$$Q_1 \cup Q_2 = \varphi^{\mathcal{B}}(\overline{d}, \overline{x}) \cup \psi^{\mathcal{B}}(\overline{d}, \overline{x}).$$

Moreover, since $\varphi^{\mathcal{B}}(\overline{d}, \overline{x}) \cup \psi^{\mathcal{B}}(\overline{d}, \overline{x})$ is $\Delta^0_\alpha(\mathcal{B})$ (actually, it is computable), Q_1 and Q_2 are $\Delta^0_\alpha(\mathcal{B})$. Therefore, Q_1 is the required separator.

Exercise: For Theorem 10.12, prove that $(1) \Rightarrow (2)$.

Problem 5 *Find syntactical conditions on a computable structure $\mathcal{A}$, with added relations R and S, guaranteeing that in any copy $\mathcal{B}$, if the image of R is c.e. relative to $\mathcal{B}$, then so is the image of S.*

Example: Let V be an infinite dimensional vector space over Q, let R and S be subspaces such that $R \subsetneq S$, R has infinite dimension, and S has finite dimension over R and infinite co-dimension. If R is c.e. relative to V, then so is S.

The reason is that S is defined by a formula of a special form. Let $\overline{c}$ be a basis for S over R. Let $(\tau_n(\overline{c}))_{n \in \omega}$ be a computable list of the terms naming linear combinations of the elements of $\overline{c}$. Then S is defined by the formula

$$\varphi(\overline{c}, x) = \exists u \, \exists v \, (x = u + v \ \& \ R(u) \ \& \ \bigvee_n v = \tau_n(\overline{c})).$$

If F is an isomorphism from $\mathcal{A}$ onto $\mathcal{B}$, then we can apply a uniform effective procedure to convert an enumeration of the atomic diagram of $\mathcal{B}$ and the relation R into an enumeration of S. Therefore, $F(R)$ is enumeration reducible to $\mathcal{B}$ and $F(S)$.

The definability in the example above illustrates the general pattern.

10.3. IMAGES OF A PAIR OF RELATIONS

Proposition 10.13 *Let $\mathcal{A}$ be a computable structure, and let R and S be further relations on $\mathcal{A}$. Then the following are equivalent:*

(1) for all $\mathcal{B} \cong \mathcal{A}$, if the image of R is c.e. relative to $\mathcal{B}$, then so is the image of S,

(2) S is definable in $(\mathcal{A}, R)$ by a computable Σ_1 formula $\varphi(\bar{c}, \bar{x})$, with a finite tuple of parameters $\bar{c}$, in which R appears only positively.

Proof sketch: It is easy to see that (2) implies (1). We suppose (1) and prove (2). For simplicity, we suppose that R and S are unary. We consider two cases, according to whether $\mathcal{A}$ is trivial. Recall (from Chapter 3) that a structure $\mathcal{A}$ is trivial if its automorphism group includes all permutations that fix some finite tuple $\bar{c}$.

Case 1: Suppose that $\mathcal{A}$ is trivial.

If R is finite or co-finite, then S is also finite or co-finite. Otherwise, there would be automorphisms of $\mathcal{A}$ that leave R invariant and give the image of S arbitrary Turing degree. Suppose that R is infinite and co-infinite. Without loss, we may suppose that R is computable—there is a computable copy of $\mathcal{A}$ in which the image of R is computable. The part of S in R is finite or co-finite relative to R, and the part of S in $\neg R$ is finite or co-finite relative to $\neg R$. Otherwise, there would be automorphisms of $\mathcal{A}$ that leave R invariant and give the image of S arbitrary Turing degree.

It cannot be the case that S differs finitely from $\neg R$, for then there would be an automorphism of $\mathcal{A}$ in which the image of R is c.e. but not computable, and the image of S would not be c.e. So, S is finite, or S is co-finite, or S differs only finitely from R. In the first two cases, S is definable in $\mathcal{A}$ by an open formula with finitely many parameters. In the third case, S is definable in $(\mathcal{A}, R)$ by an open formula with finitely many parameters and only a positive instance of R.

Case 2: Suppose $\mathcal{A}$ is non-trivial.

Claim: For any set X and any isomorphism F from $\mathcal{A}$ onto a copy $\mathcal{B}$ such that $\mathcal{B} \leq_T X$, if $F(R)$ is c.e. relative to X, then so is $F(S)$.

Proof of Claim: By Theorem 3.21, there is an isomorphism g from $\mathcal{B}$ onto some $\mathcal{C}$ such that $\mathcal{C} \equiv_T X$ and $g \leq_T X$. Let $G = g \circ F$. Since $F(R)$ is c.e. in X and g is computable in X, $G(R)$ is also c.e. in X, or in $\mathcal{C}$. Then by (1), $G(S)$ must be c.e. in $\mathcal{C}$, or in X. Since g^{-1} is computable in X, $F(S)$ is c.e. in X. This proves the claim.

Let L be the language of $\mathcal{A}$. For simplicity, suppose R and S are unary. Our plan is to produce a generic set X and a generic isomorphism F from $\mathcal{A}$ onto a copy $\mathcal{B}$ such that $\mathcal{B} \leq_T X$ and $F(R)$ is c.e. relative to X. Let B be an infinite computable set of constants, for the universe of $\mathcal{B}$. Let P be the set of atomic sentences in the language $L \cup$ B. Let $P' = P \cup \{Sb : b \in B\}$—the result of adding to P the atomic sentences of the form Sb for $b \in B$. Let C be an infinite computable set disjoint from $P \cup B$.

We determine a computable tree of height 2. There are infinitely many elements at level 1, and each of these has infinitely many successors at level 2. Formally, the elements of the tree are sequences of length 1 or 2, but here it will be convenient to forget this. We think of the tree as a structure with two unary relations, for the levels, and a binary relation, successor. Let B be the set of elements at level 1, and let C be the set of elements at level 2.

We shall take X to be a subset of $P \cup C$, and we determine a $1-1$ function F from $\mathcal{A}$ onto $\mathcal{B}$ such that if $\mathcal{B}$ is the structure induced by F, then $X \cap P$ is the set of atomic sentences in $D(\mathcal{B})$. Then $\mathcal{B} \leq_T X$. To guarantee that $F(R)$ is c.e. relative to X, we make sure that for all $b \in B$,

$$b \in F(R) \Leftrightarrow (\exists b' \notin X)\,(b' \text{ is a successor of } b).$$

We may think of S as a predicate symbol and expand $\mathcal{B}$ to $\mathcal{B}'$, where $S^{\mathcal{B}'} = F(S)$. We may also think of the set of atomic sentences of $D(\mathcal{B}')$ as a structure for the propositional language P'. Then for each e, we have a natural computable infinitary sentence

$$\bigwedge_b (Sb \leftrightarrow \varphi_e^D(b)\downarrow)$$

with the meaning $F(S) = W_e^{\mathcal{B}}$.

Let $\mathcal{F}$ be as in earlier proofs, the set of finite $1-1$ functions from B to $\mathcal{A}$. The forcing conditions are pairs (p, σ), where $p \in \mathcal{F}$, σ is a finite partial function from B' to $2 = \{0, 1\}$, and the following conditions are satisfied:

(a) if $\sigma(x) = 0$ and b is the predecessor of x, then $p(b) \in R$,
(b) if $p(b) \in R$, then for some successor x of b, $\sigma(x) = 0$.

We use the natural partial ordering, where

$$(p, \sigma) \subseteq (q, \tau) \Leftrightarrow p \subseteq q \;\&\; \sigma \subseteq \tau.$$

We define forcing in the obvious way, and the usual lemmas (Truth-and-Forcing, etc.) all hold. Take a complete forcing sequence $((p_n, \sigma_n))_{n \in \omega}$, deciding all of the computable infinitary sentences in P', and entering additional dense sets so that each element of $\mathcal{A}$ is eventually in $ran(p_n)$ and each element of C is eventually in $dom(\sigma_n)$. Then $\cup_n p_n$ is a $1-1$ function from B onto $\mathcal{A}$. Let F be the inverse, and let $\mathcal{B}$ be the structure induced on B. Now, the set of atomic sentences of true in $\mathcal{B}$ is a subset of P. Also, $\cup_n \sigma_n$ is the characteristic function of a subset of C. Let X be the union of the positive part of $D(\mathcal{B})$ with this set. As planned, $\mathcal{B} \leq_T X$ and $F(R)$ is c.e. relative to X.

By assumption (1), we have $F(S) = W_e^{\mathcal{B}}$, for some e. By the Truth-and-Forcing Lemma, some forcing condition (p, σ) forces this statement. We extract the definition of S as follows. Note that $a \in S$ if and only if

$$(\exists (q, \tau) \supseteq (p, \sigma))\,[q(b) = a \;\&\; (q, \tau) \Vdash \varphi_e^D(b)\downarrow].$$

Say p takes $\overline{d}$ to $\overline{c}$. Suppose (q, τ) is a forcing condition such that

10.4. ISOMORPHISMS

(i) $(q, \tau) \supseteq (p, \sigma)$,
(ii) $(q, \tau) \Vdash \varphi_e^D(b) \downarrow$,
(iii) q takes $\overline{d}, b, \overline{b}_1$ to to $\overline{c}, a, \overline{a}_1$.

Let $\psi(\overline{d}, b, \overline{b}_1)$ be the conjunction of the sentences from $D(\mathcal{B})$ made true by q and used in the computation, and let $\rho(\overline{d}, b, \overline{b}_1)$ be the conjunction of the positive atomic sentences involving R made true by q in $(\mathcal{A}, R)$. Note that if $q' \supseteq p$, where $dom(q') = dom(q)$ and q' makes $(\psi(\overline{d}, b, \overline{b}_1) \ \& \ \rho(\overline{d}, b, \overline{b}_1))$ true in $(\mathcal{A}, R)$, then (q', τ) is a forcing condition, and we have
(i) $(q', \tau) \supseteq (p, \sigma)$, and
(ii) $(q', \tau) \Vdash \varphi_e^D(b) \downarrow$.
The set of formulas

$$\exists \overline{u} \, (\, \psi(\overline{c}, x, \overline{u}) \ \& \ \rho(\overline{c}, x, \overline{u}) \,)$$

that arise in this way is computably enumerable, and the disjunction of the set is the required definition of S.

It is possible to formulate a much more general version of Problem 5, in which the single relation R is replaced by a family of relations, and we look for definability conditions guaranteeing that in any copy $\mathcal{B}$ of $\mathcal{A}$ in which the images of the relations in this family are $\Sigma_\beta^0(\mathcal{B})$, for various specified β, the image of the relation S is $\Sigma_\alpha^0(\mathcal{B})$. The more general problem is solved in [10].

10.4 Isomorphisms

The next two problems concern isomorphisms from a given computable structure $\mathcal{A}$ to various copies.

Problem 6 *Find syntactical conditions on a computable structure $\mathcal{A}$ guaranteeing that for all copies $\mathcal{B}$, some isomorphism F from $\mathcal{A}$ onto $\mathcal{B}$ is $\Delta_\alpha^0(\mathcal{B})$.*

Problem 7 *Find syntactical conditions on a computable structure $\mathcal{A}$ guaranteeing that for all copies $\mathcal{B}$, all isomorphisms from $\mathcal{A}$ onto $\mathcal{B}$ are $\Delta_\alpha^0(\mathcal{B})$.*

The syntactical conditions are suggested by the following examples.

Example 1: Let $\mathcal{A}$ be an infinite dimensional vector space over the rationals. For any copy $\mathcal{B}$, some isomorphism from $\mathcal{A}$ onto $\mathcal{B}$ is $\Delta_2^0(\mathcal{B})$.

The reason is that $\mathcal{A}$ has a c.e. Scott family consisting of computable Π_1 formulas $\varphi(\overline{x})$, with no parameters. Using $\Delta_2^0(\mathcal{B})$, we can enumerate the finite partial isomorphisms p which take a tuple $\overline{a}$ in $\mathcal{A}$ to a tuple $\overline{b}$ in $\mathcal{B}$ satisfying the same formula of Φ. This is a back-and-forth family, and using it, we can build an isomorphism that is $\Delta_2^0(\mathcal{B})$

Example 2: Let $\mathcal{A}$ be a computable ordering which is the shuffle sum of finite orderings of type $n + 1$ for all $n \in \omega$. For any copy $\mathcal{B}$, some isomorphism from $\mathcal{A}$ onto $\mathcal{B}$ is $\Delta_3^0(\mathcal{B})$.

The reason is that $\mathcal{A}$ has a c.e. Scott family Φ consisting of finitary Σ_3 formulas $\varphi(x)$ (with no parameters).

Example 3: Let $\mathcal{A} = (Z, S)$ (the integers with the successor function). For any copy $\mathcal{B}$, all isomorphisms from $\mathcal{A}$ onto $\mathcal{B}$ are computable relative to $\mathcal{B}$.

The reason is that $\mathcal{A}$ has a c.e. defining family Φ, consisting of finitary existential formulas $\varphi(0, x)$, with parameter 0, where $\varphi(0, x)$ says, for some k, that x is the kth successor of 0, or that 0 is the kth successor of x. We can argue as in Example 1 that some isomorphism from $\mathcal{A}$ onto a copy $\mathcal{B}$ is computable in $(\mathcal{B})$. Moreover, we can start with an arbitrary image of 0. Since there is a unique isomorphism with a given image of 0, all isomorphisms are computable in $\mathcal{B}$.

Example 4: Let $\mathcal{A}$ be the ordered field of real algebraic numbers. For any copy $\mathcal{B}$, all isomorphisms from $\mathcal{A}$ onto $\mathcal{B}$ are computable relative to $\mathcal{B}$.

The reason is that there is a c.e. defining family consisting of finitary existential formulas $\varphi(x)$, with no parameters, where $\varphi(x)$ says that x is the kth root of some polynomial $p(x)$ with integer coefficients.

In Section 10.1, we indicated that computable copies of a structure, if they exist, are the ones that interest us most, and we admitted that the results on the image of a relation in arbitrary copies of a given structure were proved later than the corresponding results for computable copies. However, the results for arbitrary copies are cleaner, and the proofs are simpler. They should have been proved first. The same is true here, for the results on isomorphisms. The results on Problems 6 and 7 were proved later than the corresponding results for computable copies, but it would have been useful to prove these results first.

We begin with the result on Problem 6. The syntactical conditions were isolated, in the context of computable structures, by Goncharov [53] in the case where $\alpha = 1$ and by Ash [2] for arbitrary computable ordinals α.

Theorem 10.14 *For a computable structure $\mathcal{A}$, the following are equivalent:*

(1) $\mathcal{A}$ has a c.e. Scott family, consisting of computable Σ_α formulas $\varphi(\overline{c}, \overline{x})$ with a fixed tuple of parameters $\overline{c}$,

(2) $\mathcal{A}$ has a Σ_α^0 Scott family, consisting of computable Σ_α formulas $\varphi(\overline{c}, \overline{x})$ with a fixed tuple of parameters $\overline{c}$,

(3) for any $\mathcal{B} \cong \mathcal{A}$, there is an isomorphism F from $\mathcal{A}$ onto $\mathcal{B}$ such that F is $\Delta_\alpha^0(\mathcal{B})$.

Sketch of proof: Trivially, (1) $\Rightarrow$ (2).

(2) $\Rightarrow$ (3) Take $\overline{d}$ such that that $(\mathcal{A}, \overline{c}) \cong (\mathcal{B}, \overline{d})$. The construction of the isomorphism F proceeds by stages. At stage s, we have a finite part F_s of F, where if F_s maps $\overline{a}$ to $\overline{b}$, then there exists $\varphi(\overline{c}, \overline{x}) \in \Phi$ such that

$$\mathcal{A} \models \varphi(\overline{c}, \overline{a}) \text{ and } \mathcal{B} \models \varphi(\overline{d}, \overline{b}) \ .$$

10.4. ISOMORPHISMS

At stage $s + 1$, we take the first element $a' \in \mathcal{A}$ not in $\bar{a}$, and we determine an image b', or we take the first element $b' \in \mathcal{B}$ not in $\bar{b}$ and we determine a preimage a'. Using $\Delta_\alpha^0(\mathcal{B})$, we search for a formula $\psi(\bar{c}, \bar{x}, y) \in \Phi$ and an element b' (or a') such that

$$\mathcal{A} \models \psi(\bar{c}, \bar{a}, a') \text{ and } \mathcal{B} \models \psi(\bar{d}, \bar{b}, b') .$$

We let $F_{s+1} \supseteq F_s$, where $F_{s+1}(a') = b'$.

(3) $\Rightarrow$ (1) We determine a generic copy $\mathcal{B}$ of $\mathcal{A}$. The forcing conditions are as for Theorem 10.1. Among the computable infinitary formulas decided by our c.f.s. are the formulas ψ_e, saying that $\varphi_e^{\Delta_\alpha^0(\mathcal{B})}$ is an isomorphism from $\mathcal{A}$ onto $\mathcal{B}$—where $\mathcal{B}$ is the generic copy. There are computable Σ_α formulas which we denote by $\varphi_e^{D_\alpha}(\bar{a}) = \bar{b}$, saying that $\varphi_e^{\Delta_\alpha^0(\mathcal{B})}$ maps $\bar{a}$ to $\bar{b}$. If $\mathcal{B}$ is the generic copy, then for some e, $\varphi_e^{\Delta_\alpha^0(\mathcal{B})}$ is an isomorphism. Therefore, there is a forcing condition p such that $p \Vdash \psi_e$. Say p takes $\bar{d}$ to $\bar{c}$. By the definition of forcing, for each $\bar{a}$ in $\mathcal{A}$, there exist $q \supseteq p$, $\bar{b}$, and $\bar{a}^*$ such that $q(\bar{b}) = \bar{a}$, and

$$q \Vdash \varphi_e^{D_\alpha}(\bar{a}^*) = \bar{b} .$$

Moreover, for any such $\bar{a}^*$,

$$(\mathcal{A}, \bar{a}) \cong (\mathcal{A}, \bar{a}^*) ,$$

since q^{-1} extends to an isomorphism F from $\mathcal{A}$ onto a generic $\mathcal{C}$ such that $F(\bar{a}) = \bar{b}$ and $\varphi_e^{\Delta_\alpha^0}$ is an isomorphism from $\mathcal{A}$ onto $\mathcal{C}$ taking $\bar{a}^*$ to $\bar{b}$.

Given $\bar{a}^*$ and $\bar{b}$ (of the same length), and $\bar{b}_1$ (of arbitrary length), we can find a computable Σ_α formula $Force_{\bar{d}, \bar{b}, \bar{b}_1; \varphi_e^{D_\alpha}(\bar{a}^*) = \bar{b}}(\bar{c}, \bar{x}, \bar{u}_1)$, in the language of $\mathcal{A}$, such that for q mapping $\bar{d}, \bar{b}, \bar{b}_1$ to $\bar{c}, \bar{a}, \bar{a}_1$,

$$q \Vdash \varphi_e^{D_\alpha}(\bar{a}^*) = \bar{b} \Leftrightarrow \mathcal{A} \models Force_{\bar{d}, \bar{b}, \bar{b}_1, \varphi_e^{D_\alpha}(\bar{a}^*) = \bar{b}}(\bar{c}, \bar{a}, \bar{a}_1) .$$

For each tuple $\bar{a}^*$, we can find a computable Σ_α formula $\varphi_{\bar{a}^*}(\bar{c}, \bar{x})$, logically equivalent to the disjunction, over the appropriate tuples $\bar{b}, \bar{b}_1$, of the formulas

$$\exists \bar{u}_1 \, Force_{\bar{d}, \bar{b}, \bar{b}_1, \varphi_e^{D_\alpha}(\bar{a}^*) = \bar{b}}(\bar{c}, \bar{x}, \bar{u}_1) .$$

From what we have said already, it should be clear that the set of these formulas $\varphi_{\bar{a}^*}(\bar{c}, \bar{x})$ is the required Scott family.

There is some terminology associated with Theorem 10.14. We say that $\mathcal{A}$ is *relatively* Δ_α^0 *categorical* if for all $\mathcal{B} \cong \mathcal{A}$, some isomorphism from $\mathcal{A}$ onto $\mathcal{B}$ is $\Delta_\alpha^0(\mathcal{B})$. A *formally* Σ_α^0 *Scott family* is a c.e. Scott family Φ consisting of computable Σ_α formulas. Theorem 10.14 says that $\mathcal{A}$ is relatively Δ_α^0 categorical if and only if it has a formally Σ_α^0 Scott family.

If a computable structure $\mathcal{A}$ has no formally Σ_1^0, or formally c.e. Scott family, then by Theorem 10.14, there exists $\mathcal{B} \cong \mathcal{A}$ with no isomorphism $F \leq_T \mathcal{B}$ from $\mathcal{A}$ onto $\mathcal{B}$. The next result strengthens this, saying that there are many such $\mathcal{B}$.

Theorem 10.15 (McCoy) *Suppose $\mathcal{A}$ is a computable structure. If $\mathcal{A}$ has no formally c.e. Scott family, then*
 (a) for each $n > 1$, there exist $\mathcal{B}_i \cong \mathcal{A}$, for $i = 1,\ldots,n$, such that for $i < j$, there is no isomorphism F from $\mathcal{B}_i$ onto $\mathcal{B}_j$ such that $F \leq_T \mathcal{B}_1,\ldots,\mathcal{B}_n$.
 (b) there is an infinite sequence $(\mathcal{B}_i)_{i<\omega}$ such that $\mathcal{B}_i \cong \mathcal{A}$ for all i, and for $i < j$, there is no isomorphism F from $\mathcal{B}_i$ onto $\mathcal{B}_j$ such that $F \leq_T (\mathcal{B}_i)_{i<\omega}$.

Goncharov defined the *computable dimension* of a structure $\mathcal{A}$ to be the number of equivalence classes of computable copies under the relation of computable isomorphism. There are (specially constructed) examples having computable dimension n for all finite $n \geq 1$. Goncharov and Ventsov asked whether there was a meaningful notion of relative computable dimension. Theorem 10.15 says that there is not.

The second author had announced Part (a), but the proof is trickier than she thought. McCoy [104] worked out the details of the proof for (a), and he also proved (b).

We turn to Problem 7. Again the syntactical conditions were first isolated, in the context of computable structures, by Goncharov [53] in the case where $\alpha = 1$ and by Ash [1], [3] for arbitrary computable ordinals α.

Theorem 10.16 *For a computable structure $\mathcal{A}$, the following are equivalent:*
 (1) $\mathcal{A}$ has a c.e. defining family Φ, consisting of computable Σ_α formulas with a fixed tuple of parameters,
 (2) $\mathcal{A}$ has a Σ_α^0 defining family Φ, consisting of computable Σ_α formulas with a fixed tuple of parameters,
 (3) for any isomorphism F from $\mathcal{A}$ onto a copy $\mathcal{B}$, F is $\Delta_\alpha^0(\mathcal{B})$.

Sketch of proof: Trivially, $(1) \Rightarrow (2)$.

$(2) \Rightarrow (3)$ Suppose F maps $\bar{c}$ to $\bar{d}$. To compute $F(a)$, using $\Delta_\alpha^0(\mathcal{B})$, we search for $\varphi(\bar{c},x) \in \Phi$ and b such that $\mathcal{A} \models \varphi(\bar{c},a)$ and $\mathcal{B} \models \varphi(\bar{d},b)$. Then $F(a) = b$.

$(3) \Rightarrow (1)$ We determine a generic copy of $\mathcal{A}$ as in the proof Theorem 10.1. Among the formulas decided by our c.f.s. are the computable Σ_α formulas $\varphi_e^{D_n}(a) = b$. Let $q \in \mathcal{D}_e$ if and only if for some $a \in ran(q)$, $q \Vdash \varphi_e^{D_n}(a) \uparrow$ or $q \Vdash \varphi_e^{D_n}(a) = b$, where for some $b' \neq b$, $q(b') = a$. We do not claim that the sets $\mathcal{D}_e$ are all dense, but we choose the c.f.s. to enter as many as we can.

As before, having fixed a c.f.s., we obtain a one-one function $F = (\cup_n p_n)^{-1}$ from $\mathcal{A}$ onto an infinite computable set of constants B. The generic structure $\mathcal{B}$ is that induced by F, so that F is an isomorphism from $\mathcal{A}$ onto $\mathcal{B}$. Now, F must be $\Delta_\alpha^0(\mathcal{B})$, say $F = \varphi_e^{\Delta_\alpha^0(\mathcal{B})}$. For this e, our c.f.s. fails to enter $\mathcal{D}_e$. Therefore, $\mathcal{D}_e$ is not dense. There exists p such that for all $p' \supseteq p$ and all a, there exists $q \supseteq p'$ such that $q \Vdash \varphi_e^{D_n}(a) = b$, and for any such q, $q(b) = a$. Say p maps $\bar{d}$ to $\bar{c}$.

Given a, b and $\bar{b}_1$, we can find a computable infinitary Σ_α formula in the language of $\mathcal{A}$, $Force_{\bar{d},b,\bar{b}_1,\varphi_e^{D_n}(a)=b}(\bar{c},x,\bar{u}_1)$, such that for q mapping $\bar{d},b,\bar{b}_1$ to $\bar{c},a,\bar{a}_1$,

$$q \Vdash \varphi_e^{D_n}(a) = b \Leftrightarrow \mathcal{A} \models Force_{\bar{d},b,\bar{b}_1,\varphi_e^{D_n}(a)=b}(\bar{c},a,\bar{a}_1).$$

10.5. EXPANSIONS

For each a, we can find a computable Σ_α formula $\varphi_a(\bar{c}, x)$ which is logically equivalent to the disjunction over $b, \bar{b}_1$ of the formulas

$$\exists \bar{u}\, Force_{\bar{d}, b, \bar{b}_1, \varphi_c^{p_\alpha}(a)=b}(\bar{c}, x, \bar{u}) \ .$$

The set of these formulas $\varphi_a(\bar{c}, x)$ is the required defining family Φ.

There is some terminology associated with Theorem 10.16. We say that $\mathcal{A}$ is *relatively* Δ_α^0 *stable* if for all $\mathcal{B} \cong \mathcal{A}$, all isomorphisms from $\mathcal{A}$ onto $\mathcal{B}$ are $\Delta_\alpha^0(\mathcal{B})$. A *formally* Σ_α^0 *defining family* is a c.e. defining family Φ consisting of computable Σ_α formulas. Theorem 10.16 says that $\mathcal{A}$ is relatively Δ_α^0 stable if and only if it has a formally Σ_α^0 defining family.

10.5 Expansions

The following problem is the subject of [8]

Problem 8 *Let $\mathcal{A}$ be a computable structure, and let ψ be a computable infinitary sentence involving a new relation symbol. Give conditions guaranteeing that for all $\mathcal{B} \cong \mathcal{A}$, there is some R such that $(\mathcal{B}, R) \models \psi$ and R is computable, or c.e. relative to $\mathcal{B}$.*

Example 1: Suppose $\mathcal{A}$ consists of two disjoint copies of a given structure, and ψ says, of a new binary relation symbol F, that it is an isomorphism between the two. If the original structure is relatively computably categorical, then for any $\mathcal{B} \cong \mathcal{A}$, there is an isomorphism F between the two parts such that $F \leq_T \mathcal{B}$, so

$$(\mathcal{B}, F) \models \psi \ .$$

If the original structure is not relatively computably categorical, then there is some $\mathcal{B} \cong \mathcal{A}$ with no isomorphism F between the two parts such that $F \leq_T \mathcal{B}$

Example 2: Let $\mathcal{A}$ be an infinite dimensional vector space over the rationals, and let ψ say, of a new unary relation symbol, that it is an infinite linearly independent set. By a result of Metakides and Nerode [117], there is actually a computable copy of $\mathcal{A}$ with no infinite c.e. linearly independent set. We shall mention this again in Chapter 12.

In [8], there are general results for the case where ψ is a computable Π_2 sentence. The sentences in the examples above are both of this form.

10.6 Sets computable in all copies

Problem 9 *For a given computable structure $\mathcal{A}$, characterize the sets S that are c.e. relative to $\mathcal{B}$ for all $\mathcal{B} \cong \mathcal{A}$.*

Problem 10 *For a given computable structure $\mathcal{A}$, characterize the sets S that are computable relative to $\mathcal{B}$ for all $\mathcal{B} \cong \mathcal{A}$.*

For Problem 9, the solution, from [81] and [6], involves the notion of *enumeration reducibility*, defined in Chapter 1.

Theorem 10.17 *For a computable structure $\mathcal{A}$, and a set S, the following are equivalent:*
(1) for all $\mathcal{B} \cong \mathcal{A}$, S is c.e. relative to $\mathcal{B}$,
(2) for some tuple $\bar{a}$ in $\mathcal{A}$, S is enumeration reducible to the existential type of $\bar{a}$ (i.e., the set of finitary Σ_1 formulas true of $\bar{a}$).

Proof: (2) $\Rightarrow$ (1) Suppose S is enumeration reducible to the existential type of $\bar{a}$. If $\mathcal{B} \cong \mathcal{A}$, then enumerating the atomic diagram of $\mathcal{B}$, we can enumerate the Σ_1 (or existential) type of $\bar{a}$, so we can enumerate S.

(1) $\Rightarrow$ (2) Suppose that for all $\mathcal{B} \cong \mathcal{A}$, S is c.e. relative to $\mathcal{B}$. We form a generic $\mathcal{B}$, using the same forcing conditions as in 10.1. For some p and e, we have $p \Vdash S = W_e^{\mathcal{B}}$. This means that

$$n \in S \Leftrightarrow (\exists q \supseteq p)\, q \Vdash \varphi_e^{\mathcal{B}}(n)\downarrow\ .$$

Say p maps $\bar{b}$ to $\bar{a}$. For each n, we can find a computable Σ_1 formula $\psi_n(\bar{x})$ saying that if p' maps $\bar{b}$ to $\bar{x}$, then $(\exists q \supseteq p')\, q \Vdash \varphi_e^{\mathcal{B}}(n)\downarrow$. Note that

$$n \in S \Leftrightarrow \mathcal{A} \models \psi_n$$

Given an enumeration of the existential type of $\bar{a}$, we can watch for the disjuncts of $\psi_n(\bar{x})$, for various n, and we can enumerate S.

The solution to Problem 9 yields the solution to Problem 10.

Corollary 10.18 *For a computable structure $\mathcal{A}$ and a set S, the following are equivalent:*
(1) for all $\mathcal{B} \cong \mathcal{A}$, S is computable relative to $\mathcal{B}$,
(2) $S \oplus \neg S$ is enumeration reducible to the existential type of some tuple $\bar{a}$ in $\mathcal{A}$.

Benedict and McCoy [24] considered the following problem.

Problem 11 *Let $\mathcal{A}$ be a computable structure, and let F be a function on $\mathcal{A}$. Give conditions guaranteeing that if $(\mathcal{B}, G) \cong (\mathcal{A}, F)$, then for all sets S, $G^{-1}(S)$ is computable relative to $(\mathcal{B}, S)$.*

Benedict proved the following intriguing result.

Proposition 10.19 (Benedict) *For a total function F on the structure $\mathcal{N}$, F is definable in $\mathcal{N}$ if and only if for all sets S, $F^{-1}(S)$ is definable in $(\mathcal{N}, S)$.*

McCoy generalized Benedict's result.

10.7. COPIES OF AN ARBITRARY STRUCTURE

Theorem 10.20 (McCoy) *Let $\mathcal{A}$ be a computable structure. For a total function F on $\mathcal{A}$, the following are equivalent:*
  (1) for all $(\mathcal{B}, G) \cong (\mathcal{A}, F)$, and for all sets $S \subseteq \mathcal{B}$, $G^{-1}(S)$ is arithmetical relative to $(\mathcal{B}, S)$,
  (2) F is relatively intrinsically Σ_n^0 on $\mathcal{A}$, for some $n < \omega$.

Remark: If F is total and relatively Σ_n^0 on $\mathcal{A}$, then it is relatively Δ_n^0 on $\mathcal{A}$. It follows that for all $(\mathcal{B}, G) \cong (\mathcal{A}, F)$ and all $S \subseteq \mathcal{B}$, $G^{-1}(S)$ is Σ_n^0 relative to $(\mathcal{B}, S)$.

Exercise: Prove Proposition 10.19 using Theorem 10.20.

Exercise: Prove Theorem 10.20.

10.7 Copies of an arbitrary structure

Throughout this chapter, we have been considering arbitrary copies of a *computable* structure. We could start with an *arbitrary* structure. For the results in this setting, the computable infinitary formulas do not suffice. Recall from Chapter 7, the X-*computable* infinitary formulas, for an arbitrary set X. These are defined in exactly the same way as the computable infinitary formulas except that the disjunctions and conjunctions are c.e. relative to X.

Here is the natural generalization of Theorem 10.1.

Theorem 10.21 *For an arbitrary structure $\mathcal{A}$ and further relation R, the following are equivalent:*
  (1) in all copies $\mathcal{B}$ of $\mathcal{A}$, the image of R is $\Sigma_\alpha^0(\mathcal{A}, \mathcal{B})$,
  (2) R is definable in $\mathcal{A}$ by a $\mathcal{A}$-computable Σ_α formula $\varphi(\overline{c}, \overline{x})$.

Exercise: Prove Theorem 10.21.

Here is the natural generalization of Theorem 10.14.

Theorem 10.22 *For an arbitrary structure $\mathcal{A}$, the following are equivalent:*
  (1) for all $\mathcal{B} \cong \mathcal{A}$, some isomorphism from $\mathcal{A}$ onto $\mathcal{B}$ is $\Delta_\alpha^0(\mathcal{A}, \mathcal{B})$,
  (2) $\mathcal{A}$ has a Scott family Φ, c.e. relative to $\mathcal{A}$, made up of $\mathcal{A}$-computable Σ_α formulas.

Exercise: Prove Theorem 10.22.

Here is the natural generalization of Theorem 10.16.

Theorem 10.23 *For an arbitrary structure $\mathcal{A}$, the following are equivalent:*
  (1) for all $\mathcal{B} \cong \mathcal{A}$, all isomorphisms from $\mathcal{A}$ onto $\mathcal{B}$ are $\Delta_\alpha^0(\mathcal{A}, \mathcal{B})$,
  (2) $\mathcal{A}$ has a defining family Φ, c.e. relative to $\mathcal{A}$, made up of $\mathcal{A}$-computable Σ_α formulas.

Exercise: Prove Theorem 10.23.

In later chapters, we shall give analogues of many of the results of this chapter for computable, rather than arbitrary, copies of a given structure $\mathcal{A}$. In Chapters 11 and 16, we give results on images of a relation. In Chapters 12 and 17, we give results on isomorphisms.

Chapter 11

The Ash-Nerode Theorem

Let $\mathcal{A}$ be a computable structure, with a further relation R. Ash and Nerode [19] considered the following two problems.

Problem 1 *When is there a computable copy of $\mathcal{A}$ in which the image of R is not computable?*

Problem 2 *When is there a computable copy of $\mathcal{A}$ in which the image of R is not c.e.?*

There is some terminology associated with these problems. We say that R is *intrinsically computable on* $\mathcal{A}$ if, in all computable copies of $\mathcal{A}$, the image of R is computable. Similarly, R is *intrinsically c.e. on* $\mathcal{A}$ if, in all computable copies of $\mathcal{A}$, the image of R is c.e.

In this chapter, we give the results from [19] on Problems 1 and 2, and we give results of Davey [33] and others on some related problems. These results were obtained earlier than the corresponding results from Chapter 10, on relations that are *relatively intrinsically computable, relatively intrinsically c.e.*, etc. Where forcing was the main technique used to prove the results in Chapter 10, the main technique here, and also in Chapter 12, is the finite-injury priority. See [144] for a discussion of the method and various applications to pure computability theory.

11.1 Simple examples

Before proving the results of Ash and Nerode, we give some simple examples.

Example 1: Let $\mathcal{A} = (\omega, <)$, and let S be the successor relation. Then S is not intrinsically c.e. on $\mathcal{A}$, while $\neg S$ is relatively intrinsically c.e. on $\mathcal{A}$.

Proof sketch: In Chapter 10, we observed that $\neg S$ is relatively intrinsically c.e. on $\mathcal{A}$. To show that S is not intrinsically c.e. on $\mathcal{A}$, we build a computable linear ordering $\mathcal{B}$ of type ω in which the successor relation is not c.e. We use a finite-injury priority construction. The requirements are as follows.

R_e: W_e is not the successor relation on $\mathcal{B}$

At each stage, we determine a finite substructure of $\mathcal{B}$—an initial segment of the ordering. The strategy for a single requirement is to add two new elements b_1, b_2 at the end of the ordering, and to make b_2 the successor of b_1 until the pair (b_1, b_2) appears in W_e. This may never happen, but if it does, then we add an element between b_1 and b_2.

Example 2: Let $\mathcal{A}$ be an ordering of type $\omega + \omega^*$, and let I be the initial segment of type ω. Then
 (a) I is not intrinsically c.e. on $\mathcal{A}$,
 (b) I is relatively intrinsically Δ^0_2 on $\mathcal{A}$.

Proof sketch: In Chapter 10, we observed that (b) holds. To prove (a), we build a computable copy $\mathcal{B}$ of type $\omega + \omega^*$ in which the initial segment of type ω is not c.e. We proceed in stages. At each stage, we determine a finite ordering, with an initial segment designated for I.

The requirements are as follows.

R_e: $W_e \neq I$

The strategy for a single requirement R_e is to add a witness b to the ordering, locating it at the end of the current I. We keep b in I until it appears in W_e. This may never happen, but if it does, and there has been no action on earlier requirements since we designated b, then we shift b (and any elements to the right of it) into $\neg I$. If there is action on an earlier requirement, involving a witness d which lies to the left of b, then we abandon b as a witness for R_e. We add a new witness b', locating it at the end of the current I. We act at most once on R_e after the last action on earlier requirements. Therefore, each requirement is met.

We shall return to Example 2 later in this chapter, and we shall give a stronger version of (a).

11.2 Results of Ash and Nerode

We are ready to give the general results on Problems 1 and 2. We begin with Problem 2. In Chapter 10, we saw that if a relation R is formally Σ^0_1, or formally c.e., on $\mathcal{A}$, then it is relatively intrinsically c.e. on $\mathcal{A}$. It follows that R is intrinsically c.e. on $\mathcal{A}$. If R is not formally c.e. on $\mathcal{A}$, then it is not relatively intrinsically c.e. on $\mathcal{A}$. This means that there is an isomorphism F from $\mathcal{A}$ onto a copy $\mathcal{B}$, not necessarily computable, such that $F(R)$ is not c.e. relative to $\mathcal{B}$.

11.2. RESULTS OF ASH AND NERODE

Here we show that under an added effectiveness condition, on just the one copy $\mathcal{A}$, we can take $\mathcal{B}$ to be computable.

Effectiveness condition: Suppose that for a tuple $\bar{c}$ in $\mathcal{A}$ and a finitary existential formula $\varphi(\bar{c}, \bar{x})$, we can decide whether there exists $\bar{a} \notin R$ such that $\mathcal{A} \models \varphi(\bar{c}, \bar{a})$.

Remark: If the existential diagram of $(\mathcal{A}, R)$ is decidable, then the effectiveness condition holds.

Theorem 11.1 (Ash-Nerode) *Let $\mathcal{A}$ be a computable structure, and let R be a further relation on $\mathcal{A}$. If the effectiveness condition holds, then the following are equivalent:*
(1) R is intrinsically c.e. on $\mathcal{A}$,
(2) R is formally Σ_1^0 on $\mathcal{A}$.

Proof: We must show that if R is intrinsically c.e. on $\mathcal{A}$, then it is definable by a computable Σ_1 formula. For simplicity, we suppose that R is unary. We show that either R is definable by a computable Σ_1 formula, or else there is a computable copy of $\mathcal{A}$ in which the image of R is not c.e. We consider two cases.

Case 1: Suppose there is a tuple $\bar{c}$ such that for each $a \in R$, there is a finitary existential formula $\psi(\bar{c}, x)$ true of a and not true of any element of $\neg R$.

By the effectiveness condition, for each a, we can find such a $\psi_a(\bar{c}, x)$. Then in this case, R is defined by the formula

$$\bigvee_{a \in R} \psi_a(\bar{c}, x) ,$$

which is computable Σ_1.

Case 2: Suppose that for each tuple $\bar{c}$, there exists $a \in R$ such that any finitary existential formula $\varphi(\bar{c}, x)$ true of a is also true of some element of $\neg R$.

In this case, we construct an isomorphism F from $\mathcal{A}$ onto a computable $\mathcal{B}$, satisfying the following requirements.

R_e: $F(R) \neq W_e$

Let B be an infinite computable set of constants, for the universe of $\mathcal{B}$. At each stage, we tentatively determine a finite part of F^{-1}, and we determine a finite part of $D(\mathcal{B})$, once and for all. The strategy for R_e is as follows: We choose a witnessing constant b not yet in W_e, and we map it to the first element of R not in $ran(p)$. If b later enters W_e, then we try to change the function so as to map b into $\neg R$, preserving what we have enumerated into $D(\mathcal{B})$.

By the effectiveness condition, we can effectively determine whether it is possible to change the function in this way. If not, then we abandon b as a witness. We map a new b', not yet in W_e, to the next element of R. If b' appears in W_e, then again we try to change the function so as to map b' into $\neg R$, preserving $D(\mathcal{B})$. We continue in this way until we have a witnessing constant that is mapped to an element of R and never appears in W_e, or appears in W_e and is then mapped to an element of $\neg R$. We must succeed when our witnessing constant is mapped to the element a in the hypothesis, if not before.

We have already said more about this particular finite-injury priority construction than we did about earlier ones, but are not ready to leave it. Rather, we shall give an even more formal treatment. The reason for doing this is that in Chapter 16, we shall extend Theorem 11.1 to higher levels in the hyperarithmetical hierarchy, where the difficulty of the construction forces us to be formal, and we want at least some aspects of the later proof to look familiar.

Let $\mathcal{F}$ be the set of finite $1-1$ functions p from initial segments of B into $\mathcal{A}$. We use letters p, q, r, possibly with decorations, to denote elements of $\mathcal{F}$. For $p \in \mathcal{F}$ such that $dom(p)$ consists of the first n elements of B, let $E(p)$ be the set of basic sentences (atomic sentences or negations) φ such that the constants appearing in φ are in in $dom(p)$, φ has Gödel number less than n, and p makes φ true in $\mathcal{A}$.

We record our action at stage s, attempting the first few requirements, in a finite sequence of elements of $\mathcal{F}$ and witnessing constants. At stage 0, the sequence is just $p_0 = \emptyset$. Suppose at stage s, the sequence is $\ell_s = p_0 b_0 p_1 b_1 \ldots b_{r-1} p_r$, where

(a) $p_n \in \mathcal{F}$,

(b) $p_0 = \emptyset$,

(d) $dom(p_n)$ includes the first n elements of B and $ran(p_n)$ includes the first n elements of $\mathcal{A}$, and

(e) $p_{n+1}(b_n) \in R \Leftrightarrow b_n \notin W_{n,s}$.

We describe the action—the change in the sequence ℓ_s—at stage $s+1$. First, suppose there is no $n < r$ such that $b_n \in W_{n,s+1} - W_{n,s}$. Let b be the first constant not in $dom(p_r)$. Take the first $a \notin ran(p_r)$ such that

$$a \in R \Leftrightarrow b \notin W_{r,s+1},$$

and let $q \supseteq p_r$ where $q(b) = a$. Then the stage $s+1$ sequence is $\ell_{s+1} = p_0 b_0 \ldots p_r b q$. Now, suppose that for some first $n < r$, $b_n \in W_{n,s+1} - W_{n,s}$. We search for $a \notin ran(p_n)$ and $b \notin dom(p_n)$ such that one of the following holds:

(i) $p_r(b) = a$ and $a \in R \Leftrightarrow b \notin W_{n,s+1}$,

(ii) $p_r(b) = a$, $a \in R$, $b \in W_{n,s+1}$, and there exists $q \supseteq p_n$ such that $q(b) \notin R$ and $E(p_r) \subseteq E(q)$,

(iii) $a \notin ran(p_r)$, $b \notin dom(p_r)$, and $a \in R \Leftrightarrow b \notin W_{n,s+1}$.

We will find such an a and b, since for $\bar{c} = ran(p_n)$, there exists $a \in R$ that is free over $\bar{c}$, in the sense that any existential formula $\varphi(\bar{c}, x)$ true of a is true of some $a' \notin R$. If (i) holds, then the stage $s + 1$ sequence is $p_0 b_0 \ldots p_n b p_r$. If (ii) holds, then the stage $s+1$ sequence is $p_0 b_0 \ldots p_n bq$. If (iii) holds, then the stage $s + 1$ sequence is $p_0 b_0 \ldots p_n bq$, where $q \supseteq p$ and $q(b) = a$.

We act only finitely many times on each requirement. For requirement R_e, after the last action on earlier requirements, when we have determined p_e, we may try several witnesses, but we cannot go on forever without mapping some b to a that is free over $ran(p_e)$, then the witness b will not change again, and the function will change at most once more.

The result on Problem 2 yields a result on Problem 1.

Corollary 11.2 *Let $\mathcal{A}$ be a computable structure, and let R be a further relation on $\mathcal{A}$. Suppose that both R and $\neg R$ satisfy the effectiveness condition. Then the following are equivalent:*
(1) R and $\neg R$ are both formally Σ_1^0 on $\mathcal{A}$,
(2) R is intrinsically computable on $\mathcal{A}$.

Proof: By Corollary 10.7, if R and $\neg R$ are both formally Σ_1^0 on $\mathcal{A}$, then R is relatively intrinsically computable on $\mathcal{A}$. Suppose R is not definable in this way, and the effectiveness condition is satisfied, then by Theorem 11.1, there is a computable copy of $\mathcal{A}$ in which the image of R is not c.e. The same is true for $\neg R$. In either case, the image of R is not computable.

Remark: Theorem 11.1 would look nicer without the effectiveness condition. However, Manasse [114] constructed examples showing that it cannot be dropped.

11.3 Applications

Here we give a thorough analysis of the intrinsically computable relations on vector spaces and algebraically closed fields.

11.3.1 Vector spaces

Let $\mathcal{A}$ be an infinite vector space over a computable field F. (See the Appendix for a discussion of these structures.) There are a couple of important facts to bear in mind. First, we have effective elimination of quantifiers. That is, given an arbitrary finitary formula, we can effectively determine a finitary quantifier-free formula equivalent to the given formula, over the theory of $\mathcal{A}$. It follows that if $\mathcal{A}$ is computable, then it is decidable. Second, the formula $x = x$ is strongly minimal in $Th(\mathcal{A})$, so for any finitary formula $\varphi(\overline{c}, x)$, with just x free, the set $\varphi^{\mathcal{A}}(\overline{c}, x)$ is either finite or co-finite. If $\mathcal{A}$ is computable, then we can search for n such that one of $\varphi^{\mathcal{A}}(\overline{c}, x)$, $\neg\varphi^{\mathcal{A}}(\overline{c}, x)$ has size less than n. Thus, we can effectively determine whether $\varphi^{\mathcal{A}}(\overline{c}, x)$ is finite, and if so, then we can find all of the elements.

Proposition 11.3 *Let $\mathcal{A}$ be a computable vector space (over a computable field F).*

(a) If $\mathcal{A}$ has finite dimension, then every computable relation is relatively intrinsically computable on $\mathcal{A}$.

(b) If $\mathcal{A}$ has infinite dimension, then a unary relation R is intrinsically computable on $\mathcal{A}$ if and only if it is either finite or co-finite.

Proof sketch: (a) Suppose $\mathcal{A}$ has finite dimension, with basis $\bar{c}$. There is a c.e. defining family, made up of atomic formulas $x = \tau(\bar{c})$—expressing x as a linear combination of elements of $\bar{c}$. It follows that if R is computable, then there are computable Σ_1 definitons for both R and $\neg R$.

(b) Suppose $\mathcal{A}$ has infinite dimension. If R is either finite or co-finite, then R and $\neg R$ are both defined trivially by finitary open formulas, so R is intrinsically computable. Suppose that R is both infinite and co-infinite. We may suppose that R is computable, or it is not intrinsically computable. If $\varphi^{\mathcal{A}}(\bar{c}, x)$ is finite, then we can determine all the elements, and test them one-by-one, to determine whether any are in R (or $\neg R$). If $\varphi^{\mathcal{A}}(\bar{c}, x)$ is co-finite, then there must be elements in both R and $\neg R$, because both sets are infinite.

We are in a position to apply Corollary 11.2. Suppose R is intrinsically computable, hoping for a contradiction. By Corollary 11.2, there are computable Σ_1 formulas defining R and $\neg R$, with parameters in some finite tuple $\bar{c}$. Each of the (finitary) disjuncts in the defining formulas is satisfied by only finitely many elements. We may suppose that each disjunct has the form $x = \tau(\bar{c})$. Then R and $\neg R$ are both contained in the linear span of $\bar{c}$, contradicting the fact that $\mathcal{A}$ has infinite dimension.

11.3.2 Algebraically closed fields

Algebraically closed fields have many of the same properties as vector spaces (see the Appendix). We have effective elimination of quantifiers, so if an algebraically closed field $\mathcal{A}$ is computable, then it is decidable. Also, $x = x$ is strongly minimal in $Th(\mathcal{A})$, so for any finitary formula $\varphi(\bar{c}, x)$, $\varphi^{\mathcal{A}}(\bar{c}, x)$ is either finite or co-finite.

Proposition 11.4 *Let $\mathcal{A}$ be a computable algebraically closed field, and let R be a unary relation on $\mathcal{A}$.*

(a) If $\mathcal{A}$ has finite transcendance degree, then R is intrinsically computable on $\mathcal{A}$ if and only if R is finite or co-finite, or else R and $\neg R$ are each the union of the solution sets for some c.e. set of polynomial equations $p(\bar{c}, x) = 0$, for some fixed finite tuple $\bar{c}$.

(b) If $\mathcal{A}$ has infinite transcendance degree, then R is intrinsically computable on $\mathcal{A}$ if and only if it is finite or co-finite.

Proof sketch: If R is finite or co-finite, then, trivially, R is relatively intrinsically computable on $\mathcal{A}$. Suppose that R is infinite and co-infinite. We may suppose that R is computable; otherwise, it is not intrinsically computable. For a finitary formula $\varphi(\bar{c}, x)$, we can determine whether there are elements of $\varphi^{\mathcal{A}}(\bar{c}, x)$ in R (or $\neg R$). We are in a position to apply Corollary 11.2.

(a) If each of R and $\neg R$ is the the union of the solution sets for a c.e. set of equations $p(\bar{c}, x) = 0$, with fixed parameters $\bar{c}$, then we have computable Σ_1 definitions. Therefore, R is relatively intrinsically computable. Now, suppose we have computable Σ_1 definitions for R and $\neg R$, with parameters $\bar{c}$. Since neither R nor $\neg R$ is co-finite, each disjunct is satisfied by only finitely many elements. For a given disjunct $\alpha(\bar{c}, x)$, we can find the elements satisfying it. Moreover, for each element a, we can find a polynomial $p(\bar{c}, x)$ such that

$$\mathcal{A} \models p(\bar{c}, a) = 0 \ \& \ \forall x\, (\, p(\bar{c}, x) = 0 \to \alpha(\bar{c}, x)\,)\ .$$

Thus, R and $\neg R$ are c.e. unions of the required form.

(b) Suppose R is intrinsically computable, hoping for a contradiction. There are computable Σ_1 formulas defining R and $\neg R$, with parameters in some tuple $\bar{c}$. Since neither R nor $\neg R$ is co-finite, each disjunct is satisfied by only finitely many elements. Therefore, R and $\neg R$ are both subsets of the algebraic closure of $\bar{c}$. This contradicts our assumption that $\mathcal{A}$ has infinite transcendance degree.

11.4 Expansions

The Ash-Nerode Theorem was inspired by results on a different sort of question.

Problem 3 *For a computable structure $\mathcal{A}$ and a computable Π_2 sentence ψ, involving a new relation symbol R in addition to the symbols from the language of $\mathcal{A}$, give conditions guaranteeing that some computable copy of $\mathcal{A}$ has no expansion to a model of ψ in which the interpretation of R is c.e.*

In Chapter 10, we mentioned the variant of Problem 3 for arbitrary copies of a given structure. In [9], [8], and [18], there are general results on computable and c.e. expansions, and on expansions that are relatively computable, or relatively c.e. We give some special cases, due to Metakides and Nerode [117].

Theorem 11.5 *(a) There is a computable infinite-dimensional vector space V, over the rationals, such that V has no infinite c.e. linearly independent subset.*

(b) In any characteristic, there is a computable algebraically closed field, of infinite transcendance degree, with no infinite c.e. algebraically independent subset.

(c) There is a real closed field with infinitely many algebraically independent elements but with no infinite algebraically independent subset. (See the Appendix for some basic information about real closed fields.)

Exercise: Prove Theorem 11.5, imitating the proof of Theorem 11.1, but substituting the following requirements.

R_e: W_e is not an infinite algebraically independent set.

11.5 Results of Harizanov

Harizanov [61] considered the possible degrees of the image of a given relation under isomorphisms on a given structure. The setting is as for Problems 1 and 2—we have a computable structure $\mathcal{A}$, with a further relation R. Harizanov showed that the conditions from Corollary 11.2, for making the image of R not computable, are sufficient to give it arbitrary c.e. degree.

Theorem 11.6 (Harizanov) *Let $\mathcal{A}$ be a structure, and let R be a further relation on $\mathcal{A}$. Suppose that one of R, $\neg R$ is not definable in $\mathcal{A}$ by a computable Σ_1 formula. Suppose also that the existential diagram of $(\mathcal{A}, R)$ is computable (this can be weakened, as in Theorem 11.1). Then for any c.e. set S, there is an isomorphism F from $\mathcal{A}$ onto a computable copy $\mathcal{B}$ such that $F(R) \equiv_T S$.*

Sketch of proof: Suppose R is not definable by any computable Σ_1 formula. Then for any tuple $\bar{c}$, we can find $a \in R$ such that any existential formula true of $\bar{c}, a$ is true of $\bar{c}a'$, for some $a' \notin R$. Let B be an infinite computable set of constants. We determine a $1-1$ function F from B onto $\mathcal{A}$, and let $\mathcal{B}$ be the structure induced by F on B. We make $F \leq_T S$ and $S \leq_T F(R)$. This implies that $F(R) \equiv_T S$.

We have the following requirements:

R_n: Code the fact that $n \in S$, if true.

The construction proceeds in stages. We describe the strategy for requirement R_n. Suppose that at stage s, we have a finite partial $1-1$ function p from B to $\mathcal{A}$, taking care of earlier requirements. Let $\bar{c} = ran(p)$. If n has already appeared in S, we choose new a witness $b \in$ B and map it to some new $a \notin R$. Suppose that n has not yet appeared in S. We choose $a \in R$ such that any existential formula $\varphi(\bar{c}, x)$ true of a is true of some $a' \notin R$. We also choose a new witness $b \in$ B, and we map b to a. If at some later stage, some $x \leq n$ enters S, then we change the function, taking b to some $a' \notin R$, and preserving what we have enumerated into $D(\mathcal{B})$. In this case, we discharge any witnesses that we may have chosen for $m > n$, and begin choosing new, larger ones.

Using S, we can determine the final sequence of witnesses and the corresponding sequence of partial functions. From this, it follows that $F \leq_T S$. Using $F(R)$, we can recover, for $n = 0, 1, 2, \ldots$, the final witness b_n for n, and we can determine $\chi_S(n)$, since

$$n \in S \Leftrightarrow b_n \in F(R) \ .$$

Harizanov also considered the following problem.

Problem 4 *When is there a computable copy of $\mathcal{A}$ in which the image of R is c.e. and not computable?*

By the easy direction of Theorem 10.9, if $\neg R$ is definable in $(\mathcal{A}, R)$ by a computable Σ_1 formula in which R occurs only positively, then in all copies $\mathcal{B}$ of $\mathcal{A}$ in which the image of R is c.e., it is computable. Harizanov showed that under a suitable effectiveness condition, the converse holds. Moreover, the conditions for making the image of R c.e. and not computable are sufficient to make it c.e. and of arbitrary c.e. degree.

Theorem 11.7 (Harizanov) *Let $\mathcal{A}$ be a structure with a further relation R. Suppose that the existential diagram of $(\mathcal{A}, R)$ is computable (this effectiveness condition can be weakened). Finally, suppose that $\neg R$ is not defined by any computable Σ_1 formula in which R occurs only positively. Then for any c.e. set C, there is an isomorphism F from $\mathcal{A}$ onto a computable copy $\mathcal{B}$ such that $F(R)$ is c.e. and $F(R) \equiv_T C$.*

Exercise: Prove Theorem 11.7.

11.6 Ershov's hierarchy

The following problems are considered in [13]. The setting is the same as for Problems 1–4.

Problem 5 *(a) When is there a computable copy of $\mathcal{A}$ in which the image of R is not d-c.e.?*

(b) When is there a computable copy of $\mathcal{A}$ in which the image of R is not α-c.e.?

We say that R is *intrinsically d-c.e.*, or *intrinsically α-c.e.* on $\mathcal{A}$ if in all computable copies of $\mathcal{A}$, the image of R is d-c.e., or α-c.e.

By Theorem 10.10, if R is definable in $\mathcal{A}$ by a formula of the form

$$\varphi(\overline{c}, \overline{x}) \ \& \ \neg\psi(\overline{c}, \overline{x}) \ ,$$

where φ and ψ are computable Σ_1 formulas, then R is *relatively* intrinsically d-c.e. on $\mathcal{A}$, so it is intrinsically d-c.e. on $\mathcal{A}$. If R is not definable by any such formula, then there is a copy $\mathcal{B}$ in which the image of R is not d-c.e. relative to $\mathcal{B}$. Under some effectiveness conditions, we can take $\mathcal{B}$ to be computable.

We define a notion of freeness, one of many. A tuple $\overline{a}$ is *1-free over* $\overline{c}$ if every existential formula $\varphi(\overline{c}, \overline{x})$ true of $\overline{a}$ is also true of some $\overline{a}'$ such that

$$\overline{a} \in R \ \Leftrightarrow \ \overline{a}' \notin R \ .$$

We say that $\overline{a}$ is *2-free over* $\overline{c}$ if every existential formula $\varphi(\overline{c}, \overline{x})$ true of $\overline{a}$ is also true of some $\overline{a}'$ such that

$$\overline{a} \in R \ \Leftrightarrow \ \overline{a}' \notin R$$

and $\overline{a}'$ is 1-free over $\overline{c}$. The result below says that if no $\overline{a} \in R$ is 2-free over $\overline{c}$, then, assuming some effectiveness, R has a definition of the required form.

Theorem 11.8 *Let $\mathcal{A}$ be a computable structure and let R be a further computable relation on $\mathcal{A}$. Suppose that for all tuples $\bar{c}$, we can effectively determine which tuples are 1-free over $\bar{c}$, and we can effectively find existential formulas witnessing the lack of 2-freeness, or 1-freeness. If no tuple in R is 2-free over $\bar{c}$, then R has a definition of the form*

$$\varphi(\bar{c},\bar{x}) \;\&\; \neg\psi(\bar{c},\bar{x}) ,$$

where $\varphi(\bar{c},\bar{x})$ and $\psi(\bar{c},\bar{x})$ are computable Σ_1.

Proof: For simplicity, we suppose that R is unary. For $a \in R$, let $\varphi_a(\bar{c},x)$ be an existential formula witnessing the fact that a is not 2-free over $\bar{c}$, and let $\varphi(\bar{c},x)$ be the disjunction of these formulas. For $a' \in \neg R$ satisfying $\varphi(\bar{c},x)$, let $\psi_{a'}(\bar{c},x)$ be an existential formula witnessing the fact that a' is not 1-free over $\bar{c}$, and let $\psi(\bar{c},x)$ be the disjunction of these formulas. Then

$$\varphi(\bar{c},x) \;\&\; \neg\psi(\bar{c},x)$$

is the desired definition of R.

The next result says that if, for each $\bar{c}$, there is some $\bar{a} \in R$ such that $\bar{a}$ is 2-free over $\bar{c}$, then with some effectiveness, $\mathcal{A}$ has a computable copy in which the image of R is not d-c.e.

Theorem 11.9 *Suppose $\mathcal{A}$ is a computable structure, and let R be a further relation on $\mathcal{A}$. Suppose that for each tuple $\bar{c}$, we can find $a \in R$ such that a is 2-free over $\bar{c}$. Suppose also that for all tuples $\bar{c}$, we can effectively determine which elements are 1-free over $\bar{c}$. Then there is an isomorphism F from $\mathcal{A}$ onto a computable $\mathcal{B}$ such that $F(R)$ is not d-c.e.*

Proof sketch: Let B be an infinite computable set of constants, for the universe of $\mathcal{B}$. For simplicity, suppose R is unary. We have requirements

$$R_{\langle e_1,e_2\rangle}: F(R) \neq W_{e_1} - W_{e_2}.$$

At each stage, we have enumerated finitely many sentences into $D(\mathcal{B})$, and we have tentatively determined a finite partial $1-1$ function p from B to $\mathcal{A}$, making these sentences true. The strategy for a given requirement $R_{\langle e_1,e_2\rangle}$ is as follows. Say we have p taking care of earlier requirements. We let $q \supseteq p$ map a new constant b to some $a \in R$ such that a is 2-free over $ran(p)$. We extend q until b enters W_{e_1}. Then we take $q' \supseteq p$ mapping b to some $a' \notin R$ such that a' is 1-free over $ran(p)$. We preserve the truth of sentences enumerated into $D(\mathcal{B})$. We extend q' until b enters W_{e_2}. Then we replace our extension of q' by $q'' \supseteq p$ mapping b to some $a'' \in R$, again preserving the truth of sentences enumerated into the diagram.

We can extend the two results above to show that, under some effectiveness conditions, the image of R in computable copies of $\mathcal{A}$ is α-c.e. if and only if R

11.6. ERSHOV'S HIERARCHY

has a definition of the form given in Theorem 10.11. We extend the definition of freeness (proceeding by induction). For any tuples $\bar{c}$ and $\bar{a}$, where the length of $\bar{a}$ matches the arity of R, we say that a is β-free over $\bar{c}$ if for any existential formula $\varphi(\bar{c},\bar{x})$ true of $\bar{a}$ and any $\gamma < \beta$, there is some $\bar{a}'$ satisfying $\varphi(\bar{c},\bar{x})$ such that $\bar{a}'$ is γ-free over $\bar{c}$ and $\bar{a}'$ and $\bar{a}$ are on opposite sides of R; i.e., $\bar{a}' \in R$ if and only if $\bar{a} \notin R$.

Here is the extension of Theorem 11.8.

Theorem 11.10 *Let α be a computable ordinal. Suppose $\mathcal{A}$ is a computable structure, and let R be a further computable relation on $\mathcal{A}$. Suppose that for the tuple $\bar{c}$, no $\bar{a} \in R$ is α-free over $\bar{c}$, and for each further tuple $\bar{a}$, we can find an existential formula witnessing the failure. Also, suppose that for any $\beta < \alpha$, we can effectively determine whether $\bar{a}$ is β-free over $\bar{c}$, and, if not, then we can find an existential formula witnessing this. Then there are computable sequences of computable Σ_1 formulas $(\varphi_\beta(\bar{c},\bar{x}))_{\beta \leq \alpha}$, $(\psi_\beta(\bar{c},\bar{x}))_{\beta \leq \alpha}$ such that*

(1) for all tuples $\bar{a}$, and all $\beta \leq \alpha$, if

$$\mathcal{A} \models (\varphi_\beta(\bar{c},\bar{a}) \ \& \ \psi_\beta(\bar{c},\bar{a}))$$

then for some $\gamma < \beta$,

$$\mathcal{A} \models (\varphi_\gamma(\bar{c},\bar{x}) \vee \psi_\gamma(\bar{c},\bar{x})) ,$$

(2) R is defined by the formula

$$\bigvee_{\beta < \alpha}(\varphi_\beta(\bar{c},\bar{x}) \ \& \ \neg \bigvee_{\gamma < \beta} \psi_\beta(\bar{c},\bar{x})) .$$

Proof sketch: For simplicity, we suppose that R is unary. For each $a \in R$ and each $\beta \leq \alpha$ such that a is not β-free over $\bar{c}$, let $\varphi_{\beta,a}(\bar{c},x)$ be the effectively determined existential formula witnessing this—$\varphi_{\beta,a}(\bar{c},x)$ is true of a, and for some $\gamma < \beta$, it is not true of any $a' \notin R$ such that a' is γ-free over $\bar{c}$. Let $\varphi_\beta(\bar{c},x)$ be the disjunction of the formulas $\varphi_{\beta,a}(\bar{c},x)$. Similarly, for each $a \notin R$ and each $\beta < \alpha$ such that a is not β-free over $\bar{c}$, let $\psi_{\beta,a}(\bar{c},x)$ be the effectively determined existential formula witnessing this, and let $\psi_\beta(\bar{c},x)$ be the disjunction of the formulas $\psi_{\beta,a}(\bar{c},x)$. It is not difficult to verify that these formulas satisfy (1) and (2).

Here is the extension of Theorem 11.9.

Theorem 11.11 *Let α be a computable ordinal. Suppose $\mathcal{A}$ is a computable structure and let R be a further computable relation on $\mathcal{A}$. Suppose that for each tuple $\bar{c}$ in $\mathcal{A}$, we can find $\bar{a} \in R$ such that $\bar{a}$ is α-free over $\bar{c}$, and for $\beta < \alpha$, we can recognize β-freeness. Then there is an isomorphism F from $\mathcal{A}$ onto a computable copy $\mathcal{B}$ such that $F(R)$ is not α-c.e.*

The proof of Theorem 11.11 is like that for Theorem 11.9. The requirements are as follows.

$R_{(e,e')}$: Suppose e is the index of a function $g : \omega \times \omega \to \{0,1\}$ and e' is the index of a function $o : \omega \times \omega \to \{\beta : \beta \leq \alpha\}$, such that
 (i) $g(x, 0) = 0$,
 (ii) $o(x, 0) = a$,
 (iii) $o(x, s+1) \leq o(x, s)$,
 (iv) $g(x, s+1) \neq g(x, s) \Rightarrow o(x, s+1) < o(x, s)$.
Then for some b,
$$\lim_{s \to \infty} g(b, s) \neq \chi_{F(R)}(b) \ .$$

We have seen that in a linear ordering of type $\omega + \omega^*$, the initial segment of type ω is relatively intrinsically Δ^0_2, and not intrinsically c.e. Using Theorem 11.11, we can show more.

Proposition 11.12 *In a linear ordering of type $\omega + \omega^*$, the initial segment of type ω is not intrinsically α-c.e. for any computable ordinal α.*

Proof: There is a computable ordering $\mathcal{A}$ of type $\omega + \omega^*$ in which we can effectively determine the sizes of the intervals to the left and right of any given element. We can easily describe freeness. For any tuple $\bar{c}$ and any ordinal $\beta \geq 1$, an element a is β-free over $\bar{c}$ just in case a lies closer to the "cut" than any element of $\bar{c}$; that is, a is to the right of any element of $\bar{c}$ in the initial segment of type ω, and to the left of any element of $\bar{c}$ in the terminal segment of type ω^*. We are in a position to apply Theorem 11.11. For each computable ordinal α, we get a computable ordering of type $\omega + \omega^*$ in which the initial segment of type ω is not α-c.e.

Remark: We did not mention ordinal notation above, but of course the theorem depends on notation. Fixing a path in O with a notation for α, and identifying ordinals with their notations on this path, we have a family of α-c.e. sets, and we obtain a computable ordering of type $\omega + \omega^*$ in which the initial segment of type ω is none of these.

11.7 Pairs of relations

We vary the basic setting slightly. Let $\mathcal{A}$ be a computable structure, and let R and S be a pair of further relations on $\mathcal{A}$. Davey [33] considered the following problem.

Problem 6 *When is there a computable copy of $\mathcal{A}$ in which the images of R and S have no computable separator?*

Note that if $S = \neg R$, then Problem 6 is the same as Problem 1. We say that R and S are *intrinsically computably separable on $\mathcal{A}$* if in all computable copies of $\mathcal{A}$, the images of R and S have a computable separator. In Chapter 10, we considered the corresponding relative notions.

11.7. PAIRS OF RELATIONS

Theorem 11.13 (Davey) *Let $\mathcal{A}$ be a computable structure, and let R and S be computable relations on $\mathcal{A}$, of the same arity. Suppose that given a finitary existential formula in the language of $\mathcal{A}$, we can decide whether there exists $\bar{a} \in R$ such that $\mathcal{A} \models \varphi(\bar{c}, \bar{a})$, and similarly, for S replacing R. Then the following are equivalent:*

(1) R and S are intrinsically computably separable on $\mathcal{A}$,

(2) there exist computable Σ_1 formulas $\varphi(\bar{c}, \bar{x})$ and $\psi(\bar{c}, \bar{x})$ such that

   (i) $\mathcal{A} \models \forall \bar{x} (\varphi(\bar{c}, \bar{x}) \vee \psi(\bar{c}, \bar{x}))$,

   (ii) $\varphi(\bar{c}, \bar{x})$ is satisfied by all $\bar{a} \in R$ and no $\bar{a} \in S$, while $\psi(\bar{c}, \bar{x})$ is satisfied by all $\bar{a} \in S$ and no $\bar{a} \in R$.

Proof: (2) $\Rightarrow$ (1) This follows from Theorem 10.12, letting $\alpha = 1$.

(1) $\Rightarrow$ (2) For simplicity, we take R and S to be unary relations. Suppose there is some tuple $\bar{c}$ such that for each element a, there is a finitary existential formula $\varphi_a(\bar{c}, x)$ is true of a and not true of elements of *both* R and S. By the effectiveness assumptions, we can find such a formula $\varphi_a(\bar{c}, x)$. Let $\varphi(\bar{c}, x)$ be the disjunction of the formulas $\varphi_a(\bar{c}, x)$ not satisfied by any $b \in S$, and let $\psi(\bar{c}, x)$ be the disjunction of the formulas $\varphi_a(\bar{c}, x)$ not satisfied by any $b \in R$. Then $\varphi(\bar{c}, x)$ and $\psi(\bar{c}, x)$ are computable Σ_1 formulas with the features required for (2).

Suppose that for each tuple $\bar{c}$, there in an element a that is *free* over $\bar{c}$ in the sense that any existential formula $\varphi(\bar{c}, x)$ true of a is true of elements of both R and S. We shall produce an isomorphism F from $\mathcal{A}$ onto a computable copy $\mathcal{B}$, such that $F(R)$ and $F(S)$ have no computable separator. Let B be an infinite computable set of constants, for the universe of $\mathcal{B}$. Let $\mathcal{F}$ be the set of finite $1-1$ functions from B to $\mathcal{A}$. We have the following requirements:

$R_{\langle e, e' \rangle}$: If $\mathrm{B} \subseteq W_e \cup W_{e'}$, then either $F(R) \cap W_{e'} \neq \emptyset$ or $F(S) \cap W_e \neq \emptyset$.

The strategy for $R_{\langle e, e' \rangle}$ is as follows. Take p satisfying the earlier requirements. Let a be the first element of $\mathcal{A}$ not in $ran(p)$, and map b to a. Continue extending p and enumerating sentences into $D(\mathcal{B})$ until b appears in $W_e \cup W_{e'}$. If this happens when we have $q \supseteq p$, then we try to replace q by some $q' \supseteq p$ such that either $q'(b) \in S$ and $b \in W_e$, or $q'(b) \in R$ and $b \in W_{e'}$. If we are unable to do this, then we try using the next element a', instead of a. We cannot continue indefinitely; we are certain to succeed by the time we come to an element that is free over $ran(p)$.

Remark: Theorem 11.13 implies Corollary 11.2—take S to be $\neg R$.

In the next two problems, as in Problem 6, $\mathcal{A}$ is a computable structure with a pair of added relations R and S.

Problem 7 *When is there a computable copy of $\mathcal{A}$ in which the image of R is c.e. and the image of S is not?*

Problem 8 *When is there a computable copy of $\mathcal{A}$ in which the images of R and S are c.e. and neither is computable relative to the other?*

Note that if $S = \neg R$, then Problem 7 is the same as Problem 4. Problem 7 is considered in [11], along with further generalizations involving more relations at various levels. We can see what the appropriate syntactical conditions should be by looking at Proposition 10.13.

Theorem 11.14 *Let $\mathcal{A}$ be a structure, with further relations R and S. Suppose the existential diagram of $(\mathcal{A}, R, S)$ is computable. Then the following are equivalent:*

(1) for all computable copies of $\mathcal{A}$ in which the image of R is computably enumerable, the image of S is also computably enumerable,

(2) S is definable in $(\mathcal{A}, R)$ by a computable Σ_1 formula in which R occurs only positively.

Exercise: Prove Theorem 11.14.

The priority method was originally developed by Friedberg [51] and Muchnik [124], independently, to prove the existence of computably enumerable sets A and B such that A $\not\leq_T$ B and B $\not\leq_T$ A. Problem 8 puts this in a model-theoretic setting.

Example: Let $\mathcal{A}$ be an infinite dimensional vector space over a computable field, where $\mathcal{A}$ is direct sum of infinite dimensional subspaces R and S. Note that $\neg R(x)$ if and only if

$$\exists u \, \exists v \, (\, R(u) \,\&\, S(v) \,\&\, v \neq 0 \,\&\, x = u + v \,) \ .$$

From this, it follows that in a computable copy of $\mathcal{A}$ in which the images of R and S are both c.e., R must be computable (S must also be computable).

If, in $(\mathcal{A}, R, S)$, $\neg R$ is definable by a computable Σ_1 formula in which R occurs only positively, or $\neg S$ is definable by a computable Σ_1 formula in which S occurs only positively, then in a computable copy, the images of R and S cannot be c.e. and independent. Under a suitable effectiveness condition, the converse holds, see [14] and [118] for this result and some extensions to higher levels.

Theorem 11.15 *Let $\mathcal{A}$ be a structure, with further relations R and S. Suppose that the existential diagram of $(\mathcal{A}, R, S)$ is computable. Then the following are equivalent:*

(1) for any computable copy of $\mathcal{A}$ in which the images of R and S are both c.e., one is computable in the other,

(2) in $(\mathcal{A}, R, S)$, either $\neg R$ is definable by a computable Σ_1 formula in which R occurs only positively or else $\neg S$ is definable by a computable Σ_1 formula in which S occurs only positively.

11.7. PAIRS OF RELATIONS

Exercise: Prove Theorem 11.15

Remark: We may obtain the original Friedberg-Muchnik Theorem from Theorem 11.15 by taking $\mathcal{A}$ to be ω, and letting R and S be, for example, the set of even numbers, and the set of numbers divisible by 3.

The constructions in pure computability theory are often carried out on the natural numbers, with no structure. It is an intriguing line of research to add underlying structure, to determine what definability conditions are needed for the construction, and to interpret the results in various natural examples. There has been some work besides the result given above. Hird, in particular, isolated a notion of "quasi-simplicity", which is related to simplicity, but turns out to be more natural in the context of structures such as vector spaces. See [66], [37], and [17]. Harizanov [62] has investigated further notions. This seems a very worthwhile type of investigation. One potential benefit is the increased understanding of what makes the constructions in computability theory work. Another benefit is in seeing what familiar notions mean in mathematical settings, and what constructions from computability tell us about ordinary mathematical structures.

Chapter 12

Computable categoricity and stability

It is interesting to imagine moving to a world in which the only objects present are the computable ones, and then examining familiar mathematical objects. Here we consider structures and isomorphisms between them. A pair of computable structures, isomorphic in the real world, need not be isomorphic in the computable world.

Goncharov [53] considered the following two problems.

Problem 1 *For a computable structure $\mathcal{A}$, when is there a computable copy $\mathcal{B}$ with no computable isomorphism from $\mathcal{A}$ onto $\mathcal{B}$?*

Problem 2 *When does $\mathcal{A}$ have a computable copy $\mathcal{B}$ with a non-computable isomorphism from $\mathcal{A}$ onto $\mathcal{B}$?*

There is some terminology associated with these problems. We say that a computable structure $\mathcal{A}$ is *computably categorical* if for every computable copy $\mathcal{B}$, some isomorphism from $\mathcal{A}$ onto $\mathcal{B}$ is computable. We say that $\mathcal{A}$ is *computably stable* if for every computable copy $\mathcal{B}$, every isomorphism from $\mathcal{A}$ onto $\mathcal{B}$ is computable. In Chapter 10, we considered notions of *relative computable categoricity* and *relative computable stability*. We showed that $\mathcal{A}$ is relatively computably categorical if and only if it has a formally c.e. Scott family (consisting of finitary existential formulas). We showed that $\mathcal{A}$ is relatively computably stable if and only if it has a formally c.e. defining family (also consisting of finitary existential formulas).

12.1 Simple examples

Before giving the general results, we mention a couple of examples.

Example 1: Let $\mathcal{A} = (\omega, <)$. Then $\mathcal{A}$ is not computably categorical.

To see why, first note that the successor relation is computable in the standard copy $\mathcal{A}$. In Chapter 11, we saw that there is a computable copy $\mathcal{B}$ in which the successor relation is not c.e. It follows that there is no computable isomorphism from $\mathcal{A}$ onto $\mathcal{B}$.

Example 2: Let $\mathcal{A} = (Q, <)$. Then $\mathcal{A}$ is relatively computably categorical, and therefore, computably categorical, but it is not computably stable.

Relative computable categoricity follows from the usual back-and-forth argument, or from the existence of a formally c.e. Scott family (there is one made up of finitary open formulas, with no parameters). Computable stability fails, since for any copy, there are $2^{\aleph_0}$ isomorphisms.

12.2 Relations between notions

Computable stability obviously implies computable categoricity. The precise relationship between the two notions is given in the result below.

Proposition 12.1 *Let $\mathcal{A}$ be a computable structure. Then $\mathcal{A}$ is computably stable if and only if for some tuple $\bar{c}$, $(\mathcal{A}, \bar{c})$ is computably categorical and rigid.*

Proof: First, suppose $\mathcal{A}$ is computably stable. Then there are only countably many automorphisms of $\mathcal{A}$, so by Theorem 6.13, there exists $\bar{c}$ such that $(\mathcal{A}, \bar{c})$ is rigid. Clearly, $(\mathcal{A}, \bar{c})$ is computably categorical. Now, suppose that $(\mathcal{A}, \bar{c})$ is computably categorical and rigid. For any isomorphism F from $\mathcal{A}$ onto a computable structure $\mathcal{B}$, if F maps $\bar{c}$ to $\bar{d}$, then F is the unique isomorphism from $(\mathcal{A}, \bar{c})$ onto $(\mathcal{B}, \bar{d})$. Since there is a computable isomorphism, F must be computable.

The next result shows that computable stability can be defined in terms of intrinsic computability of relations (see [19]).

Proposition 12.2 (Ash-Nerode) *For any computable structure $\mathcal{A}$, the following are equivalent:*
(1) $\mathcal{A}$ is computably stable,
(2) every computable relation R on $\mathcal{A}$ is intrinsically computable on $\mathcal{A}$.

Proof: (1) $\Rightarrow$ (2) If $\mathcal{A} \cong_F \mathcal{B}$, where $\mathcal{B}$ is computable, then, assuming (1), F must be computable. It follows that $F(R)$ is computable.

(2) $\Rightarrow$ (1) We may suppose that $\mathcal{A}$ has universe ω. The successor relation S on ω is computable, and the structure $(\mathcal{A}, S)$ has a formally c.e. defining family, with 0 as a parameter. Therefore, $(\mathcal{A}, S)$ is relatively computably stable. Let $\mathcal{A} \cong_F \mathcal{B}$, where $\mathcal{B}$ is computable. Assuming (2), F is computable. Then $F(S)$ is computable. Now, F is also an isomorphism from $(\mathcal{A}, S)$ onto $(\mathcal{B}, F(S))$. Since $(\mathcal{A}, S)$ is computably stable, F must be computable.

12.3 Computable categoricity

We turn to Problem 1 and Goncharov's result on computable categoricity [53]. The definability conditions are the same as for relative computable categoricity, but there is an additional effectiveness condition. The result below was proved first, before the corresponding result from Chapter 10 (Theorem 10.14).

Theorem 12.3 (Goncharov) *Let $\mathcal{A}$ be a computable structure whose finitary Π_2 diagram is computable. Then the following are equivalent:*
(1) $\mathcal{A}$ is computably categorical,
(2) $\mathcal{A}$ has a formally c.e. Scott family.

Remark: Recall that a Scott family for a structure is a set of infinitary formulas, with a fixed finite tuple of parameters, defining all of the orbits (under automorphisms) of tuples in the structure. A Scott family Φ is said to be *formally* Σ_1^0, or *formally c.e.* if the set Φ is c.e. and the elements of Φ are computable Σ_1 formulas. Taking the individual disjuncts of the formulas on Φ, we obtain another c.e. Scott family, made up of finitary existential formulas. We may assume that our formally c.e., Scott families have this form.

Proof of Theorem 12.3: The fact that (2) $\Rightarrow$ (1) follows from Theorem 10.14. We show that (1) $\Rightarrow$ (2). We give a finite-injury priority construction which either succeeds (meaning that all of the requirements are met), in which case (1) is false, or fails, in which case (2) is true. Let B be an infinite computable set of constants (for the universe of $\mathcal{B}$). As in earlier proofs, we let $\mathcal{B}$ be the structure induced on B by a $1-1$ function F from $\mathcal{A}$ onto B.

Here are the requirements that we attempt to satisfy:

R_e: φ_e is not an isomorphism from $\mathcal{A}$ onto $\mathcal{B}$.

We consider finite partial $1-1$ functions p from B to $\mathcal{A}$—possible parts of F^{-1}. The strategy for a single requirement R_e is as follows. Let p be the part of F^{-1} being preserved for higher priority requirements, say p maps $\overline{d}$ to $\overline{c}$. We try to guarantee that if φ_e is a total $1-1$ function from A onto $\mathcal{B}$, then for some $\overline{b}$, the pre-images of $\overline{d}, \overline{b}$ under φ_e and F do not satisfy the same existential formulas.

Suppose that at stage s, we have $q \supseteq p$ mapping $\overline{d}, \overline{b}, \overline{b}_1$ to $\overline{c}, \overline{a}, \overline{a}_1$, and we see that φ_e takes $\overline{c}^*, \overline{a}^*$ to $\overline{d}, \overline{b}$. Let $\delta(\overline{d}, \overline{b}, \overline{b}_1)$ be the conjunction of the sentences enumerated into the diagram by this stage, and suppose that q makes $\delta(\overline{d}, \overline{b}, \overline{b}_1)$ true in $\mathcal{A}$. If

$$\mathcal{A} \models \neg \exists \overline{v}\, \delta(\overline{c}^*, \overline{a}^*, \overline{v}) ,$$

then q takes care of the requirement. Suppose that

$$\mathcal{A} \models \exists \overline{v}\, \delta(\overline{c}^*, \overline{a}^*, \overline{v}) .$$

Let $\bar{u}, \bar{x}$ be variables corresponding to $\bar{d}, \bar{b}$. If among the first s existential formulas, we find $\psi(\bar{u}, \bar{x})$ such that

$$\mathcal{A} \models \psi(\bar{c}, \bar{a}) \leftrightarrow \neg\psi(\bar{c}^*, \bar{a}^*) \; ,$$

then again q takes care of the requirement. Suppose this does not happen. It may be that for one of the first s existential formulas ψ, some of the tuples satisfying $\exists \bar{v}\, \delta(\bar{c}, \bar{x}, \bar{v})$, in $\mathcal{A}$ satisfy $\psi(\bar{c}, \bar{x})$ and some do not. Suppose we have $\bar{a}'$ and $\bar{a}'_1$ such that

$$\mathcal{A} \models \delta(\bar{c}, \bar{a}', \bar{a}'_1)$$

and

$$\mathcal{A} \models \psi(\bar{c}, \bar{a}') \leftrightarrow \neg\psi(\bar{c}^*, \bar{a}^*) \; .$$

Then we can satisfy the requirement by replacing q by $q' \supseteq p$, where q' maps $\bar{d}, \bar{b}, \bar{b}_1$ to $\bar{c}, \bar{a}', \bar{a}'_1$.

Let $\varphi(\bar{d}, \bar{b}, \bar{b}_2)$ be the first atomic sentence not decided in δ, where $\bar{b}_2 \supseteq \bar{b}_1$. If for

$$\psi(\bar{u}, \bar{x}) = \exists \bar{v}'\, (\, \delta(\bar{u}, \bar{x}, \bar{v}') \,\&\, \varphi(\bar{u}, \bar{x}, \bar{v}') \,) \; ,$$

some of the tuples satisfying $\exists \bar{v}\, \delta(\bar{c}, \bar{x}, \bar{v})$ satisfy $\psi(\bar{c}, \bar{x})$ and some do not, or if the same is true with $\neg\varphi$ replacing φ, then as above, we can satisfy the requirement.

We have described several conditions that would allow us to act on requirement R_e. If none of these holds, then we are not in a position to act on the requirement at stage s. We choose $q \supseteq p$, acting on the first requirement that we can. We extend q, if necessary, to include in the domain the constants from $\varphi(\bar{d}, \bar{b}, \bar{b}_2)$, so that q makes the sentence either true or false in $\mathcal{A}$, and we enumerate $\varphi(\bar{d}, \bar{b}, \bar{b}_2)$ or its negation into $D(\mathcal{B})$, accordingly.

Suppose that for each $e' < e$, $R_{e'}$ is eventually satisfied, but R_e never is. The failure to satisfy R_e can be made to yield a formally c.e. Scott family, in the following way. Say by stage s_0, all action on earlier requirements has been completed, we have p mapping $\bar{d}$ to $\bar{c}$, and φ_{e,s_0} has range $\bar{d}$—we have seen the computations by stage s_0. For each $s > s_0$, let $\varphi_{e,s}$ map $\bar{c}^*, \bar{a}^*_s$ to $\bar{d}, \bar{b}_s$, and let $\delta_s(\bar{d}, \bar{b}_s, \bar{b}'_s)$ be the conjunction of the sentences enumerated into the diagram by stage s.

Claim 1: For any tuple $\bar{a}$, if $\bar{a}$ satisfies $\exists \bar{v}\, \delta_s(\bar{c}, \bar{x}, \bar{v})$ and $t > s$, then there exists $\bar{a}'$ such that $\bar{a}, \bar{a}'$ satisfies $\exists \bar{v}'\, \delta_t(\bar{c}, \bar{x}, \bar{x}', \bar{v}')$.

The claim is easily proved, by induction on t.

Now, we are ready to describe the Scott family Φ. For each tuple $\bar{b}$ from B, we determine a formula $\psi_{\bar{b}}(\bar{c}, \bar{x})$ as follows. Take the first stage $s > s_0$ at which we see that the range of $\varphi_{e,s}$ includes the elements of $\bar{d}, \bar{b}$, and our function $q \supseteq p$ is defined at least on these elements. Suppose q maps $\bar{d}, \bar{b}, \bar{b}_1$ to $\bar{c}, \bar{a}, \bar{a}_1$, and let $\delta(\bar{d}, \bar{b}, \bar{b}_1)$ be the conjunction of the formulas enumerated into $D(\mathcal{B})$ by this stage. Let

$$\psi_{\bar{b}}(\bar{c}, \bar{x}) = \exists \bar{v}\, \delta(\bar{c}, \bar{x}, \bar{v}) \; .$$

Let Φ consist of the formulas $\psi_{\bar{b}}(\bar{c}, \bar{x})$.

Claim 2: Φ is a Scott family.

Proof of Claim: We apply Proposition 6.10. We must show that each tuple satisfies some formula in Φ, and if two tuples satisfy the same formula in Φ, then they satisfy exactly the same existential formulas $\psi(\bar{c}, \bar{x})$. Since R_e is not satisfied, φ_e maps $\mathcal{A}$ $1-1$ onto B. For a given $\bar{a}$, suppose F maps $\bar{c}, \bar{a}$ to $\bar{d}, \bar{b}$. Then $\bar{a}$ must satisfy $\psi_{\bar{b}}(\bar{c}, \bar{x})$. At the stage when $\psi_{\bar{b}}(\bar{c}, \bar{x})$ is determined (in the way described above, witnessed by what we have enumerated into the diagram), we may have q taking $\bar{b}$ to something other than $\bar{a}$. Still, $\bar{a}$ must satisfy $\psi_{\bar{b}}(\bar{c}, \bar{x})$, since when we change the function, we preserve what has been enumerated into the diagram.

Let $\psi(\bar{c}, \bar{x})$ be a finitary existential formula satisfied by some tuple that also satisfies $\psi_{\bar{b}}(\bar{c}, \bar{x})$. Suppose that $\psi_{\bar{b}}(\bar{c}, \bar{x})$ is determined at stage s, and take $t \geq s$ such that ψ is among the first t existential formulas. The fact that we were unable to use $\psi(\bar{c}, \bar{x})$ at stage t to satisfy R_e means that if some tuple satisfying $\exists \bar{v}' \delta_t(\bar{c}, \bar{x}, \bar{x}', \bar{v}')$ satisfies $\psi(\bar{c}, \bar{x})$, then all do. By Claim 1,

$$\mathcal{A} \models \forall \bar{x} \, (\, \psi_{\bar{b}}(\bar{c}, \bar{x}) \to \exists \bar{x}' \, \exists \bar{v}' \, \delta_t(\bar{c}, \bar{x}, \bar{x}', \bar{v}') \,) \ .$$

Therefore,

$$\mathcal{A} \models \forall x \, (\, \psi_{\bar{b}}(\bar{c}, \bar{x}) \to \psi(\bar{c}, \bar{x}) \,) \ .$$

This completes the proof of Claim 2, which was all that remained to prove Theorem 12.3.

In Theorem 12.3, we assumed that the finitary Π_2 diagram of $\mathcal{A}$ is computable. The theorem would look nicer without this assumption, but we cannot drop it. In [54], there is an example of a structure that is computably categorical but has no formally c.e. Scott family.

Theorem 12.3 can be modified to give a further result of Goncharov [53], related to work of Nurtazin [129], with infinitely many copies having no computable isomorphisms. The result was proved well before the related result of McCoy in Chapter 10 (Theorem 10.15).

Theorem 12.4 (Goncharov) *Let $\mathcal{A}$ be a structure whose finitary Π_2 diagram is computable. If $\mathcal{A}$ is not computably categorical, then there is an infinite family of computable copies of $\mathcal{A}$, no two of which are computably isomorphic.*

Exercise: Prove Theorem 12.4

12.4 Decidable structures

The results that we have given for computable structures yield some results on decidable structures. The first result concerns prime models. Recall from Chapter 3 that for a complete theory T, a (countable) model is prime if and only if it is atomic; i.e., the model realizes only principal types.

Corollary 12.5 *If $\mathcal{A}$ is decidable, then the following are equivalent:*

(1) for all decidable $\mathcal{B} \cong \mathcal{A}$, there is a computable isomorphism from $\mathcal{A}$ onto $\mathcal{B}$,

(2) for some tuple $\bar{c}$, $(\mathcal{A}, \bar{c})$ is prime, and the set of generators for complete types realized in this structure is computable.

Proof: Let $\mathcal{A}^*$ be the expansion of $\mathcal{A}$ with added relations $\varphi^{\mathcal{A}}$ for all formulas φ in the language of $\mathcal{A}$. Computable copies of $\mathcal{A}^*$ correspond to decidable copies of $\mathcal{A}$.

(2) $\Rightarrow$ (1) If $\mathcal{A}$ satisfies (2), then we easily obtain a formally c.e. Scott family Φ^* for $\mathcal{A}^*$. By Theorem 12.3, for any computable copy of $\mathcal{A}^*$, and hence, for any decidable copy of $\mathcal{A}$, there is a computable isomorphism. Therefore, we have (1).

(1) $\Rightarrow$ (2) If $\mathcal{A}$ satisfies (1), then $\mathcal{A}^*$ is computably categorical. Then by Theorem 12.3, $\mathcal{A}^*$ has a formally c.e. Scott family Φ^*. Now, Φ^* consists of finitary existential formulas in the language of $\mathcal{A}^*$, with a finite tuple of parameters, say $\bar{c}$. We convert Φ^* into a formally c.e. Scott family Φ for $\mathcal{A}$, in the obvious way, replacing occurrences of symbols for the added relations by their definitions. Now, Φ consists of finitary formulas in the language of $\mathcal{A}$, with parameters $\bar{c}$. We show that $\mathcal{A}$ satisfies (2). It is easy to see that $(\mathcal{A}, \bar{c})$ is prime. To decide whether a formula $\gamma(\bar{c}, \bar{x})$ generates a complete type, we first check whether

$$\mathcal{A} \models \exists \bar{x}\, \gamma(\bar{c}, \bar{x}) \ .$$

If so, then we locate $\psi(\bar{c}, \bar{x}) \in \Phi$ such that

$$\mathcal{A} \models \exists \bar{x}\, (\gamma(\bar{c}, \bar{x})\ \&\ \psi(\bar{c}, \bar{x}))\ ,$$

and we check whether

$$\mathcal{A} \models \forall \bar{x}\, (\gamma(\bar{c}, \bar{x}) \to \psi(\bar{c}, \bar{x}))\ .$$

If so, then $\gamma(\bar{c}, \bar{x})$ generates a complete type.

Recall that a prime model is homogeneous.

Exercise: Prove Corollary 12.5 from the more general result of Goncharov and Peretyat'kin on decidable homogeneous structures (Theorem 9.27).

Using Corollary 12.5, we obtain the following criterion for computable categoricity of decidable prime models.

Corollary 12.6 *If T is a complete theory with a decidable prime model, then the following are equivalent:*

(1) any two decidable prime models of T are computably isomorphic,

(2) the set of generators of complete types realized in the prime model of T is computable.

12.5. COMPUTABLE STABILITY

Proof: Let $\mathcal{A}$ be a decidable prime model of T.

(2) $\Rightarrow$ (1) By Corollary 12.5, for any decidable copy of $\mathcal{A}$, there is a computable isomorphism. If $\mathcal{B}_1$ and $\mathcal{B}_2$ are two decidable copies, we have computable functions f_1, f_2 such that $\mathcal{A} \cong_{f_i} \mathcal{B}_i$. Then $f_2 \circ f_1^{-1}$ is a computable isomorphism from $\mathcal{B}_1$ onto $\mathcal{B}_2$.

(1) $\Rightarrow$ (2) For all decidable copies $\mathcal{B}$ of $\mathcal{A}$, there is a computable isomorphism from $\mathcal{A}$ onto $\mathcal{B}$. By Corollary 12.5, there is a tuple $\bar{c}$ such that the set of generators for complete types realized in $(\mathcal{A}, \bar{c})$ is computable. Suppose that $\varphi(\bar{u})$ generates the complete type realized by the tuple $\bar{c}$. To decide whether a formula $\gamma(\bar{x})$ generates a complete type, we apply the following procedure. First, we check whether $\exists \bar{x}\, \gamma(\bar{x}) \in T$. If so, then we locate a formula $\psi(\bar{c}, \bar{x})$, generating a complete type realized in $(\mathcal{A}, \bar{c})$, such that

$$\exists \bar{u}\, \exists \bar{x}\, (\,\varphi(\bar{u})\ \&\ \psi(\bar{u}, \bar{x})\ \&\ \gamma(\bar{x})\,) \in T\ ,$$

and we check whether

$$\forall \bar{x}\, (\,\gamma(\bar{x}) \to \exists \bar{u}\,[\,\varphi(\bar{u})\ \&\ \psi(\bar{u}, \bar{x})\,]\,) \in T\ .$$

12.5 Computable stability

We now turn to Problem 2 and Goncharov's result on computable stability [53]. The definability conditions are the same as for relative computable stability, but there is an additional effectiveness condition.

Remark: Recall that a defining family for a structure is a set of infinitary formulas defining all of the elements. In Chapter 10, we said that a defining family Φ is *formally* Σ_1^0 if the set Φ is c.e., and the elements of Φ are computable Σ_1 formulas. Taking disjuncts from these formulas, we obtain a c.e. defining family consisting of finitary existential formulas. We may assume that our formally Σ_1^0, or formally c.e., defining families have this form.

Theorem 12.7 (Goncharov) *Let $\mathcal{A}$ be a structure whose existential diagram is computable. Then the following are equivalent:*
(1) $\mathcal{A}$ is computably stable,
(2) $\mathcal{A}$ has a formally c.e. defining family.

Proof: The fact that (2) $\Rightarrow$ (1) follows from Theorem 10.16. We show that (1) $\Rightarrow$ (2). Suppose that for some tuple $\bar{c}$, each element of $\mathcal{A}$ is defined by an existential formula $\psi(\bar{c}, x)$ with parameters $\bar{c}$. Since the existential diagram is computable, for each a, we can find an existential formula $\varphi(\bar{c}, x)$ such that

$$\mathcal{A} \models \varphi(\bar{c}, a)\ \&\ \neg \exists x\, (\,\varphi(\bar{c}, x)\ \&\ x \neq a\,)\ .$$

The set Φ made up of these formulas $\varphi(\bar{c}, x)$ is a formally c.e. defining family. Therefore, we have (2).

Now, suppose that for each $\bar{c}$, there is some element a not definable in $\mathcal{A}$ by any existential formula $\varphi(\bar{c}, x)$ with parameters $\bar{c}$. In this case, we produce, using a finite-injury priority construction, a computable $\mathcal{B} \cong \mathcal{A}$, with a noncomputable isomorphism F from $\mathcal{A}$ onto $\mathcal{B}$.

Let B be an infinite computable set of constants (for the universe of $\mathcal{B}$). We have the following requirements.

$$R_e \colon \varphi_e \neq F^{-1}.$$

The strategy for a single requirement R_e is as follows. Let p be the part of F^{-1} being preserved for earlier requirements, where p maps $\bar{d}$ to $\bar{c}$. We take $q_0 \supseteq p$ mapping some b_0 to the first $a_0 \notin ran(p)$. We extend q_0, enumerating sentences into $D(\mathcal{B})$, until we find that $\varphi_e(b_0) = a_0$. This may never happen, in which case, q_0 takes care of the requirement. Suppose we find that $\varphi_e(b_0) = a_0$, at a stage when we have q mapping $\bar{d}, b_0, \bar{b}$ to $\bar{c}, a_0, \bar{a}$, and the conjunction of the sentences enumerated into $D(\mathcal{B})$ is $\delta(\bar{d}, b_0, \bar{b})$. If there exist $a_0' \neq a_0$ and $\bar{a}'$ such that

$$\mathcal{A} \models \delta(\bar{c}, a_0', \bar{a}'),$$

then we replace q by q' taking $\bar{d}, b_0, \bar{b}$ to $\bar{c}, a_0', \bar{a}'$. If there do not exist such a_0' and $\bar{a}'$ (i.e., if a_0 is defined by $\psi(\bar{c}, x) = \exists \bar{u}\delta(\bar{c}, x, \bar{u})$), then we stop trying to use a_0 to satisfy the requirement, and instead consider the next element of $\mathcal{A}$, say a_1. We take $q_1 \supseteq q$ mapping some b_1 to a_1, and we extend q_1 until we see that $\varphi_e(b_1) = a_1$. We continue in this way. Since there is an element that is not definable by any existential formula $\psi(\bar{c}, x)$, we will eventually come to some a_n that never needs replacing. This completes the proof of Theorem 12.7.

Remark: It is possible to prove Theorem 12.7 from Theorem 11.1. We need a definition to describe the appropriate combination of structures. If $\mathcal{A}_1$ and $\mathcal{A}_2$ are disjoint structures for the same language, then the *cardinal sum*, denoted by $\mathcal{A}_1 \oplus \mathcal{A}_2$, is the structure $(|\mathcal{A}_1| \cup |\mathcal{A}_2|, \mathcal{A}_1, \mathcal{A}_2)$, including, for each $i = 1, 2$, the relations and operations of $\mathcal{A}_i$, plus a relation for the universe.

Proposition 12.8 *Suppose $\mathcal{A}$ is a computable structure. Let $\mathcal{A}^* = \mathcal{A}_1 \oplus \mathcal{A}_2$, where $\mathcal{A}_1$ and $\mathcal{A}_2$ are disjoint computable copies of $\mathcal{A}$, and let F be an isomorphism from $\mathcal{A}_1$ onto $\mathcal{A}_2$. Then $\mathcal{A}$ is computably stable if and only if F is intrinsically c.e. on $\mathcal{A}^*$.*

Proof: Since F is a total function on $|\mathcal{A}_1|$, F is intrinsically c.e. if and only if it is intrinsically computable. Suppose first that $\mathcal{A}$ is computably stable. Suppose $\mathcal{B}_1, \mathcal{B}_2$ are computable copies of $\mathcal{A}$, and let G be an isomorphism from $\mathcal{B}_1$ onto $\mathcal{B}_2$. We show that G is computable. Let F_1 be an isomorphism from $\mathcal{A}$ onto $\mathcal{B}_1$, and let $F_2 = G \circ F_1$. Then F_2 is an isomorphism from $\mathcal{A}$ onto $\mathcal{B}_2$. Since $\mathcal{A}$ is computably stable, F_1 and F_2 are computable. Then $G = F_2 \circ F_1^{-1}$ is computable. It follows that F is intrinsically computable on $\mathcal{A}^*$.

Now, suppose that F is intrinsically computable on $\mathcal{A}^*$. Let G be an isomorphism from $\mathcal{A}$ onto a computable copy $\mathcal{B}$. If $\mathcal{A}$ and $\mathcal{B}$ are disjoint, then there is an isomorphism from $\mathcal{A}^*$ onto $\mathcal{A} \oplus \mathcal{B}$ under which the image of F is G. Therefore, G is computable. If $\mathcal{A}$ and $\mathcal{B}$ are not disjoint, then we have computable isomorphisms h_1, h_2 from $\mathcal{A}$, $\mathcal{B}$ onto disjoint computable structures $\mathcal{A}'$, $\mathcal{B}'$. Composing, we obtain an isomorphism $G' = h_2 \circ G \circ h_1^{-1}$ from $\mathcal{A}'$ onto $\mathcal{B}'$. By the reasoning above, G' is computable, and then $G = h_2^{-1} \circ G' \circ h_1$ is also computable. This completes the proof.

In Theorem 12.7, we assumed that the existential diagram is decidable. The theorem would look nicer without this assumption, but we cannot drop it. After Theorem 12.3, we mentioned an example from [54] of a structure $\mathcal{A}$ that is computably categorical but has no formally c.e. Scott family. The structure is actually rigid. Therefore, it is computably stable. It does not have a formally c.e. defining family, since that could be converted into a formally c.e. Scott family.

12.6 Computable dimension

Goncharov defined the *computable dimension* of $\mathcal{A}$ to be the number of equivalence classes of computable copies of $\mathcal{A}$ under the equivalence relation of computable isomorphism. The definition is not vacuous. Goncharov [54], [57], constructed examples of computable structures having computable dimension n, for all finite n. There are other examples, with further special features.

We state (without proof) a basic lemma of Goncharov on existence of a family of sets whose enumerations have special properties, and we show how this yields an example of a structure of computable dimension 2.

Recall that for a countable family $\mathcal{S}$ of subsets of ω, an *enumeration* is a binary relation R such that if $R_i = \{x : (i, x) \in R\}$, then $\mathcal{S} = \{R_i : i \in \omega\}$. The number i is called an R-*index* of the set R_i. An enumeration R is *univalent* if no set has two different indices. If R and R' are two univalent enumerations of the same family, then there is a unique permutation f of ω such that for all i, $R'_i = R_{f(i)}$—$f(i)$ is the R-index of the set whose R'-index is i. If f is computable, then we say that R and R' are *computably equivalent*. The following lemma is from [52].

Lemma 12.9 (Goncharov) *There is a countable family $\mathcal{S}$ of subsets of ω with two univalent enumerations R and R', both c.e., such that*

(1) R and R' are not computably equivalent,

(2) every univalent c.e. enumeration of $\mathcal{S}$ is computably equivalent either to R or to R'.

Using Lemma 12.9, we can prove the following.

Theorem 12.10 (Goncharov) *There exists a structure $\mathcal{A}$ of computable dimension 2.*

Proof: Let $\mathcal{S}$ be the family of sets from Lemma 12.9. The construction of the corresponding structure $\mathcal{A}$ differs only slightly from the one in Proposition 9.22. We form a directed graph with a single connected component for each set $X \in \mathcal{S}$ (instead of infinitely many). As before, the connected component corresponding to X has one special element a such that $a \to a$. In addition, for each $n \in X$, there is a cycle of length $n + 1$, of the form

$$a \to a_0 \to \ldots \to a_{n-1} \to a \;.$$

The computable copies of $\mathcal{A}$ correspond to c.e., univalent enumerations of $\mathcal{S}$. Moreover, we can pass effectively from an index for a copy $\mathcal{B}$ of $\mathcal{A}$ to a c.e. index for a univalent enumeration R of $\mathcal{S}$ such that from the R-index n of a set X, we can find the special element b in the connected component of $\mathcal{B}$ that codes the set X.

Let $\mathcal{B}_1, \mathcal{B}_2$ be computable copies of $\mathcal{A}$, and let R_1, R_2 be the corresponding c.e., univalent enumerations of $\mathcal{S}$. If R_1 and R_2 are equivalent, via f, then we can define a computable isomorphism between $\mathcal{B}_1$ and $\mathcal{B}_2$. (We use f to match the special elements, and the remaining elements of a connected component are all definable from the special one by finitary existential formulas.) Conversely, if there is a computable isomorphism F from $\mathcal{B}_1$ onto $\mathcal{B}_2$, then F maps special elements to special elements, so, using F, we can find the R_2-index of the set with a given R_1-index.

There has been quite a lot of work on structures with finite computable dimension. Cholak, Goncharov, Khoussainov, and Shore constructed examples of computably categorical structures that acquire computable dimension $n \geq 2$ after naming a constant [28]. There are further examples with other special features, due to Kudinov [91], Khoussainov and Shore [78], Hirschfeldt, Khoussainov, Slinko, and Shore [68], and Hirshfeldt by himself [67].

We shall not discuss this body of work, although it is interesting. The results do not support the thesis of this book—the claim that definability can account for bounds on complexity which persist under isomorphism. Definability does *not* account for the fact that these structures have computable dimension $n \geq 2$.

In Chapter 10, we relativized the notions of computable categoricity, stability, etc. Theorem 10.15 says that the notion of computable dimension does not relativize to anything interesting—it is always either 1 or ∞. Even though there are counterexamples, the behavior that we expect from everyday computable structures is the same—either there is a formally c.e. Scott family and the computable dimension is 1, or there is no formally c.e. Scott family, and the computable dimension is ∞.

12.7 One or infinitely many

For many familiar kinds of structures, the conclusion of Theorem 12.4 holds, even when the hypothesis fails. The next two results illustrate this. The result below, for linear orderings, is from [40], [41].

12.7. ONE OR INFINITELY MANY

Theorem 12.11 (Dzgoev-Goncharov) *If $\mathcal{A}$ is a linear ordering, then $\mathcal{A}$ has computable dimension 1 if the successor relation is finite, and ∞ otherwise.*

Proof: First, suppose $\mathcal{A}$ is an ordering in which the successor relation is finite. Let $\bar{a}$ include the elements with immediate successors or immediate predecessors, and the first and last elements, if any. Then $\bar{a}$ partitions the ordering into open intervals which are all either empty or dense. From this, it follows that $\mathcal{A}$ has a formally c.e. Scott family with parameters in $\bar{a}$.

Now, suppose $\mathcal{A}$ is an ordering in which the successor relation is infinite. We show that for any finite collection of computable copies $\mathcal{B}_1, \ldots, \mathcal{B}_n$, there is another computable copy $\mathcal{C}$ not computably isomorphic to any $\mathcal{B}_k$. We could do the same for any infinite collection of copies $(\mathcal{B}_k)_{k\in\omega}$ that is uniformly computable.

As usual, we take an infinite computable set of constants, call it C this time, for the universe of $\mathcal{C}$. We determine a $1-1$ function F from $\mathcal{A}$ onto C, and we let $\mathcal{C}$ be the induced structure. We consider finite partial $1-1$ functions from C to $\mathcal{A}$, and when we tentatively decide that p is a part of F^{-1}, we permanently decide the ordering on $dom(p)$.

We have the following requirements.

$R_{\langle e,k\rangle}$: $\mathcal{B}_k \not\cong_{\varphi_e} \mathcal{C}$

The strategy for $R_{\langle e,k\rangle}$ is as follows. Let p be a finite partial function taking care of earlier requirements. Then $ran(p)$ partitions $\mathcal{A}$ into finitely many intervals. If some infinite interval in $\mathcal{A}$ corresponds under $\varphi_e^{-1} \circ p^{-1}$ to an interval in $\mathcal{B}_k$, where this contains some pair b, b' in the successor relation, and φ_e maps b, b' to points c, c' in C, then we extend p, mapping c, c' to a pair a, a' that is not in the successor relation. Of course, we can only guess which intervals in $\mathcal{A}$ are infinite, and we can only guess successors in $\mathcal{B}_k$.

At stage s, we determine the ordering $\mathcal{C}_s$ on the first few constants in C. We have a tentative isomorphism p_s between $\mathcal{C}_s$ and a finite substructure of $\mathcal{A}$. We guess that b' is the successor of b in $\mathcal{B}_k$ if none of the first s constants lies between them. We use $\varphi_{e,s}$ as the stage s approximation of φ_e. We guess that an interval in $\mathcal{A}$ is infinite if the sth constant from the universe of $\mathcal{A}$ lies in that interval.

Requirement $R_{\langle e,k\rangle}$ *admits action* at stage $s+1$ on an interval (a, a') with endpoints a and a' in $ran(p)$ if the sth constant of $\mathcal{A}$ lies in the interval, and
 (1) φ_e^{-1} takes $p^{-1}(a), p^{-1}(a')$ to a pair b, b' in $\mathcal{B}_k$,
 (2) for the first pair d, d' such that $b \leq d < d' \leq b'$ and d' appears to be the successor of d in $\mathcal{B}_k$, φ_e takes d, d' to some c, c' in $\mathcal{C}_s$,
 (3) φ_e preserves order and c' is the successor of c in $\mathcal{C}_s$.

To act, we let $\mathcal{C}_{s+1}$ be the result of adding an element c^* between c and c', and we take $p_{s+1} \supseteq p$, where p_{s+1} preserves order on the interval $p^{-1}(a), p^{-1}(a')$ and p_{s+1} agrees with p_s outside this interval. Requirement $R_{\langle e,k\rangle}$ *requires action* at stage $s+1$ if there is some interval on which action is admitted, and there is no other interval on which we took action earlier, and on which the successor relation $\mathcal{B}_k$ remains unchanged.

At stage s, we act on the first requirement $R_{\langle e,k\rangle}$ that requires action. If no requirement requires action, then $p_{s+1} \supseteq p_s$, and $\mathcal{C}_{s+1}$ is the result of adding a new element to $\mathcal{C}_s$ to match the next element of $\mathcal{A}_{s+1}$. Since we preserve the ordering, the structure $\mathcal{C} = \cup_s \mathcal{C}_s$ is clearly computable.

Claim: Each requirement $R_{\langle e,k\rangle}$ will be acted on only finitely many times and will eventually be satisfied.

Proof of Claim: If p takes care of previous requirements, then at least one of the intervals with endpoints in $ran(p)$ is infinite and contains a pair of successors. At some stage, we will have correctly guessed a pair of successors in some interval, and the next element of $\mathcal{A}$ will turn out to lie in this interval. Then, if the requirement is not already satisfied in some trivial way, we are in a position to act on it for the last time.

For each n, let $s(n)$ be the first s such that after stage s, there is no action on $R_{\langle e,k\rangle}$, for $\langle e,k\rangle \leq n$. Then $\cup_n p_{s(n)}$ is an isomorphism from $\mathcal{C}$ onto $\mathcal{A}$.

Corollary 12.12 *A linear ordering $\mathcal{A}$ is computably stable if and only if it is finite.*

Proof: If $\mathcal{A}$ is finite, it is trivially computably stable. Suppose $\mathcal{A}$ is computably stable. Then it is computably categorical. By Theorem 12.11, the successor relation on $\mathcal{A}$ is finite. Then either $\mathcal{A}$ is finite, or else it has a dense interval. If $\mathcal{A}$ has a dense interval, then $(\mathcal{A}, \bar{c})$ is not rigid for any tuple $\bar{c}$. Then by Proposition 12.1, $\mathcal{A}$ is not computably stable, a contradiction. Therefore, $\mathcal{A}$ must be finite.

The next result, from [52], [41], is for Boolean algebras. It is similar to Theorem 12.11.

Theorem 12.13 (Goncharov) *A Boolean algebra has computable dimension 1 if there are only finitely many atoms, and ∞ otherwise.*

Proof: First, suppose $\mathcal{A}$ is a Boolean algebra with only finitely many atoms. Let $\bar{a}$ be a tuple including all of the atoms. Then $\mathcal{A}$ has a formally c.e. Scott famil consisting of finitary open formulas with parameters $\bar{a}$.

Now, suppose $\mathcal{A}$ is a computable Boolean algebra with infinitely many atoms. As above, to show that $\mathcal{A}$ has computable dimension ∞, we show that for any finite collection of computable copies $\mathcal{B}_1, \ldots, \mathcal{B}_n$, or any infinite, uniformly computable collectiion of copies $(\mathcal{B}_k)_{k \in \omega}$, there is another computable copy $\mathcal{C}$ that is not computably isomorphic to any $\mathcal{B}_k$. We have the following requirements.

$R_{\langle e,k\rangle}$: $\mathcal{B}_k \not\cong_{\varphi_e} \mathcal{C}$

The strategy for $R_{\langle e,k\rangle}$ is as follows. Suppose we have a finite $1-1$ function p taking care of earlier requirements. We suppose that $ran(p)$ is a subalgebra of $\mathcal{A}$. Let $a_1,\ldots,a_r$ be the atoms of this finite subalgebra, where $p(c_i) = a_i$. Suppose φ_e threatens to be an isomorphism, mapping c_i to b_i in $\mathcal{B}_k$. For some i, b_i bounds an atom and is infinite (i.e., it is not the join of finitely many atoms) in $\mathcal{B}_k$. We do not know which elements of $\mathcal{A}$ are infinite, and we can only guess at the atoms of $\mathcal{B}_k$. For each i, we take the first $b'_i \leq b_i$, if any, that appears to be an atom in $\mathcal{B}_k$, and if φ_e maps b'_i to $c'_i \leq c_i$, then we try to extend p to q, preserving the algebra, and mapping c'_i to a some a'_i that is not an atom of $\mathcal{A}$.

If we act, believing that b'_i is an atom, and later we find that b'_i is not an atom, then we may need to act again. The requirement will be satisfied by the time we have found and acted on the first atom $b'_i \leq b_i$, for some i such that a_i is actually infinite.

Corollary 12.14 *A Boolean algebra $\mathcal{A}$ is computably stable if and only if it is finite.*

Proof: If $\mathcal{A}$ is finite, then it is computably stable. Suppose $\mathcal{A}$ is computably stable. Then it is computably categorical. By Theorem 12.13, there are only finitely many atoms. If $\mathcal{A}$ is infinite, then $(\mathcal{A}, \bar{c})$ is not rigid, for any tuple $\bar{c}$. Then by Proposition 12.1, $\mathcal{A}$ is not computably stable, a contradiction. Therefore, $\mathcal{A}$ is finite.

Goncharov [56] proved results on Abelian groups, similar to the results on linear orderings and Boolean algebras.

12.8 Quotient structures

We could consider categoricity and stability in the context of copies that are not arbitrary, and not computable, but somewhere in between—c.e. quotient structures, and Γ-structures. For results on Γ-structures, see [11]. Here we give a result on stability among c.e. quotient structures, due to Love [100]. Let $\mathcal{A}$ be a computable structure. An isomorphism F from $\mathcal{A}$ onto a quotient structure $\mathcal{B}/\sim$ is *computable* if the relation $\{(a,b) : F(a) = b/\sim\}$ is c.e. This means that given a, we can search effectively for a pair (a,b) such that $F(a) = b/\sim$. Since $F(a)$ is defined, the search will halt, and we know that b is one name for $F(a)$. We say that $\mathcal{A}$ is *computably stable, among c.e. quotient structures* provided that for all c.e. quotient structures $\mathcal{B}/\sim$ isomorphic to $\mathcal{A}$, all isomorphisms are computable in the sense that we have just defined.

Theorem 12.15 (Love) *Let $\mathcal{A}$ be a structure whose finitary Σ_1 diagram is computable. Then the following are equivalent:*

(1) $\mathcal{A}$ is computably stable among c.e. quotient structures,

(2) $\mathcal{A}$ has a c.e. defining family Φ, consisting of finitary Σ_1 formulas $\psi(\bar{c}, x)$, with fixed parameters $\bar{c}$, in which $=$ occurs only positively.

Note: Love stated Theorem 12.15 only for quotient *algebras*, but there is no added difficulty in passing to relational structures.

Sketch of proof: We suppose that the language of $\mathcal{A}$ is relational.

(2) $\Rightarrow$ (1) Let $\bar{c} = (c_1, \ldots, c_n)$. Suppose that F is an isomorphism from $\mathcal{A}$ onto a c.e. quotient structure $\mathcal{B}/\sim$, where F maps c_i to $d_i/\sim$. The fact that the relation
$$R = \{(a,b) : F(a) = b/\sim\}$$
is c.e. follows from Theorem 7.5, using the uniformity. (It is also easy to see from first principles.) Given $a \in \mathcal{A}$ and $b \in \mathcal{B}$, we have an effective procedure which halts just in case $F(a) = b/\sim$. We first locate a formula $\psi(\bar{c}, x)$ in Φ such that
$$\mathcal{A} \models \psi(\bar{c}, a) .$$
Then, enumerating the diagram of $\mathcal{B}$ and the congruence relation $\sim$, we search for evidence that
$$\mathcal{B}/\sim \; \models \psi(\bar{d}, b) .$$

(1) $\Rightarrow$ (2) Suppose that for some tuple $\bar{c}$ in $\mathcal{A}$, each element a is definable by a finitary Σ_1 formula $\psi(\bar{c}, x)$ in which $=$ appears only positively. For each a, we can search until we find $\psi(\bar{c}, x)$, of the proper form, such that
$$\mathcal{A} \models \psi(\bar{c}, a) \; \& \; \neg \exists x \, (\, \psi(\bar{c}, x) \; \& \; x \neq a \,) .$$
Therefore, we have the required defining family.

Now, suppose that for each $\bar{c}$, there is some a not definable by a finitary Σ_1 formula $\psi(\bar{c}, x)$ in which $=$ appears only positively. Then we construct computable $\mathcal{B}$ and a c.e. congruence relation $\sim$ such that $\mathcal{A} \cong_F \mathcal{B}/\sim$, where $\{\langle a,b \rangle : F(a) = b/\sim\}$ is not c.e. Let B be an infinite computable set of constants, for the universe of $\mathcal{B}$. As in other constructions, we work with finite $1-1$ functions p from B to $\mathcal{A}$—possible parts of F^{-1}. When we choose p, we enumerate into $D(\mathcal{B})$ basic sentences not involving $=$ that p makes true in $\mathcal{A}$. We also enumerate into $\sim$ any pairs in $\mathcal{B}$ that are mapped by p to the same element of $\mathcal{A}$.

We have the following requirements.

R_e: $W_e \neq \{\langle a,b \rangle : a \in \mathcal{A} \; \& \; F(a) = b/\sim\}$

The strategy for R_e is as follows. Suppose p is being preserved for earlier requirements. We extend p, mapping some b to the first a not in $ran(p)$. If $\langle a,b \rangle$ eventually appears in W_e, then we act as follows. Let $\delta(\bar{d}, b, \bar{b}_1)$ be the conjunction of the basic sentences not involving $=$ that have been enumerated into $D(\mathcal{B})$, together with the sentences $b_1 = b_2$, where (b_1, b_2) has been enumerated into $\sim$. If
$$\psi(\bar{c}, x) = \exists \bar{u} \, \delta(\bar{c}, x, \bar{u})$$
defines a, then we replace a by the next a' not in $ran(p)$, and try again. If not, say a' satisfies $\psi(\bar{c}, x)$, where $a' \neq a$, then we change the function, mapping b to

12.8. QUOTIENT STRUCTURES

a'. For b' such that $p(b') = a'$, we add appropriate pairs to $\sim$. In any case, we enumerate into $D(\mathcal{B})$ further basic sentences not involving $=$, where these are made true by p.

Chapter 13

n-systems

13.1 Introduction

In a talk at the Kleene symposium, in 1978, Morley stated the following result (he acknowledged that it had been discovered independently by a number of other mathematicians).

Theorem 13.1 *Recursion theory is very hard.*

Many of the results and problems in computability theory (recursion theory) have statements which can be readily understood. It is the proofs which are hard, especially certain priority constructions. We have already given several priority constructions—all relatively simple. Roughly speaking, the object of a priority construction is to enumerate a set (or tuple of sets), so as to satisfy a list of requirements, using information which can only be approximated. The strategies for different requirements may conflict, and in acting for one requirement, we may "injure" another. There is a system of "priorities" to determine which requirements are protected and which may be injured at a given stage.

In a *finite-injury* priority construction, each requirement is injured at most finitely often, and the information needed to see exactly how the requirements are satisfied is Δ_2^0. There are *infinite-injury* priority constructions, in which some requirements may be injured infinitely often, and the information needed to see how the requirements are satisfied is Δ_3^0. There are also Δ_4^0 constructions, Δ_ω^0 constructions, etc.

We shall prove a general "metatheorem" with a list of conditions guaranteeing the success of a priority construction. The metatheorem is due to Ash [1], [3], [4]. In the present chapter, we give a version of the metatheorem for Δ_n^0 constructions from [1], incorporating simplifications and improvements from [3].

The motivation for the metatheorem, both the finite version given in this chapter, and the transfinite version given in the next chapter, was to bring within reach (of Ash and Ewan Barker, who was writing his Master's thesis

213

under Ash's supervision) certain constructions that seemed too complicated to attempt directly. The motivating applications of Ash and Barker will be given in Chapters 16 and 17. There are by now a number of further applications.

There have been other attempts to systematize the writing up of priority constructions of different kinds, in particular, by Lempp and Lerman [95], [96], and by Shoenfield [142]. The different approaches are successful in different ways. Ash's metatheorem has the advantage of being relatively easy to use, where it applies. Shoenfield's approach applies to certain important kinds of constructions. Lempp and Lerman's approach applies to a much broader class of constructions. It is sufficiently complicated to have scared some possible users, but this may change.

Groszek, Mytilinaios, and Slaman [59], [60], [125] Kontostathis [89], Chong and Yang [30], [31] and others have analyzed priority constructions with a different goal. Instead of trying to provide a way to write up new constructions taking advantage of what has already been done, they trying to classify constructions according the amount of induction needed to carry them out. Although the goals are different, the work is similar to that of Ash, Lempp, Lerman, and Shoenfield, in that it requires picking out the important features of various constructions, and describing them abstractly.

It seems to us very worthwhile for a researcher who has thought through a difficult priority construction, to record in some abstract way the reasons for the success of the construction. If the same features appear again, it should not be necessary to rediscover the strategies for dealing with them, and when new features appear, they should be easily recognized as new, so that effort can be concentrated on finding appropriate new strategies.

13.2 Statement of the metatheorem

Before we can state the metatheorem, we must formulate in a more precise mathematical way the object of a priority construction. The setting is as follows. An *alternating tree* on sets L and U is a tree P consisting of non-empty finite "alternating sequences"

$$\sigma = \ell_0 u_1 \ell_1 u_2 \ell_2 \ldots ,$$

where $\ell_i \in L$ and $u_i \in U$. We suppose that P has no terminal nodes (i.e., every sequence in P has a proper extension in P).

An *instruction function* for P is a function q from the set of sequences in P of odd length (those with last term in L) to U, such that if $q(\sigma) = u$, then $\sigma u \in P$. A *run* of (P, q) is a path

$$\pi = \ell_0 u_1 \ell_1 u_2 \ell_2 \ldots$$

such that q gives the terms in U; i.e., for all n,

$$u_{n+1} = q(\ell_0 u_1 \ell_1, \ldots, u_n \ell_n) .$$

13.2. STATEMENT OF THE METATHEOREM

An *enumeration function* on L is a function E assigning a finite set to each element of L. We may extend the enumeration function E from elements of L to paths $\pi = \ell_0 u_1 \ell_1 u_2 \ell_2 \ldots$ through P, letting $E(\pi) = \cup_{i \in \omega} E(\ell_i)$.

We are ready to describe the object of a Δ_n^0 construction. Let L and U be c.e. sets, let P be a c.e. "alternating tree" on L and U, in which all of the sequences start with a fixed $\hat{\ell} \in L$, let E be a partial computable enumeration function, defined on L, and let q be a Δ_n^0 instruction function on P.

Object: Produce a run π of (P, q) such that $E(\pi)$ is c.e. (while π is Δ_n^0).

The instruction function q provides the high-level information for satisfying the requirements. The other given objects are all c.e. or partial computable. We do not expect to obtain a *computable* run π, since the terms in U are given by the instruction function q, which is Δ_n^0. Still, we hope to make $E(\pi)$ c.e. The conditions guaranteeing that we can do this involve further c.e. binary relations $\leq_k$ on L.

An *n-system* is a structure of the form

$$(L, U, \hat{\ell}, P, E, (\leq_m)_{m<n}),$$

where L and U are c.e. sets, P is a c.e. alternating tree on L and U in which all of the sequences start with $\hat{\ell}$, E is a partial computable enumeration function defined on L, and for each $m < n$, $\leq_m$ is a c.e. binary relation on L such that the following conditions hold:

(1) $\leq_m$ is reflexive and transitive,
(2) for $k < m$, $\ell \leq_m \ell' \Rightarrow \ell \leq_k \ell'$,
(3) $\ell \leq_0 \ell' \Rightarrow E(\ell) \subseteq E(\ell')$,
(4) if $\sigma u \in P$, where σ ends in ℓ^0 (an element of L), and

$$\ell^0 \leq_{m_0} \ell^1 \leq_{m_1} \ldots \leq_{m_{r-1}} \ell^r,$$

for $n > m_0 > m_1 > \ldots > m_r$, then there exists ℓ^* such that $\sigma u \ell^* \in P$ and $\ell^i \leq_{m_i} \ell^*$ for all $i \leq r$.

Here is the statement of the metatheorem for n-systems.

Theorem 13.2 *Let $(L, U, \hat{\ell}, P, E, (\leq_m)_{m<n})$ be an n-system, and let q be a Δ_n^0 instruction function for P. Then there is a Δ_n^0 run π of (P, q) such that $E(\pi)$ is c.e. Moreover, from a Δ_n^0 index for q, together with c.e. and computable indices for the components of the n-system, we can effectively determine a Δ_n^0 index for π, and a c.e. index for $E(\pi)$.*

Remark: Above, we assumed that the tree P has no terminal nodes. This assumption has been made redundant by later assumptions. The existence of an instruction function implies that there is no terminal node with last term in L, and Condition (4) implies that there is no terminal node with last term in U.

13.3 Some examples

Before proving Theorem 13.2, we give some relatively simple examples of its use. The result below was already proved in Chapter 9, using a finite-injury priority construction.

Example 1 (Theorem 9.7) *If $\mathcal{A}$ is a Δ_2^0 linear ordering, then $\mathcal{A}$ is isomorphic to a c.e. quotient structure (in the language with min replacing $<$).*

Proof using Theorem 13.2: The effect of the change in language is that we count as a congruence relation any equivalence relation in which the equivalence classes are intervals. Let $\mathcal{A}$ be a Δ_2^0 linear ordering. We may suppose that the universe of $\mathcal{A}$ is an infinite computable set A of constants. We let B be another infinite computable set of constants, for the universe of $\mathcal{B}$.

We determine a Δ_2^0 function F from B onto A such that the equivalence relation $\sim$, given by
$$b \sim b' \Leftrightarrow F(b) = F(b')$$
is c.e. We also determine a computable linear ordering of B, such that $\sim$ is a congruence relation on the resulting structure $\mathcal{B}$, and F induces an isomorphism from $\mathcal{B}/\sim$ onto $\mathcal{A}$

We shall define a 2-system
$$(L, U, \hat{\ell}, P, E, \leq_0, \leq_1)$$
and a Δ_2^0 instruction function q. Let U consist of the linear orderings on finite subsets of the universe of A. Let $\mathcal{F}$ consist of the finite partial functions from B to A. Each $p \in \mathcal{F}$ induces an equivalence relation $\sim_p$ on its domain, where
$$b \sim_p b' \Leftrightarrow p(b) = p(b') \,.$$

Let L consist of the triples $\ell = (p, u, v)$, where $u \in U$, v is a linear ordering on a finite subset of B, and p is a finite partial function from v onto u inducing an isomorphism from $v/\sim_p$ onto u. For $\ell = (p, u, v)$ in L, let $E(\ell)$ be the set of atomic sentences true in $(v, \sim_p)$—these sentences have form $b < b'$, $b = b'$, or $b \sim_p b'$. Let $\hat{\ell} = (\emptyset, \emptyset, \emptyset)$.

For $\ell = (p, u, v)$ and $\ell' = (p', u', v')$ in L, let
$$\ell \leq_0 \ell' \Leftrightarrow E(\ell) \subseteq E(\ell')$$
(i.e., $v \subseteq v'$ and $b \sim_p b' \Rightarrow b \sim_{p'} b'$), and
$$\ell \leq_1 \ell' \Leftrightarrow (p \subseteq p' \ \& \ u \subseteq u' \ \& \ v \subseteq v') \,.$$

Let P consist of the finite sequences
$$\sigma = \ell_0 u_1 \ell_1 u_2 \ell_2 \ldots$$
such that $\ell_k \in L$, $u_k \in U$, and the following hold:

13.3. SOME EXAMPLES

(a) $\ell_0 = \hat{\ell}$,
(b) u_k is an ordering of the first k constants from the universe of $\mathcal{A}$ ($k \geq 1$),
(c) ℓ_k has the form (p_k, u_k, v_k), where v_k includes the first k constants from B,
(d) $\ell_k \leq_1 \ell_{k+1}$—this implies that $u_k \subseteq u_{k+1}$,

We have defined all of the ingredients of the 2-system. To show that we actually have a 2-system, we must verify the four conditions. Conditions (1), (2), and (3) are clear. Toward Condition (4), let $\sigma u \in P$, where σ ends in $\ell^0 = (p_0, u_0, v_0)$, and suppose $\ell^0 \leq_1 \ell^1$, where $\ell^1 = (p_1, u_1, v_1)$. We must show that there exists $\ell^* = (p^*, u, v^*)$ such that $\sigma u \ell^* \in P$, $\ell^0 \leq_1 \ell^*$, and $\ell^1 \leq_0 \ell^*$.

First, we collapse extra equivalence classes, defining p^* on v_1 so that all equivalence classes correspond to elements of u_0, provided this is not empty. If $p_1(b) \in u_0$, then $p^*(b) = p_1(b)$. If $p_1(b) \notin u_0$, then $p^*(b) = p_1(b')$, where b' is the closest to b on the left such that $p_1(b') \in u_0$, or the closest on the right, if there is no such b' on the left, but there is on the right.

Next, we add an equivalence class corresponding to the new element a in $u - u_0$ (there is only one). For the first new constant $b \in B$, we let $p^*(b) = a$, and we let v^* be the extension of v_1 with b located to the right of any b' such that
$$u \models p^*(b') < a,$$
and to the left of any b' such that
$$u \models a < p^*(b').$$
In the trivial case where $u_0 = \emptyset$, we collapse v_1 to a single equivalence class, adding the first constant b, if necessary, to form v^*, and we let p^* map all of the elments to a. In the even more trivial case where $v_1 = \emptyset$ as well, then we start an equivalence class, using the first constant b, and we let p^* map b to a. It is not difficult to see that $\ell^* = (p^*, u, v^*)$ has the desired features. Therefore, we have a 2-system.

Next, we define the instruction function q. For $\sigma \in P$ of length $2n+1$, let $q(\sigma)$ be the restriction of the ordering $\mathcal{A}$ to the first $n+1$ constants. Now, we are in a position to apply Theorem 13.2. We obtain a run
$$\pi = \hat{\ell} u_1 \ell_1 u_2 \ell_2 \ldots$$
of (P, q) such that $E(\pi)$ is c.e. Say $\ell_k = (p_k, u_k, v_k)$. Then
$$F = \cup_k p_k$$
is a function from B onto $\mathcal{A}$, $\cup_k v_k$ is a linear ordering $\mathcal{B}$ with universe B, and
$$\sim \; = \; \cup_k \sim_{p_k}$$
is a congruence relation on $\mathcal{B}$ such that $b \sim b' \Leftrightarrow F(b) = F(b')$. Moreover, F induces an isomorphism from $\mathcal{B}/\sim$ onto $\mathcal{A}$. The fact that $E(\pi)$ is c.e. implies that the atomic diagram of $\mathcal{B}$ is c.e. Since $\mathcal{B}$ is a linear ordering, the fact that the atomic diagram is c.e. implies that it is actually computable. The congruence relation $\sim$ is c.e. This completes the proof.

The next example is Watnik's Theorem. Like the first example, it was proved in Chapter 9.

Example 2 (Theorem 9.10) *If $\mathcal{A}$ is a Δ^0_3 linear ordering, then $(\omega^* + \omega) \cdot \mathcal{A}$ has a computable copy.*

Proof using Theorem 13.2: We suppose that $\mathcal{A}$ has computable universe A, and we let B be another infinite computable set of constants, for the universe of $\mathcal{B}$. We produce a Δ^0_3 function F from B onto A, yielding an equivalence relation $\sim$. We also determine a computable ordering on B such that $\sim$ is a congruence relation on the resulting $\mathcal{B}$, and F induces an isomorphism from $\mathcal{B}/\sim$ onto $\mathcal{A}$.

We define a 3-system

$$(L, U, \hat{\ell}, P, E, \leq_0, \leq_1, \leq_2)$$

and a Δ^0_3 instruction function q. Let U, L and $\hat{\ell}$ be as in the previous proof. The other objects, E, $\leq_i$, P, are not the same. For $\ell = (p, u, v) \in L$, let $E(\ell)$ be the set of basic sentences true in the ordering v. We turn to the relations $\leq_k$. Suppose $\ell = (p, u, v)$ and $\ell' = (p', u', v')$ are elements of L.

$\leq_0$: We let $\ell \leq_0 \ell'$ if and only if $v \subseteq v'$ (i.e., the ordering is preserved).

$\leq_1$: We let $\ell \leq_1 \ell'$ if and only if
  (i) $\ell \leq_0 \ell'$,
  (ii) if $b \sim_p b'$, then $b \sim_{p'} b'$, and
  (iii) if b' is the successor of b in v, then the same is true in v'
(i.e., the ordering is preserved, equivalence is preserved from left to right, and within equivalence classes, successor is preserved from left to right).

$\leq_2$: We let $\ell \leq_2 \ell'$ if and only if $\ell \leq_1 \ell'$ & $p \subseteq p'$.

Now, we define the tree P. This consists of the finite sequences $\ell_0 u_1 \ell_1 u_2 \ell_2 \ldots$ such that $\ell_k \in L$, $u_k \in U$, and the following conditions hold:

(a) $\ell_0 = \hat{\ell}$,

(b) u_k is an ordering on the first k constants from A,

(c) ℓ_k has the form (p_k, u_k, v_k), with second component u_k, and v_k includes the first k elements of B,

(d) $\ell_k \leq_2 \ell_{k+1}$—this implies that $u_k \subseteq u_{k+1}$,

(e) each $\sim_{p_{k+1}}$-equivalence class in v_{k+1} has *new* elements (not in v_k) at the beginning and end.

From the definition of P, it is not difficult to see the following.

Lemma 13.3 *If $\pi = \hat{\ell} u_1 \ell_1 u_2 \ell_2 \ldots$ is a path through P, then $\mathcal{A}_\pi = \cup_n u_n$ is a linear ordering with universe A, and $E(\pi)$ is the atomic diagram of an ordering $\mathcal{B}$ isomorphic to $(\omega^* + \omega) \cdot \mathcal{A}_\pi$.*

We must verify the conditions for a 3-system. As in the first example, Conditions (1)-(3) are clear. Toward Condition (4), suppose $\sigma u \in P$, where σ has length $2n + 1$, with last term in ℓ^0, and let $\ell^0 \leq_2 \ell^1 \leq_1 \ell^2$. We must show that there is some ℓ^* such that $\sigma u \ell^* \in P$, $\ell^1 \leq_1 \ell^*$, and $\ell^2 \leq_0 \ell^*$.

13.3. SOME EXAMPLES

Say $\ell^i = (u_i, p_i, v_i)$. First, take $\ell'^1 = (u_1, p'_1, v_2)$, where $p_1 \subseteq p'_1$, and the extra elements from ℓ_2—those in $v_2 - v_1$—are added to the beginning or end of neighboring equivalence classes from ℓ^1, so that $\ell^2 \leq_0 \ell'^1$ and $\ell^1 \leq_2 \ell'^1$. Note that equivalence in ℓ^2 does not imply equivalence in ℓ'^1. Equivalence classes from ℓ^1 that have been collapsed in ℓ^2 will be re-instated in ℓ'^1.

Second, take $\ell'^0 = (u_0, p'_0, v_2)$, where $p_0 \subseteq p'_0$, and the elements of $v_2 - v_0$ are added to the beginning or end of neighboring equivalence classes, such that elements equivalent in ℓ'^1 remain equivalent in ℓ'^0, with ordering and successor both preserved. Then $\ell'^1 \leq_1 \ell'^0$ and $\ell^0 \leq_2 \ell'_0$. By the definition of P, u is an extension of u_0, with one new element a. Now, let ℓ^* be (p^*, u, v^*), where v^* is an extension of v_2 with a new element at the beginning and end of each equivalence class from ℓ'^0, and with a further new element b in a new equivalence corresponding to a, and p^* is an extension of p'_0 yielding the desired equivalence classes. It should be clear that this ℓ^* has the desired properties.

We have a 3-system. We define a Δ_3^0 instruction function q such that for sequences σ of length $2n+1$, $q(\sigma)$ is the restriction of $\mathcal{A}$ to the first $n+1$ constants. We are in a position to apply Theorem 13.2. We get a Δ_3^0 run

$$\pi = \hat{\ell} u_1 \ell_1 u_2 \ell_2 \ldots$$

of (P, q) such that $E(\pi)$ is c.e. Now, $\mathcal{A} = \cup_n u_n$, and by Lemma 13.3, $E(\pi)$ is the diagram of an ordering

$$\mathcal{B} \cong (\omega^* + \omega) \cdot \mathcal{A} .$$

Since $E(\pi)$ is c.e., $\mathcal{B}$ is computable. This completes the proof.

The next example is a variant of Watnik's Theorem, given as an exercise in Chapter 9.

Example 3 (Theorem 9.11) *If $\mathcal{A}$ is a Δ_3^0 linear ordering, then $\omega \cdot \mathcal{A}$ has a computable copy.*

Sketch of proof, using Theorem 13.2: There are two cases to consider, depending on whether $\mathcal{A}$ has a first element.

Case 1: Suppose that $\mathcal{A}$ has no first element.

We define a 3-system $(L, U, P, E, \leq_0, \leq_1, \leq_2)$ and a Δ_3^0 instruction function q like the one in the previous example except for the following changes.

(1) For $\ell \leq_2 \ell'$, where $\ell = (p, u, v)$ and $\ell' = (p', u', v')$, we require that if b is first in its equivalence class in ℓ, then it is first in its equivalence class in ℓ'.

(2) For a sequence $\sigma = \hat{\ell} u_1 \ell_1 u_2 \ell_2 \ldots$ in P, u_{n+1} is an ordering on a set that includes the first $n+1$ constants and has an element to the left of all in u_n, and if $\ell_n = (p_n, u_n, v_n)$, then each equivalence class from ℓ_n acquires new elements in ℓ_{n+1}, but only at the right end.

(3) Suppose $\sigma u \in P$, where σ has length $2n+1$, with last term in ℓ^0, where $\ell^0 \leq_2 \ell^1 \leq_1 \ell^2$. There exists ℓ'^1 such that $\ell^2 \leq_0 \ell'^1$ and $\ell^1 \leq_2 \ell'^1$. The extra

elements from ℓ^2 are added to the nearest equivalence class on the left in ℓ'^1, or else they go into a single new first equivalence class. There exists ℓ'^0 such that $\ell^0 \leq_2 \ell'^0$ and $\ell'^1 \leq_1 \ell'^0$. The equivalence classes in ℓ'^0 come from ℓ^0 except for a single new first class. The new elements of an equivalence class from ℓ'^1 all come at the end.

We have a 3-system. We define a Δ_3^0 instruction function q such that for all sequences $\sigma = \hat{\ell} u_1 \ell_1 \ldots u_n \ell_n$ in P, where u_n is a substructure of $\mathcal{A}$, $q(\sigma)$ is a finite substructure u_{n+1} of $\mathcal{A}$, including the first $k+1$ constants from A, plus the first constant that, in $\mathcal{A}$, lies to the left of all elements of u_n. The rest of the construction is as before. We obtain a computable copy of $\omega \cdot \mathcal{A}$.

Case 2: Suppose that $\mathcal{A}$ has a first element.

Then $\omega^* + \mathcal{A}$ does not have a first element. Let $\mathcal{C}$ be computable, where $\mathcal{C} \cong \omega^* + \mathcal{A}$. Then $\omega \cdot \mathcal{C}$ also has a computable copy $\mathcal{B}$. Say b is the first element of the copy of ω corresponding to the first element of $\mathcal{A}$. Taking the restriction of $\mathcal{B}$ to the set $\{d : \mathcal{B} \models b \leq d\}$, we have a computable copy of $\omega \cdot \mathcal{A}$.

13.4 Proof of the metatheorem

We shall prove Theorem 13.2 by induction, using two lemmas. The first lemma gives the theorem for 1-systems. We say that a sequence $\ell_0 u_1 \ell_1 u_2 \ldots$ *preserves* $\leq_0$ if $\ell_k \leq_0 \ell_{k+1}$ for all k (similarly, for $\leq_i$ replacing $\leq_0$).

Lemma 13.4 *Let $(L, U, \hat{\ell}, P, E, \leq_0)$ be a 1-system, and let q be a computable instruction function for P. Then there is a computable run π of (P, q) such that π preserves $\leq_0$. Moreover, from c.e. indices for L, U, P, and $\leq_0$ and a computable index for q, we can effectively determine a computable index for π and a c.e. index for $E(\pi)$.*

Proof: To obtain the computable run

$$\pi = \ell_0 u_1 \ell_1 \ldots ,$$

we alternately apply q to determine u_{s+1}, and enumerate P to locate ℓ_{s+1} such that $\hat{\ell} \ldots \ell_s u_s \ell_{s+1} \in P$. This procedure is uniform, so it is clear that from c.e. and computable indices for q and the components of the 1-system, we can determine a computable index for p and a c.e. index for $E(\pi)$.

The next lemma implies that if the theorem holds for n-systems, then it holds for $(n+1)$-systems.

Lemma 13.5 *Let $(L, U, \hat{\ell}, P, E, (\leq_m)_{m<n+1})$ be an $(n+1)$-system, and let q be a Δ_{n+1}^0 instruction function for P. There exist U^*, P^*, and q^* such that*

$$(L, U^*, \hat{\ell}, P^*, E, (\leq_m)_{m<n})$$

13.4. PROOF OF THE METATHEOREM

is an n-system, q^* is a Δ_n^0 instruction function for P^*, and for any Δ_n^0 run π^* of (P^*, q^*), there is a Δ_{n+1}^0 run π of (P, q) such that $E(\pi) = E(\pi^*)$. Moreover, given c.e. indices for L, U, P, and $\leq_m$ (for $m < n+1$), and a Δ_{n+1}^0 index for q, we can effectively determine c.e. indices for U^* and P^*, and a Δ_n^0 index for q^*. Also, given a Δ_n^0 index for π^*, we can effectively determine a Δ_{n+1}^0 index for π.

Proof: Let U^* consist of all finite sequences $c = \hat{\ell} u_1 \ell_1 \ldots \ell_k u_{k+1}$ in P, of even length, such that $\leq_n$ is preserved. Let $t(c)$ denote the last term of c that is in L; namely, ℓ_k. We define an alternating tree P^* on L and U^*, consisting of the finite sequences
$$\ell_0 c_1 \ell_1 c_2 \ell_2 \ldots ,$$
with $\ell_0 = \hat{\ell}$, such that the following hold:
  (a) $t(c_{i+1}) \leq_n \ell_i$,
  (b) $c_{i+1} \ell_{i+1} \in P$,
  (c) $t(c_{i+1}) \leq_n \ell_{i+1}$,
  (d) $\ell_i \leq_{n-1} \ell_{i+1}$.

Remark: All paths through P^* preserve $\leq_{n-1}$.

We claim that $(L, U^*, \hat{\ell}, P^*, E, (\leq_m)_{m<n})$ is an n-system. The first three conditions are clear. We must verify Condition (4). Suppose $\sigma c \in P^*$, where σ ends in ℓ^0, and
$$\ell^0 \leq_{k_0} \ell^1 \leq_{k_1} \ldots \leq_{k_{r-1}} \ell^r ,$$
for $n > k_0 > k_1 > \ldots > k_r$. We may suppose that $k_0 = n-1$. (If this is not the case initially, then we add an extra ℓ^0 at the front of the sequence, noting that by Condition (1) for the given $(n+1)$-system, $\ell^0 \leq_{n-1} \ell^0$.) Since $\sigma c \in P^*$, we have $t(c) \leq_n \ell^0$. Since the $(n+1)$-system satisfies Condition (4), there exists ℓ^* such that $c\ell^* \in P$, $t(c) \leq_n \ell^*$, and for all i, $\ell^i \leq_{k_i} \ell^*$. The three facts,
  (i) $\sigma \ell^* \in P$,
  (ii) $t(c) \leq \ell^*$, and
  (iii) $\ell^0 \leq_{n-1} \ell^*$
imply that $\sigma c \ell^* \in P^*$. Since $\ell^i \leq_{k_i} \ell^*$, for all i, this ℓ^* is the one we need to show that Condition (4) holds for the n-system.

We must define the Δ_n^0 instruction function q^* for P^*. We start with a Δ_n^0 guessing function g, approximating the given Δ_{n+1}^0 instruction function q for P. By the Limit Lemma, there is a partial Δ_n^0 function g such that for all $\sigma \in P$ with last term in L,
$$lim_{s \to \infty} g(\sigma, s) = q(\sigma) .$$

We want g with the additional features that $g(\sigma, s)$ is always defined, and if $g(\sigma, s) = u$, then $\sigma u \in P$. We can replace an arbitrary guessing function g by one with these features. (We delay using a value $u = g(\sigma, s)$ until we have seen σu in P, and if at stage s, there is no new $t \leq s$ such that $u = g(\sigma, t)$ satisfies this test, then we substitute an arbitrary u such that $\sigma u \in P$.)

We define $q^*(\rho)$ by induction on the length of ρ. Suppose ρ is an element of P^* of length $2s + 1$, and suppose that we have already defined q^* on the initial segments of ρ of length $2t + 1$, for $t < s$. If ρ is not a partial run of (P^*, q^*), then $q^*(\rho)$ is defined trivially to be the first c we find such that $\rho c \in P^*$. Now, suppose that ρ is a partial run of (P^*, q^*). We consider $\sigma \in P$, where

$$\sigma = \begin{cases} c\ell & \text{if } s > 0 \text{ and } c\ell \text{ consists of the last two terms of } \rho \\ \hat{\ell} & \text{if } s = 0 \end{cases}$$

If σ appears to be a partial run of (P, q), in that for all proper initial segments $\sigma' u'$ with $u' \in U$, we have $u' = g(\sigma', s+1)$, then we let $q^*(\rho) = \sigma u$, where $u = g(\sigma, s+1)$. Since $\sigma u \in P$ and $t(c) \leq_n \ell$, we have $\sigma u \in U^*$. If σ does not appear to be a partial run of (P, q), then we take the least initial segment $\sigma' u'$ witnessing this, where $g(\sigma', s+1) = u \neq u'$, and we let $q^*(\rho) = \sigma' u$. Again $\sigma u \in U^*$. This completes the definition of the Δ_n^0 instruction function q^*.

For a sequence $\pi = \ell_0 u_1 \ell_1 \ldots$ (finite or infinite), with alternate terms in L, let $L(\pi)$ denote the subsequence consisting of these terms. Suppose

$$\pi^* = \ell_0 c_1 \ell_1 c_2 \ell_2 \ldots$$

is a Δ_n^0 run of (P^*, q^*), and let

$$c_s = \sigma_s u_s .$$

We shall determine a Δ_{n+1}^0 run π of (P, q) such that $L(\pi)$ is a subsequence of $L(\pi^*)$. Since π^* preserves $\leq_{n-1}$, it preserves $\leq_0$, by Condition (2). Then by Condition (3), $E(\pi) = E(\pi^*)$.

Claim: There is a Δ_{n+1}^0 sequence of numbers $s(k)$ such that $\sigma_{s(k)}$ has length $2k + 1$ and for all $s > s(k)$, $\sigma_s \supseteq \sigma_{s(k)}$.

Proof of claim: Since $\sigma_1 = \ell_0$, we can take $s(0)$ to be 1. Supposing that we have determined $s(k)$, where $q(\sigma_{s(k)}) = u$, let $s' \geq s(k)$ be first such that for all $s \geq s'$, $g(\sigma_{s(k)}, s) = u$. Then $c_{s'} = \sigma_{s(k)} u$, and $\sigma_{s'+1} = \sigma_{s(k)} u \ell_{s'}$, where this has length $2k + 3$. Moreover, for $s > s' + 1$, we have $\sigma_s \supseteq \sigma_{s'+1}$. Therefore, we can take $s(k+1)$ to be $s' + 1$. This proves the claim.

Let π be the Δ_{n+1}^0 run of (P, q) having $\sigma_{s(k)}$ as the initial segment of length $2k + 1$. Clearly, $L(\pi)$ is a subsequence of $L(\pi^*)$—it is made up of the terms $\ell_{s(k+1)-1}$. Therefore, $E(\pi) = E(\pi^*)$, as required. This completes the proof of Lemma 13.5.

We are now in a position to complete the proof of Theorem 13.2, as planned. We show by induction on n that for an n-system

$$(L, U, \hat{\ell}, P, E, (\leq_m)_{m<n})$$

and a Δ_n^0 instruction function q for P, there exists a Δ_n^0 run π such that $E(\pi)$ is c.e. Moreover, from c.e. and computable indices for the components of the

13.5. LOOKING AHEAD

n-system and a Δ_n^0 index for q, we can effectively determine a Δ_n^0 index for π and a c.e. index for $E(\pi)$.

For $n = 1$, the statement follows from Lemma 13.4. Suppose the statement holds for n. Let

$$(L, U, \hat{\ell}, P, E, (\leq_m)_{m<n+1})$$

be an $(n+1)$-system, and let q be a Δ_{n+1}^0 instruction function for P. Applying Lemma 13.5, we derive an n-system

$$(L, U^*, \hat{\ell}, P*, E, (\leq_m)_{m<n})$$

and a Δ_n^0 instruction function q^*. By H.I., there is a Δ_n^0 run π^* of (P^*, q^*) such that $E(\pi^*)$ is c.e. The properties of P^* and q^* (in Lemma 13.5) guarantee that from π^*, we can derive a Δ_{n+1}^0 run π of (P, q) such that $E(\pi) = E(\pi^*)$. Therefore, $E(\pi)$ is c.e. The uniformity in Lemma 13.5 lets us pass effectively from the indices for the $(n+1)$-system and q to the indices for the n-system and q^*, and from the index for the run π^* of the n-system to the index for the run π of the $(n+1)$-system. The uniformity in the induction hypothesis lets us pass from the indices for the n-system and q^* to the indices for π^* and the c.e. index for $E(\pi^*)$.

13.5 Looking ahead

In the next chapter, we shall extend Theorem 13.2 to transfinite levels. To prepare for the proof of the more general metatheorem, we take another look at some parts of the proof that we have just given. Examining the proof of Lemma 13.5, we see that for any $m < n$, the even terms at level m (those not in L) are finite sequences of terms of the kind that appear at level $m + 1$. Thus, for a construction using a 4-system, an even term at the Δ_3^0 level is a sequence of sequences terms of the kind in the desired run, an even term at the Δ_2^0 is a sequence of sequences of terms of the kind in the desired run, and an even term at the Δ_1^0 level is a sequence of sequences of sequences of the kind in the desired run.

When we extend Theorem 13.2 to transfinite levels, we shall replace these nested sequences by simpler objects, called "pictures", which are to be "completed" by an application of Condition (4). Below, we indicate how the pictures can be extracted from the nested sequences, in the present context, where there are only finitely many levels all together.

Given an n-system

$$(L, U, \hat{\ell}, P, E, (\leq_m)_{m<n})$$

and a Δ_n^0 instruction function q, we obtain (from the proof of Theorem 13.2) a family of m-systems

$$(L, U_m, \hat{\ell}, P^m, E, (\leq_j)_{j<m})$$

with corresponding instruction functions q_m such that

(i) q_m is Δ_m^0,

(ii) a run of (P^n, q_n) is essentially a run of (P, q),

(iii) for any Δ_1^0 run π^1 of (P^1, q_1), there exist π^m, for $1 < m \leq n$, such that π^m is a Δ_m^0 run of (P^m, q_m), and $L(\pi^{m+1})$ is a subsequence of $L(\pi^m)$,

(iv) any run of (P^m, q_m) preserves $\leq_{m-1}$.

Take a family of runs π^m as in (iii), where π^m is Δ_m^0 and $L(\pi^{m+1})$ is a subsequence of $L(\pi^m)$. Then $E(\pi^m)$ is the same for all m, $E(\pi^1)$ is c.e., and π^n is the desired run of (P, q). Say that

$$\pi^m = \hat{\ell} c_1^m \ell_1^m c_2^m \ell_2^m \ldots .$$

For each m such that $1 \leq m < n$, there is a natural embedding of $L(\pi^{m+1})$ into $L(\pi^m)$. Let us trace a particular term ℓ in $L(\pi^n)$, following this chain of embeddings. Suppose that at level m, ℓ occurs as $\ell_{k_{m+1}}^m$, immediately following the sequence

$$\sigma_{k_m}^m = \hat{\ell} c_1^m \ell_1^m \ldots \ell_{k_m}^m c_{k_m+1}^m .$$

We have $\sigma_{k_n}^n \in P$, and for $1 \leq m < n$, $c_{k_m+1}^m = \sigma_{k_m}^{m+1}$. By the definition of P^m, $\ell_{k_m+1}^{m+1} \leq_m \ell_{k_m}^m$. At level 1, we have $\sigma_{k_1}^1 \ell \in P^1$. By the definition of P^1, $\ell_{k_1}^1 \leq_0 \ell$, and $\sigma_{k_2}^2 \ell \in P^2$. This means that $\ell_{k_2}^2 \leq_1 \ell$ and $\sigma_{k_3}^3 \ell \in P^3$, and so on.

In choosing ℓ, we use only a small part of the information in the sequences $\sigma_{k_m}^m$, for $1 \leq m \leq n$. It is enough to have the top sequence $\sigma_{k_n}^n$, which is an element of P, and the last terms of the sequences at lower levels, together with the levels themselves—recorded as $(m, \ell_{k_m}^m)$ for $1 \leq m < n$. Together, $\sigma_{k_n}^n$ and the sequence of pairs $(m, \ell_{k_m}^m)$, for $1 \leq m < n$, form a picture satisfying the hypotheses of Condition (4). The term ℓ is chosen to complete the picture; that is, it satisfies the conclusion of Condition (4).

13.6 Michalski's Theorem

In Chapter 14, we shall lift Theorem 13.2 to arbitrary levels in the hyperarithmetical hierarchy. There are a number of applications of Theorem 13.2 and this lifting. Some applications are very direct, while others require some cleverness, or added tricks. Sometimes, it may appear that the theorem does not apply, but it can be used if we do something special on the top, or bottom level. There are priority constructions for which Theorem 13.2 (and the lifting) do not apply. There are further metatheorems, of the same general flavor, which have been developed to handle certain of these constructions. We shall mention some of these later.

Theorem 13.2 does not handle the original priority construction—the one due to Friedberg and Muchnik. Below, we give a metatheorem due to Michalski [118], that appears to handle all finite-injury constructions. We first describe the setting. Let P be a c.e. tree, consisting of non-empty finite sequences from a c.e. set L, all starting with a fixed $\hat{\ell}$. This tree may have terminal nodes. Let E be a partial computable enumeration function on L. The object of the construction is to produce a path

$$\pi = \hat{\ell} \ell_1 \ell_2 \ldots$$

13.6. MICHALSKI'S THEOREM

such that $E(\pi) = \cup_n E(\ell_n)$ is c.e. The conditions involve an additional binary relation I on L.

Theorem 13.6 (Michalski) *Let L be a c.e. set, and let P be a c.e. tree of non-empty finite sequences from L, all starting with the same $\hat{\ell}$. Let E be a partial computable enumeration function on L, and let I be a c.e. binary relation on P satisfying the following conditions:*

(1) if $\ell_0 \ell_1 \ell_2 \ldots \ell_n \in P$, then for all $k < n$, $E(\ell_k) \subseteq E(\ell_{k+1})$,

(2) for $\sigma = \ell_0 \ell_1 \ell_2 \ldots \ell_n \in P$, either

    (a) for some ℓ_{n+1}, $\sigma \ell_{n+1} \in P$ (i.e., σ has a successor), or

    (b) for some $k < r$, the sequence $\rho = \ell_0 \ell_1 \ell_2 \ldots \ell_k$ has a successor $\rho \ell'_{k+1}$ such that $E(\ell_n) \subseteq E(\ell'_{k+1})$ and $I(\rho \ell_{k+1}, \rho \ell'_{k+1})$.

(3) for $\sigma \in P$, there is no computable sequence of successors $\tau_n = \sigma \ell_n$, such that $E(\ell_n) \subseteq E(\ell_{n+1})$ and $I(\tau_n, \tau_{n+1})$, for all n.

Then P has a path π such that $E(\pi)$ is c.e.

Proof: We extend E to P, letting $E(\sigma) = E(\ell)$, where ℓ is the last term of σ. Then we form an infinite computable sequence $(\sigma_n)_{n \in \omega}$, where $\sigma_0 = \hat{\ell}$, and σ_{n+1} is obtained from σ_n as in (2). Note that E is always preserved; i.e., $E(\sigma_n) \subseteq E(\sigma_{n+1})$ whether we are in Case (a) or Case (b). For each n, there exists $k(n)$ such that $\sigma_{k(n)}$ has length $n+1$, and for all $k > k(n)$, $\sigma_k \supseteq \sigma_k(n)$. Otherwise, we would have a subsequence of the sequence $(\sigma_n)_{n \in \omega}$ contradicting (3). Say $\sigma_{k(n)} = \hat{\ell} \ell_1 \ell_2 \ldots \ell_n$. Then $\pi = \hat{\ell} \ell_1 \ell_2 \ldots$ is a path through P. Since

$$\cup_n E(\sigma_n) = \cup_n E(\sigma_{k(n)}),$$

$E(\pi)$ is c.e. This completes the proof of Theorem 13.6.

Below we show how to use Michalski's Theorem to prove the Friedberg-Muchnik Theorem.

Theorem 13.7 (Friedberg-Muchnik) *There exists c.e. sets A and B such that $A \not\leq_T B$ and $B \not\leq_T A$.*

Proof: We have the following requirements.

R_{2e}: $\varphi_e^A \neq \chi_B$
R_{2e+1}: $\varphi_e^B \neq \chi_A$

We must define L, $\hat{\ell}$, P, E, and I. Let L be the set of pairs (p, q), where $p, q \in 2^{<\omega}$. Think of p, q as possible initial segments of χ_A, χ_B, respectively. For $\ell = (p, q)$ and $\ell' = (p', q')$ in L, we write $\ell \subseteq \ell'$ if $p \subseteq p'$ and $q \subseteq q'$. Let $\hat{\ell} = (\emptyset, \emptyset)$. Let E be the function on L such that if $\ell = (p, q)$, $E(\ell)$ be the set of statements $x \in A$, for $p(x) = 1$, and $x \in B$, for $q(x) = 1$.

The tree P consists of the finite sequences $\ell_0 \ell_1 \ell_2 \ldots \ell_r$ in L, such that for $\ell_n = (p_n, q_n)$, the following conditions hold:

(1) $\ell_0 = \hat{\ell}$,

(2) $\ell_n \subseteq \ell_{n+1}$,

(3) for each $n < r$, ℓ_{n+1} "satisfies R_n through step r", where this means the following: For $n = 2e$, if k is the first number not in $dom(q_n)$, then

$$q_{n+1}(k) = \begin{cases} 1 & \text{if } \varphi_{e,n+1}^{p_{n+1}}(k) \downarrow = 0 \\ 0 & \text{otherwise} \end{cases}$$

and if $\varphi_{e,n+1}^{p_{n+1}}(k) \uparrow$, then $\varphi_{e,r}^{p_r}(k) \uparrow$. For $n = 2e + 1$, the definition is the same except that the p's and q's switch roles.

Finally, we define the binary relation I on P. Let $\sigma \in P$, where

$$\sigma = ((p_0, q_0), \ldots, (p_n, q_n)) ,$$

and let τ and τ' be a pair of successors of σ, where $\tau = \sigma(p_{n+1}, q_{n+1})$ and $\tau' = \sigma(p'_{n+1}, q'_{n+1})$. The pair (τ, τ') is in I if and only if

(i) n is even, say $n = 2e$, and for the first $k \notin dom(q_n)$, $q_{n+1}(k) = 0$ while $q'_{n+1}(k) = 1$,

(ii) n is odd, say $n = 2e + 1$, and for the first $k \notin dom(p_n)$, $p_{n+1}(k) = 0$ while $p'_{n+1}(k) = 1$.

It is not difficult to verify the conditions of Michalski's Theorem. In particular, for $\sigma \in P$, there are at most two successors as in (3)—there is certainly no infinite sequence. The theorem yields a path π through P such that $E(\pi)$ is c.e. We let A be the set of x such that the statement $x \in A$ is in $E(\pi)$, and we let B be the set of x such that the statement $x \in B$ is in $E(\pi)$. These are the desired independent c.e. sets.

Chapter 14

α-systems

In Chapter 13, we gave an abstract formulation of the object of a Δ^0_n priority construction, and we proved a metatheorem with conditions guaranteeing the success of the construction. In this chapter, we extend the metatheorem to transfinite levels.

14.1 Statement of the metatheorem

Most of the setting is as in Chapter 13. The high-level information is now given by an instruction function that is Δ^0_α instead of Δ^0_n. We define the the notion of α-system, extending that of n-system. We fix a path in Kleene's O with a notation for α, and we identify ordinals $\beta \leq \alpha$ with their unique notations along this path. Let L and U be c.e. sets, let E be a partial computable enumeration function on L, and let P be a c.e. alternating tree on L and U, made up of nonempty finite sequences, all starting with the same $\hat{\ell} \in L$. Let $\leq_\beta$, for $\beta < \alpha$, be binary relations on L, c.e. uniformly in β—this means that given β (or its notation), we can compute a c.e. index for $\leq_\beta$.

The structure
$$(L, U, \hat{\ell}, P, E, (\leq_\beta)_{\beta<\alpha})$$
is an α-system if the following conditions are satisfied:
 (1) $\leq_\beta$ is reflexive and transitive, for $\beta < \alpha$,
 (2) $\ell \leq_\gamma \ell' \Rightarrow \ell \leq_\beta \ell'$, for $\beta < \gamma < \alpha$,
 (3) if $\ell \leq_0 \ell'$, then $E(\ell) \subseteq E(\ell')$,
 (4) if $\sigma u \in P$, where σ ends in $\ell^0 \in L$, and
$$\ell^0 \leq_{\beta_0} \ell^1 \leq_{\beta_1} \ldots \leq_{\beta_{k-1}} \ell^k ,$$
for $\alpha > \beta_0 > \beta_1 > \ldots > \beta_k$, then there exists ℓ^* such that $\sigma u \ell^* \in P$, and $\ell^i \leq_{\beta_i} \ell^*$, for all $i \leq k$.

Here is the statement of the metatheorem.

Theorem 14.1 Let $(L, U, \hat{\ell}, P, E, (\leq_\beta)_{\beta<\alpha})$ be an α-system. Then for any Δ^0_α instruction function q, there is a run π of (P, q) such that $E(\pi)$ is c.e., while π itself is Δ^0_α. Moreover, from a Δ^0_α index for q, together with a computable sequence of c.e. and computable indices for the components of the α-system, we can effectively determine a Δ^0_α index for π and a c.e. index for $E(\pi)$.

14.2 Organizing the construction

To prove Theorem 14.1, we derive, from the given α-system

$$(L, U, \hat{\ell}, P, E, (\leq_\gamma)_{\gamma<\beta})$$

and Δ^0_α instruction function q, a family of β-systems

$$(L, U_\beta, \hat{\ell}, P^\beta, E, (\leq_\gamma)_{\gamma<\beta})$$

and corresponding instruction functions q_β, for $1 \leq \beta \leq \alpha$, where U_β and P^β are c.e., uniformly in β, and q_β is Δ^0_β, also uniformly in β. An element of U_β represents a situation for applying Condition (4), in which the ordinals that appear are all at least β. We define P^α and q_α to be essentially the same as P and q, and we define P^1 so that any path preserves $\leq_0$. For a sequence σ with some terms in L, $L(\sigma)$ denotes the subsequence consisting of those terms. We aim at producing a family of runs π^β of (P^β, q_β), for $1 \leq \beta \leq \alpha$, such that
  (a) π^β is Δ^0_β, uniformly in β, and
  (b) for $\beta < \gamma$, $L(\pi^\gamma)$ is a subsequence of $L(\pi^\beta)$.

Claim: This is enough to prove the theorem.

Proof of Claim: By (b) and the fact that π^1 preserves $\leq_0$, $E(\pi^1) = E(\pi^\alpha)$. By (a), π^1 is Δ^0_1. Therefore, $E(\pi^1)$ (or $E(\pi^\alpha)$) is c.e. Also by (a), π^α is Δ^0_α. Then π^α provides the desired run of (P, q).

Dealing with limit ordinals leads to some difficulties that we did not encounter in Chapter 13. First, let $\alpha = \omega$. For a Δ^0_ω instruction function q, any one value is computed with only finitely many questions to the oracle for Δ^0_ω. These questions have the form "Is $n \in \Delta^0_k$?", for various finite k. For the greatest such k, we can get the answers to all of these questions using Δ^0_k. Given an ω-system

$$(L, U, \hat{\ell}, P, E, (\leq_k)_{k<\omega}),$$

and a Δ^0_ω instruction function q, we can derive a family of n-systems

$$(L, U_n, \hat{\ell}, P^n, E, (\leq_k)_{k<n})$$

and instruction functions q_n, where U_n and P^n are c.e. uniformly in n, and q_n is Δ^0_n, also uniformly in n, such that the first few steps in a Δ^0_n run of (P^n, q_n) are essentially the first few steps in a run of (P, q)—where the oracle Δ^0_n suffices

14.2. ORGANIZING THE CONSTRUCTION

for computing the values of q—and the remaining steps approximate those in a run of (P^{n+1}, q_{n+1}).

Next, let $\alpha = \omega^2$. Given an ω^2-system

$$(L, U, \hat{\ell}, P, E, (\leq_\beta)_{\beta<\omega^2})$$

and a Δ^0_α instruction function q, we can derive a family of β-systems

$$(L, U_\beta, \hat{\ell}, P^\beta, E, (\leq_\gamma)_{\gamma<\beta})$$

and instruction functions q_β, for $1 \leq \beta \leq \omega^2$, where U_β and P^β are c.e., uniformly in β, and q_β is Δ^0_β, also uniformly in β. If $\omega \cdot n \leq \beta < \omega \cdot (n+1)$, then the first few steps in a run of (P^β, q_β) are essentially those in a run of (P, q), the next few match those in a run of $(P^{\omega \cdot (n+1)}, q_{\omega \cdot (n+1)})$, and the remaining steps approximate those in a run of $(P^{\beta+1}, q_{\beta+1})$.

For larger α, we must find our way through thickets of limit ordinals—limits of limits, limits of limits of limits, etc. We need an organization scheme so that when we define q_β, we can say exactly when q_β should match q_γ for some limit ordinal $\gamma > \beta$, and when it should approximate $q_{\beta+1}$. The next lemma is what we need to develop this organization scheme. First, we fix a computable list of the pairs (β, n), for $\beta \leq \alpha$ and $n \in \omega$, with the feature that (β, n) appears before $(\beta, n+1)$. We write

$$(\gamma, m) \ll (\beta, n)$$

if (γ, m) appears before (β, n) on the list.

Lemma 14.2 *There is a partial computable function Lim, defined on the ordinals β with $1 \leq \beta < \alpha$, such that $Lim(\beta)$ is a set of pairs (γ, k), where γ is a limit ordinal with $\alpha \geq \gamma > \beta$ and $k \in \omega$, such that the following conditions hold:*
 (a) if $(\gamma, k) \in Lim(\beta)$, then $(\gamma, k) \ll (\beta, 0)$,
 (b) for any limit ordinal γ and any n, $(\gamma, n) \in Lim(\beta)$ for all sufficiently large $\beta < \gamma$,
 (c) if $(\gamma, m) \in Lim(\beta)$ and $k < m$, then $(\gamma, k) \in Lim(\beta)$,
 (d) if $(\gamma, k) \in Lim(\beta)$ and $\beta < \delta < \gamma$, then $(\gamma, k) \in Lim(\delta)$.

Before proving the lemma, we indicate briefly how it yields the organization scheme we need. We shall define the instruction function q_β so that the first n steps in a run of (P^β, q_β) match the first n steps in a run of (P^γ, q_γ) for the pairs (γ, k) in $Lim(\beta)$, so long as the oracle Δ^0_β can compute the necessary values of the instruction function q_γ. Properties (b), (c), and (d) guarantee that for all sufficiently large $\beta < \gamma$, there will be a match; i.e., Δ^0_β will be able to compute q_γ. Property (a) guarantees that $Lim(\beta)$ is finite.

Proof of Lemma 14.2: We define a partial computable function g on the pairs (β, n), for $1 < \beta \leq \alpha$ and $n \in \omega$. If $\beta = \gamma + 1$, then $g(\beta, n) = \gamma$ for all $n \in \omega$. If β is a limit ordinal, then our notation for β yields an increasing sequence $(\beta_k)_{k \in \omega}$ with limit β. The values of $g(\beta, n)$ will form a subsequence.

We let $g(\beta, n) = \beta_k$ for the first k such that for all $(\gamma, m) \ll (\beta, n)$, the following hold:
(i) $\beta_k > \gamma$, if $\gamma < \beta$, and
(ii) $\beta_k > g(\gamma, m)$, if $g(\gamma, m) < \beta$.

Note that for any limit ordinal $\beta \leq \alpha$, $g(\beta, n)$ increases with n and has limit β.

Now, for each β such that $1 \leq \beta < \alpha$, we let

$$Lim(\beta) = \{(\gamma, m) : \gamma > \beta \,\&\, \gamma \text{ is a limit ordinal} \,\&\, g(\gamma, m) \leq \beta\} .$$

It is not difficult to verify the four properties in the statement of Lemma 14.2. We have (a) since if $(\beta, 0) \ll (\gamma, m)$, then $g(\gamma, m) > \beta$. We have (b) and (d) since $g(\gamma, m) < \gamma$ and

$$(\gamma, m) \in Lim(\beta) \Leftrightarrow g(\gamma, m) \leq \beta < \gamma .$$

We have (c) since if $k < m$ and $g(\gamma, m) \leq \beta$, then $g(\gamma, k) \leq \beta$. This completes the proof of the lemma.

14.3 Derived systems

We are ready to begin the process of deriving from the given α-system

$$(L, U, \hat{\ell}, P, E, (\leq_\gamma)_{\gamma < \alpha})$$

and Δ^0_α instruction function q, the family of β-systems

$$(L, U_\beta, \hat{\ell}, P^\beta, E, (\leq_\gamma)_{\gamma < \beta})$$

and corresponding Δ^0_β instruction functions q_β. As we said above, P^α and q_α will be essentially the same as P and q. For $1 \leq \beta < \alpha$, the definition of q^β will have "limit" steps and "approximation" steps, plus some steps of "invention", or marking time.

The sets U_β will all be subsets of a c.e. set C, where the elements of C are the pairs $c = (\sigma u; \tau)$ such that $\sigma u \in P$, σ ends in some ℓ^0 in L, and τ is a sequence $\beta_0 \ell^1 \ldots \beta_{k-1} \ell^k$, where

$$\ell^0 \leq_{\beta_0} \ell^1 \leq_{\beta_1} \ldots \leq_{\beta_{k-1}} \ell^k ,$$

for $\alpha > \beta_0 > \beta_1 > \ldots > \beta_{k-1} > 0$. We allow $\tau = \emptyset$. We refer to the elements of C as *pictures*, and the ℓ^* that we get by applying Condition (4) is said to *complete* the picture.

For a picture $c = (\sigma u; \tau)$, let $f(c) = \sigma u$, and let $s(c) = \tau$. (Here "f" stands for "first" and "s" stands for "second".) Let $\ell(c)$ denote the last element of L appearing in c. If $c = (\sigma u; \tau)$, then

$$\ell(c) = \begin{cases} \text{last term of } \tau & \text{if } \tau \neq \emptyset \\ \text{last term of } \sigma & \text{otherwise} . \end{cases}$$

14.3. DERIVED SYSTEMS

We let U^β be the set of elements of C in which the ordinals are all at least β. Note that for $c \in U_\alpha$, $s(c) = \emptyset$.

For $1 < \beta \leq \alpha$, we let P^β be the alternating tree on L and U^β consisting of the finite sequences

$$\sigma = \ell_0 c_1 \ell_1 c_2 \ell_2 \ldots$$

such that
  (i) $\ell_0 = \hat{\ell}$, and
  (ii) for all $n \geq 1$, $\ell(c_n) = \ell_{n-1}$, and ℓ_n completes the picture c_n.
For the bottom tree, P^1, we require in addition preservation of $\leq_0$. Note that for each element $\hat{\ell} c_1 \ell_1 \ldots$ of the top tree P^α, there is a corresponding element of P; namely, $\hat{\ell} u_1 \ell_1 \ldots$, where $c_k = \hat{\ell} u_1 \ell_1 \ldots u_k$.

We have now defined all of the new components of the β-systems. We must define the corresponding instruction functions q_β. We begin on the top. For $\sigma \in P^\alpha$ of odd length, we let $q_\alpha(\sigma) = (\sigma^* u; \emptyset)$, where σ^* is the element of P corresponding to σ. Clearly, if π^* is a run of (P^α, q_α), then there is a run π of (P, q) such that $\pi \leq_T \pi^*$ and $L(\pi) = L(\pi^*)$. For $1 \leq \beta < \alpha$, we shall define q_β by a somewhat complicated induction. Let P_n^β be the set of $\sigma \in P^\beta$ of length $2n + 1$. We define the restrictions $q_{\beta,n}$ of q_β to P_n^β by induction on the list of pairs. That is, we say how to compute $q_{\beta,n}$, using a Δ_β^0 oracle, and given Δ_γ^0 indices for $q_{\gamma,k}$, for the pairs $(\gamma, k) \ll (\beta, n)$. From the description, it will be clear that we can effectively compute a Δ_β^0 index for $q_{\beta,n}$.

Along with $q_{\beta,n}(\sigma)$, we define an auxilliary object σ^+. While σ^+ will depend on β and n as well as σ, we do not complicate the notation to indicate this. We say what σ^+ represents. Recall that our aim is to define q^β so that for any run π of (P^β, q_β), there is a run π^+ of $(P^{\beta+1}, q_{\beta+1})$ such that $L(\pi^+)$ is a subsequence of $L(\pi)$. Assuming that $\sigma \in P_n^\beta$ is a partial run of (P^β, q_β), σ^+ is our stage n approximation to the initial segment of the corresponding run of $(P^{\beta+1}, q_{\beta+1})$ such that $L(\sigma^+)$ is a subsequence of $L(\sigma)$.

We fix some approximation procedures. Given a $\Delta_{\beta+1}^0$ index for $q_{\beta+1,n}$, we have a Δ_β^0 approximation to $q_{\beta+1,n}$, whose value at σ and s is denoted by $q_{\beta+1,n}^s(\sigma)$. We may suppose that if $\sigma \in P_n^{\beta+1}$, then for all s, $q_{\beta+1,n}^s(\sigma)$ has some value c such that $\sigma c \in P^{\beta+1}$. Suppose γ is a limit ordinal such that $1 \leq \beta < \gamma$. For $\sigma \in P_n^\gamma$, if the Δ_β^0 oracle yields the information used in the halting computation of $q_{\gamma,k}$ on the initial segment of σ of length $2k + 1$, for all $k \leq n$, then we say that Δ_β^0 can compute $q_{\gamma,n}(\sigma)$. Note that we can determine, using a Δ_β^0 oracle, whether Δ_β^0 can compute $q_{\gamma,n}(\sigma)$.

Before giving the case-by-case definitions of $q_{\beta,n}(\sigma)$ and σ^+, we indicate, for the various kinds of steps, how $q_{\beta,n}(\sigma)$ is computed in terms of σ^+—we have already said what σ^+ represents. This overview is intended to make the clauses in the actual case-by-case definitions as predictable as possible.

Overview

First, if σ does not follow q_β, then $q_{\beta,n}(\sigma)$ will be defined trivially. Now, suppose that $\sigma \in P_n^\beta$ is a partial run of (P^β, q_β).

I. Limit steps

If there exists $(\gamma, n) \in Lim(\beta)$ such that Δ_β^0 can compute $q_{\gamma,n}(\sigma)$, then we let $q_{\beta,n}(\sigma) = q_{\gamma,n}(\sigma)$, for the first such (γ, n) (under $\ll$).

Now, suppose that there does not exist (γ, n) as required for a limit step. Let σ^+ have length $2m + 1$. Then $q_{\beta,n}(\sigma)$ will have the form

$$(f(d); s(d)\beta\ell(\sigma)),$$

where $\sigma^+ d \in P^{\beta+1}$ and $\ell(\sigma^+) \leq_\beta \ell(\sigma)$. We must specify d.

II. Approximation steps

If $(\beta + 1, m) \ll (\beta, n)$, then we let $q_{\beta,n}(\sigma) = (f(d); s(d)\beta\ell(\sigma))$, where $d = q_{\beta+1,m}^n(\sigma^+)$.

III. Invention steps

If $(\beta, n) \ll (\beta + 1, m)$, then we let $q_{\beta,n}(\sigma) = (f(d); s(d)\beta\ell(\sigma))$, where d is just the first we happen to find such that $\sigma^+ d \in P^{\beta+1}$.

It is not difficult to see (using Lemma 14.2) that if we hold to this scheme in defining q_β, then in a run of (P^β, q_β), the limit steps come first, and there are only finitely many such steps. If γ is a limit ordinal, and $\sigma \in P_n^\gamma$, where σ follows q_γ, then for all $\beta < \gamma$ such that $(\gamma, n) \in Lim(\beta)$ and Δ_β^0 can compute $q_{\gamma,n}(\sigma)$ (i.e., for all sufficiently large $\beta < \gamma$), it will be the case that $\sigma \in P^\beta$, σ follows q_β, and $q_{\beta,n}(\sigma) = q_{\gamma,n}(\sigma)$.

Above, we said that for a limit step, $q_{\beta,n}(\sigma) = q_{\gamma,n}(\sigma)$, for the *first* pair of form $(\gamma, n) \in Lim(\beta)$ such that Δ_β^0 can compute $q_{\gamma,n}(\sigma)$. Actually, we need not have said this. If there is another such pair (γ', n), where $(\gamma, n) \ll (\gamma', n)$, then
 (i) $\beta < \gamma' < \gamma$,
 (ii) $(\gamma, n) \in Lim(\gamma')$, and
 (iii) $\Delta_{\gamma'}^0$ can compute $q_{\gamma,n}(\sigma)$.
Then by the comments in the preceding paragraph, $\sigma \in P^{\gamma'}$ and $q_{\gamma',n}(\sigma) = q_\gamma(\sigma)$.

In defining $q_\beta(\sigma)$, for σ following q_β, we maintain the following three conditions:
 (1) $\ell(\sigma^+) \leq_\beta \ell(\sigma)$,
 (2) if σ is the result of limit steps only, then $\sigma^+ = \sigma$,
 (3) if ρ is the maximum initial segment of σ that results from limit steps, then σ^+ preserves $\leq_\beta$ from $\ell(\rho^+)$ on.

Case-by-case definitions

Here are the case-by case definitions of $q_{\beta,n}(\sigma)$ and σ^+, for $\sigma \in P_n^\beta$.

14.3. DERIVED SYSTEMS

Case 1: Suppose $n = 0$.

The only element of P_0^β is $\hat{\ell}$. Let $\hat{\ell}^+ = \hat{\ell}$. First, suppose that there exists $(\gamma, 0) \in Lim(\beta)$ such that Δ_β^0 can compute $q_{\gamma,0}(\hat{\ell})$. Then

$$q_{\beta,0}(\hat{\ell}) = q_{\gamma,0}(\hat{\ell}) .$$

This is a limit step. Now, suppose there is no $(\gamma, 0)$ as required for a limit step. If $(\beta + 1, 0) \ll (\beta, 0)$ and $q_{\beta+1,0}^0(\hat{\ell}) = d$, then

$$q_{\beta,0}(\hat{\ell}) = (f(d); s(d)\beta\hat{\ell}) .$$

This is an approximation step. If $(\beta, 0) \ll (\beta + 1, 0)$, then for the first d we find such that $\hat{\ell}d \in P^{\beta+1}$,

$$q_{\beta,0}(\hat{\ell}) = (f(d); s(d)\beta\hat{\ell}) .$$

This is an invention step.

Case 2: Suppose $n > 0$.

Let $\sigma \in P^{\beta_n}$, where $\sigma = \hat{\ell}c_1\ell_1 \ldots c_n\ell_n$, and for $m \leq n$, let $\sigma_k = \hat{\ell}c_1\ell_1 \ldots c_n\ell_k$. We first check whether σ follows q_β so far. If not, then $\sigma^+ = \sigma_{n-1}^+$ and $q_{\beta,n}(\sigma)$ is the first c we find such that $\sigma c \in P^\beta$. Suppose σ follows q_β. We define a tentative version of σ^+, denoted by ρ.

(i) If c_n represents a limit step, then $\rho = \sigma$.

(ii) If c_n represents an approximation step, then $\rho = \sigma_{n-1}^+ d\ell_n$, where if σ_{n-1}^+ has length $2r + 1$, then $d = q_{\beta+1,r}^{n-1}(\sigma_{n-1}^+)$.

(iii) If c_n represents an invention step, then $\rho = \sigma_{n-1}^+$.

Say that $\sigma_T^+ = \hat{\ell}d_1\ell_{i_1} \ldots d_m\ell_{i_m}$, where $1 \leq i_1 < \ldots < i_m \leq n$. We write ρ_t for the initial segment of ρ of length $2t + 1$; namely,

$$\rho_t = \hat{\ell}d_1\ell_{i_1} \ldots d_t\ell_{i_t} .$$

First, suppose that there exists $(\gamma, m) \in Lim(\beta)$ such that Δ_β^0 can compute $q_{\gamma,m}(\sigma)$. (In this case, by the comments on limit steps, we should have $\rho = \sigma$ and $m = n$.) Then $\sigma^+ = \rho$ and

$$q_{\beta,n}(\sigma) = q_{\gamma,n}(\sigma) .$$

This is a limit step.

Now, suppose that there is no (γ, m) as required for a limit step. For each k such that $1 \leq k \leq m$, we check whether $q_{\beta+1,k-1}^n(\rho_{k-1}) = d_k$. Suppose this fails for some first k. Then $\sigma^+ = \rho_{k-1}$, and if $d = q_{\beta+1,k-1}^n(\rho_{k-1})$, then

$$q_{\beta,n}(\sigma) = (f(d); s(d)\beta\ell_n) .$$

This is an approximation step. Suppose that $q_{\beta+1,k-1}^n(\rho_{k-1}) = d_k$ for all k. Then $\sigma^+ = \rho$. If $(\beta+1, m) \ll (\beta, n)$, then

$$q_{\beta,n}(\sigma) = (f(d); s(d)\beta\ell_n),$$

where $d = q_{\beta+1,m}^n(\rho)$. This is an approximation step. If $(\beta, n) \ll (\beta+1, m)$, then

$$q_{\beta,n}(\sigma) = (f(d); s(d)\beta\ell_n)$$

for the first d we find such that $\rho d \in P^{\beta+1}$. This is an invention step.

We have defined $q_{\beta,n}(\sigma)$ and σ^+ for all β and n. It is not difficult to see that the three conditions stated above are maintained.

14.4 Simultaneous runs

For $\gamma \leq \xi \leq \beta$, let π^ξ be a run of (P^ξ, q_ξ). We say that the family $(\pi^\xi)_{\gamma \leq \xi \leq \beta}$ is *coherent* if $L(\pi^\zeta)$ is a subsequence of $L(\pi^\xi)$ whenever $\xi < \zeta$. To complete the proof of Theorem 14.1, it is enough to show the existence of a coherent family of runs $(\pi^\xi)_{1 \leq \xi \leq \alpha}$ such that π^ξ is Δ_ξ^0, uniformly in ξ.

Lemma 14.3 *Let $1 \leq \gamma < \beta \leq \alpha$. Given a Δ_γ^0 index for a run π of (P^γ, q_γ), we can find a computable sequence of indices for a coherent family $(\pi^\xi)_{\gamma \leq \xi \leq \beta}$ such that π^ξ is a run of (P^ξ, q_ξ), Δ_ξ^0, uniformly in ξ, and $\pi^\gamma = \pi$. Moreover, if the initial segment of π of length $2n+1$ is the result of limit steps involving pairs (δ, k) for $\delta > \beta$, then for all ξ, the initial segment of π^ξ of length $2n+1$ matches that of π.*

Proof: We concentrate on proving the first part of the statement. The "moreover" part follows from Lemma 14.2 and the remarks on limit work preceding the case-by-case definitions of $q_{\beta,n}(\sigma)$ and σ^+. Suppose the statement holds for $\beta' < \beta$. The proof for β has two cases.

Case I: Suppose β is a successor.

By H.I., we may assume that $\beta = \gamma + 1$. Let

$$\pi = \hat{\ell} c_1 \ell_1 c_2 \ell_2 \ldots$$

be a Δ_γ^0 run of (P^γ, q_γ), and let π_k be the initial segment of length $2k+1$. We must describe a Δ_β^0 run of (P^β, q_β). We determine a Δ_β^0 sequence of numbers $(k(n))_{n \in \omega}$ such that $\pi_{k(n)}^+$ has length $2n+2$ and for all $k > k(n)$, $\pi_k^+ \supseteq \pi_{k(n)}^+$. Let k^* be first such that c_{k+1} does not represent a limit step in π. For $n \leq k^*$, we let $k(n) = n$. Suppose that we have determined $k(n)$ (where $n \geq k^*$), and let $d = q_{\gamma,n}(\pi_{k(n)}^*)$. We search (using the Δ_β^0 oracle) for the first $i > k(n)$ such that $(\beta, n) \ll (\gamma, i)$ and for all $k \geq i$, $q_{\gamma,n}^k(\pi_{k(n)}^+) = d$. Then $i+1$ serves as

14.4. SIMULTANEOUS RUNS

$k(n+1)$. Now, $\pi^+_{k(n)}$ is the initial segment of length $2n+1$ in the desired run of (P^β, q_β). This completes the proof for Case I.

Case II: Suppose β is a limit ordinal.

Let π be a Δ^0_γ run of (P^γ, q_γ), where $\gamma < \beta$. We may suppose that the sequence $(\beta_n)_{n \in \omega}$ given by our notation for β begins with γ (if this is not so initially, then we modify the sequence). By H.I., we can find sequences of indices for families $(\pi^\xi)_{\beta_n \leq \xi \leq \beta_{n+1}}$, for $n = 0, 1, 2, \ldots$, and we can combine these to form all of the desired coherent family except π^β. To obtain π^β, we form a Δ^0_β sequence of numbers $(k(n))_{n \in \omega}$ such that for all ξ such that $\beta_{k(n)} \leq \xi < \beta$, the initial segment of π^ξ of length $2n+1$ is the same. Let $k(0) = 0$. Supposing that we have $k(n)$ and letting σ_n be the initial segment of π^ξ of of length $2n+1$, for $\beta_{k(n)} \leq \xi < \beta$, take the first $k > k(n)$ such that $(\beta, n) \in Lim(\beta_k)$ and $\Delta^0_{\beta_k}$ can compute $q_{\beta,n}(\sigma_n)$. By Lemma 14.2 and the remarks on limit steps, this k serves as $k(n+1)$.

The initial segment of π^β of length $2n+1$ is that of $\pi^{\beta_{k(n)}}$. To show coherence, we must check that for $\gamma \leq \xi < \beta$, $L(\pi^\beta)$ is a subsequence of $L(\pi^x)$. By H.I., it is enough to consider ξ of the form $\beta_{k(m)}$. For each $n \leq m$, we have an embedding of $L(\pi^{\beta_{k(n+1)}})$ into $L(\pi^{\beta_{k(n)}})$. Say that

$$\pi^\beta = \hat{\ell} c_1 \ell_1 c_2 \ell_2 \ldots .$$

Note that the location of ℓ_n is the same in $\pi^{\beta_{k(n)}}$ and in π^β. Moreover, if $m \geq n$, then ℓ_n is in this same place in $L(\pi^{\beta_{k(m)}})$. If $m < n$, then, following the chosen embeddings, we can trace ℓ_n from $L(\pi^{\beta_{k(n)}})$ down to $L(\pi^{\beta_{k(m)}})$. This completes the proof for Case II.

We have proved Lemma 14.3, which was all that remained in the proof of Theorem 14.1.

Comments: The proof that we gave for Theorem 14.1 is not the original one (in [3]). First of all, the original proof involved nested finite sequences. In Chapter 13, we worked with these at finite levels. The "pictures" in the present proof extract just the essential information from the nested sequences. There are other differences between the proofs. Both involve induction, but not in the same way. The original proof did not yield an explicit sequence of steps at the computable level. The coherent family of runs gives at explicit sequence of steps at each level. In particular, the run π^1 of (P^1, q_1) gives a computable sequence of steps, showing what is enumerated into the c.e. set, which requirement receives attention, and what is injured. Finally, the present proof has shown itself to be easier to vary than the original. Several further metatheorems have been proved, by making relatively obvious changes in the definitions of the instruction functions q_β. For at least one of these metatheorems, the one in [12], earlier attempts to find the correct statement in the transfinite case, and to give a proof like that in[3], were unsuccessful.

14.5 Special (α_n)-systems

Here we describe a variant of Theorem 14.1, for limit ordinals. This variant was proved in [3], along with the original theorem. The hypotheses are somewhat weaker, so the result applies in certain situations where Theorem 14.1 does not.

Let α be a limit ordinal, and let $(\alpha_n)_{n \in \omega}$ be the sequence of ordinals picked out by our notation for α. A *special (α_n)-system* is a structure

$$(L, U, \hat{\ell}, P, E, (\leq_\beta)_{\beta < \alpha})$$

with the same properties as an ordinary α-system except that Condition (4) is replaced by the following:

(4') if $\sigma u \in P$, where σ has length $2n + 1$, ending in ℓ^0,

$$\ell^0 \leq_{\beta_0} \ell^1 \leq_{\beta_1} \ldots \leq_{\beta_{k-1}} \ell^k ,$$

and $\alpha_n > \beta_0 > \ldots > \beta_k$, then there exists ℓ^* such that $\sigma u \ell^* \in P$, and $\ell^i \leq_{\beta_i} \ell^*$, for $0 \leq i < k$.

The difference between Conditions (4) and (4') is that in Condition (4'), for $\sigma \in P$ of length $2n+1$, we consider only ordinals $\beta_0 < \alpha_n$, while in Condition (4), regardless of the length of σ, we consider arbitrary ordinals $\beta < \alpha$.

An (α_n) *instruction function* for P is an instruction function q whose restriction to sequences in P of length $2n + 1$ is $\Delta^0_{\alpha_n}$, uniformly in n. The names of both the system and the instruction function are intended to reflect the *sequence* of ordinals $(\alpha_n)_{n \in \omega}$.

Theorem 14.4 *Suppose α is a limit ordinal. Let*

$$(L, U, \hat{\ell}, P, E, (\leq_\beta)_{\beta < \alpha})$$

be a special (α_n)-system, and let q be an (α_n) instruction function for P. Then (P, q) has a run $\pi = \hat{\ell} u_1 \ell_1 \ell_2 \ldots$ such that $E(\pi)$ is c.e. Moreover, the initial segment of π of length $2n + 3$ (i.e., $\hat{\ell} u_1 \ell_1 \ldots u_{n+1} \ell_{n+1}$) is $\Delta^0_{\alpha_n}$, uniformly in n.

Proof: The proof is essentially the same as for Theorem 14.1 except that we make sure to include (α, n) in $Lim(\alpha_n)$. This will be automatic if we start with a list of pairs in which (α, n) comes before any pair (β, k) such that $\alpha_n \leq \beta < \alpha$. In a run of (P^β, q_β), where $\beta \geq \alpha_n$, the initial segment of length $2n + 3$ will be an initial segment of a run of (P, q).

To apply the metatheorems in this chapter, we must define appropriate relations $\leq_\beta$. In Chapter 15, we shall discuss the standard back-and-forth relations, and we show that for some kinds of structures, they turn out to be computably enumerable. The relations $\leq_\beta$ in most applications of the metatheorems are defined in terms of these relations. There are applications of Theorem 14.1 in Chapters 16, 17, and 18, and there is an application of Theorem 14.4 in Chapter 19.

14.6 Further metatheorems

As we said in Chapter 13, there are priority constructions for which the original metatheorem does not apply. In [12], there is a metatheorem for constructions having infinitely many requirements at each level. Using it, we can give a result on a structure $\mathcal{A}$ with added relations R_i, for $i \in \omega$, saying when there is a computable copy in which, for each i, the image of R_i is not Σ_i^0. In [11], there is a metatheorem for constructions that enumerate sets at various levels, not just at the computable level. Using it, we can give a result on a structure $\mathcal{A}$ with added relations R_i and S, saying when there is a computable copy in which, for each i, the image of R_i is Σ_i^0, and the image of S is not Σ_n^0. There is a metatheorem in [85] for constructions in which a family of sets is coded. There is a metatheorem in [86], useful for constructions in which the nth piece of high-level information is given by a special kind of limit.

Chapter 15

Back-and-forth relations

In this chapter, we define the *standard back-and-forth relations* $\leq_\beta$ between tuples from a single structure, or, more generally, from different structures. Basic model-theoretic information about these relations may be found in [23]. Results such as the Ash-Nerode Theorem, on relations that are intrinsically c.e. on a structure, and Goncharov's theorems, on structures that are computably categorical or computably stable, involve effectiveness assumptions on just one copy of the structure. In Chapters 16 and 17, when we lift these results to arbitrary levels in the hyperarithmetical hierarchy, there will be stronger effectiveness assumptions, also on just one copy. In particular, we shall suppose that the back-and-forth relations are uniformly c.e.

It is useful to calculate the back-and-forth relations in various classes of structures, so that we can apply general results on intrinsically Σ_α^0 relations, and on Δ_α^0 categoricity and stability. In addition, understanding the back-and-forth relations on a structure puts us in a position to determine the ranks r and R. The third rank that we defined, SR, is based on equivalence relations that cannot be calculated precisely from the back-and-forth relations, except at limit ordinals. We give explicit calculations of the standard back-and-forth relations for well orderings and superatomic Boolean algebras, from [2], [3].

15.1 Standard back-and-forth relations

Let K be a class of structures for a fixed language. We define the *standard back-and-forth relation* $\leq_\beta$ on pairs $(\mathcal{A}, \overline{a})$, where $\mathcal{A} \in K$ and $\overline{a}$ is a tuple in $\mathcal{A}$.

First, suppose that $\overline{a}$ in $\mathcal{A}$ and $\overline{b}$ in $\mathcal{B}$ are tuples of the same length. Then

(1) $(\mathcal{A}, \overline{a}) \leq_1 (\mathcal{B}, \overline{b})$ if and only if all finitary Σ_1 formulas true of $\overline{b}$ in $\mathcal{B}$ are true of $\overline{a}$ in $\mathcal{A}$,

(2) for $\alpha > 1$, $(\mathcal{A}, \overline{a}) \leq_\alpha (\mathcal{B}, \overline{b})$ if and only if, for each $\overline{d}$ in $\mathcal{B}$ and each $1 \leq \beta < \alpha$, there exists $\overline{c}$ in $\mathcal{A}$ such that $(\mathcal{B}, \overline{b}, \overline{d}) \leq_\beta (\mathcal{A}, \overline{a}, \overline{c})$.

Now, we extend the definition of $\leq_\beta$ to tuples of different lengths. For $\overline{a}$ in $\mathcal{A}$ and $\overline{b}$ in $\mathcal{B}$, let $(\mathcal{A}, \overline{a}) \leq_\beta (\mathcal{B}, \overline{b})$ if and only if $length(\overline{a}) \leq length(\overline{b})$

and for the initial segment $\bar{b}'$ of $\bar{b}$ such that $length(\bar{a}) = length(\bar{b}')$, we have $(\mathcal{A},\bar{a}) \leq_\beta (\mathcal{B},\bar{b}')$.

We have defined the standard back-and-forth relations $\leq_\beta$ for $\beta \geq 1$. It is convenient have in addition a *standard relation* $\leq_0$ and a *standard enumeration function* E. Let $\bar{x}$ be the sequence of the first n variables. For an n-tuple $\bar{a}$ in $\mathcal{A}$, let $E(\mathcal{A},\bar{a})$ consist of the basic formulas $\psi(\bar{x})$, with Gödel number less than n, such that $\mathcal{A} \models \psi(\bar{a})$. For $\bar{a}$ in $\mathcal{A}$ and $\bar{b}$ in $\mathcal{B}$ such that $length(\bar{a}) \leq length(\bar{b})$, let

$$(\mathcal{A},\bar{a}) \leq_0 (\mathcal{B},\bar{b}) \iff E(\mathcal{A},\bar{a}) \subseteq E(\mathcal{B},\bar{b}) \ .$$

If the language is finite, then we could consider all basic formulas $\psi(\bar{x})$ instead of just those with small Gödel number. As it is, the definition of $\leq_0$ lacks model-theoretic inevitability, but it is useful in the case of infinite languages. Moreover, the definition fits smoothly with the definitions of $\leq_\beta$ for $\beta \geq 1$, in that

$$(\mathcal{A},\bar{a}) \leq_1 (\mathcal{B},\bar{b})$$

if and only if for each tuple $\bar{d}$ in $\mathcal{B}$, there is a tuple $\bar{c}$ in $\mathcal{A}$ such that

$$(\mathcal{B},\bar{b},\bar{d}) \leq_0 (\mathcal{A},\bar{a},\bar{c}) \ .$$

Notation: We may write $\mathcal{A} \leq_\beta \mathcal{B}$ instead of $(\mathcal{A},\emptyset) \leq_\beta (\mathcal{B},\emptyset)$. If we are interested in just one structure $\mathcal{A}$, then we may write $\bar{a} \leq_\beta \bar{b}$ instead of $(\mathcal{A},\bar{a}) \leq_\beta (\mathcal{A},\bar{b})$, and $E(\bar{a})$ instead of $E(\mathcal{A},\bar{a})$.

For $\beta \geq 1$, there is a natural connection between Σ_β and Π_β formulas and the standard back-and-forth relations $\leq_\beta$.

Proposition 15.1 *Let $\mathcal{A}$ and $\mathcal{B}$ be countable structures, and suppose $\bar{a}$ and $\bar{b}$ are tuples of the same length, in $\mathcal{A}$ and $\mathcal{B}$, respectively. Then for a countable ordinal $\beta \leq 1$, the following are equivalent:*
  (1) $(\mathcal{A},\bar{a}) \leq_\beta (\mathcal{B},\bar{b})$,
  (2) the Σ_β formulas true of $\bar{b}$ in $\mathcal{B}$ are true of $\bar{a}$ in $\mathcal{A}$,
  (3) the Π_β formulas true of $\bar{a}$ in $\mathcal{A}$ are true of $\bar{b}$ in $\mathcal{B}$.

Note: The formulas here need not be computable—they are arbitrary formulas of $L_{\omega_1\omega}$, in normal form, of the appropriate complexity.

Proof: Clearly, (2) and (3) are equivalent. We prove the equivalence of (1) and (2) by induction on β. For $\beta = 1$, the statement is clear. Supposing that it holds for $1 \leq \gamma < \beta$, we prove it for β. Assuming (1), we prove (2). Let $\varphi(\bar{x})$ be a Σ_β formula such that $\mathcal{B} \models \varphi(\bar{b})$. Then $\bar{b}$ satisfies some disjunct of $\varphi(\bar{x})$, say $\exists \bar{u}\, \psi(\bar{x},\bar{u})$, where ψ is Π_γ for some γ such that $1 \leq \gamma < \beta$. Then for some $\bar{d}$ in $\mathcal{B}$, we have $\mathcal{B} \models \psi(\bar{b},\bar{d})$. By (1), there exists $\bar{c}$ in $\mathcal{A}$ such that

$$(\mathcal{B},\bar{b},\bar{d}) \leq_\gamma (\mathcal{A},\bar{a},\bar{c}) \ .$$

Then by H.I., we have $\mathcal{A} \models \psi(\bar{a},\bar{c})$, so $\mathcal{A} \models \varphi(\bar{a})$. This proves (2).

15.2. α-FRIENDLY FAMILIES

Now, assuming (2), we prove (1). Take $\bar{d}$ in $\mathcal{B}$ and $1 \leq \gamma < \beta$. By Lemma 6.6 (taking $K = \{\mathcal{A}, \mathcal{B}\}$), there is a Π_γ formula $\varphi(\bar{x}, \bar{u})$ that is satisfied, in $\mathcal{A}$ and $\mathcal{B}$, by just those tuples which satisfy all Π_γ formulas true of $\bar{b}, \bar{d}$ in $\mathcal{B}$. Now, $\exists \bar{u}\, \varphi(\bar{x}, \bar{u})$ is a Σ_β formula satisfied by $\bar{b}$ in $\mathcal{B}$. By (2), it must be satisfied by $\bar{a}$ in $\mathcal{A}$, say $\mathcal{A} \models \varphi(\bar{a}, \bar{c})$. Then by H.I.,

$$(\mathcal{B}, \bar{b}, \bar{d}) \leq_\gamma (\mathcal{A}, \bar{a}, \bar{c}) \ .$$

This proves (1).

15.2 α-friendly families

A structure $\mathcal{A}$ is said to be *α-friendly structure* if $\mathcal{A}$ is computable, and for $\beta < \alpha$, the standard back-and-forth relations $\leq_\beta$, on tuples from $\mathcal{A}$, are c.e., uniformly in β. More generally, let $K = \{\mathcal{A}_0, \mathcal{A}_1, \ldots\}$ be a finite or countably infinite family of structures. Here we may write $(i, \bar{a}) \leq_\beta (j, \bar{b})$ instead of $(\mathcal{A}_i, \bar{a}) \leq_\beta (\mathcal{A}_j, \bar{b})$. Then K is *α-friendly* if the structures $\mathcal{A}_i$ are uniformly computable, and for $\beta < \alpha$, the standard back-and-forth relations $\leq_\beta$ on the set of pairs $(i, \bar{a})$, for $\bar{a}$ in $\mathcal{A}_i$, are c.e., uniformly in β.

Remark: Any computable structure is 1-friendly. More generally, any uniformly computable family of structures is 1-friendly.

According to Lemma 6.6, for a tuple $\bar{a}$ from a structure $\mathcal{A}$ in a countable family K, there is a single Π_β formula that distinguishes the complete Π_β type realized by $\bar{a}$ in $\mathcal{A}$ from the other Π_β types realized in the structures from K. Here we are taking a Π_β type to consist of *all* Π_β formulas true of some tuple in one of the structures. The result below gives conditions on K under which the single formula can be taken to be *computable* Π_β.

Theorem 15.2 *Suppose $K = \{\mathcal{A}_0, \mathcal{A}_1, \ldots\}$ is α-friendly, and the existential diagrams of the structures $\mathcal{A}_i$ are uniformly computable. Then for any β such that $1 \leq \beta < \alpha$, any i, and any tuple $\bar{a}$ from $\mathcal{A}_i$, we can find a computable Π_β formula $\varphi^\beta_{(i,\bar{a})}(\bar{x})$ such that for all j and all $\bar{b}$ in $\mathcal{A}_j$,*

$$(i, \bar{a}) \leq_\beta (j, \bar{b}) \iff \mathcal{A}_j \models \varphi^\beta_{(i,\bar{a})}(\bar{b}) \ .$$

Proof: First, let $\varphi^1_{(i,\bar{a})}(\bar{x})$ be the conjunction of the finitary universal formulas true of $\bar{a}$ in $\mathcal{A}_i$. This clearly has the desired property. Now, let $\beta > 1$, and suppose that for $1 \leq \gamma < \beta$, we can find computable infinitary Π_γ formulas with the desired property. Let

$$C = \{(j, \bar{b}) : (i, \bar{a}) \not\leq_\beta (j, \bar{b})\} \ .$$

Note that C is co-c.e. but not necessarily computable. Using a Δ^0_2 oracle, we can associate to each $(j, \bar{b}) \in C$ some $(\bar{d}, \gamma)$ such that $\gamma < \beta$ and for all $\bar{c}$,

$$(j, \bar{b}, \bar{d}) \not\leq_\gamma (i, \bar{a}, \bar{c}) \ .$$

We have a Δ^0_2 set

$$D = \{(j,\overline{b},\overline{d},\gamma) : (j,\overline{b}) \in C \ \& \ (\overline{d},\gamma) \text{ corresponds to } (j,\overline{b})\} \ .$$

The formula

$$\bigwedge_{(j,\overline{b},\overline{d},\gamma) \in D} \forall \overline{u} \, \neg \varphi^\gamma_{(j,\overline{b},\overline{d})}(\overline{x},\overline{u})$$

has the meaning that we want for $\varphi^\beta_{(i,\overline{a})}(\overline{x})$. Moreover, this formula is logically equivalent to a computable Π_β formula, since by Proposition 7.12, we can replace the Δ^0_2 conjunction by one that is c.e.

15.3 Examples

It may seem implausible that the relations $\leq_\beta$ could ever be c.e. We cannot, in finitely many steps, examine all Π_β formulas, determine whether they are true of the first of a pair of tuples, and for those that are true, determine whether they are also true of the second tuple. Nonetheless, we can calculate the standard back-and-forth relations for certain familiar structures and classes of structures. Using these calculations, we show the existence of some α-friendly structures and families.

15.3.1 Arithmetic

In the standard model of arithmetic, $\mathcal{N}$, every element is definable by a finitary open formula. Hence, if $\overline{a}$ and $\overline{b}$ are tuples from $\mathcal{N}$, of the same length, then for all $\beta \geq 1$,

$$\overline{a} \leq_\beta \overline{b} \Leftrightarrow \overline{a} = \overline{b} \ .$$

It follows that $\mathcal{N}$ is α-friendly, for all computable ordinals α.

15.3.2 Vector spaces

For any complete theory of vector spaces over a given field, we have quantifier elimination, so the back-and-forth relations collapse (see the Appendix). Therefore, we have the following.

Lemma 15.3 *Suppose $\mathcal{A}$ is a vector space over a field F, and $\overline{a}$ and $\overline{b}$ are tuples from $\mathcal{A}$, of the same length. Then the following are equivalent:*

(1) the open formulas true of $\overline{a}$ are true of $\overline{b}$, i.e., each a_i is independent of, or expressible as a particular linear combination of $\{a_j : j < i\}$ just in case the same is true of b_i and $\{b_j : j < i\}$,

(2) there is an automorphism of $\mathcal{A}$ taking $\overline{a}$ to $\overline{b}$.

(3) $\overline{a} \leq_\beta \overline{b}$, for some (or for all) $\beta \geq 1$.

15.3. EXAMPLES

Note that in vector spaces over a *finite* field, we can say that the dimension is at least n, using a finitary existential formula. In vector spaces over an *infinite* field, saying that the dimension is at least n requires a computable Σ_2 formula. Using Lemma 15.3, we get the existence of α-friendly vector spaces.

Proposition 15.4 *Let F be a computable field. For all computable ordinals α, there is an α-friendly vector space over F in each countable dimension, finite or (countably) infinite.*

Proof: Let $\mathcal{A}$ be a vector space over F, of the desired dimension, with a computable basis. For any tuple $\bar{a}$ from $\mathcal{A}$, we can express the elements a_i in terms of this basis. Having done this, we can systematically decide, for each a_i, whether it is independent of $\{a_j : j < i\}$, and if not, then we can express a_i uniquely as a linear combination of a previously determined maximal independent subset of $\{a_j : j < i\}$. From this, it is clear that for tuples $\bar{a}$ and $\bar{b}$ in $\mathcal{A}$, of the same length, we can determine whether they satisfy the same open formulas. Therefore, by Lemma 15.3, $\mathcal{A}$ is α-friendly.

Above, we considered single vector spaces. Now, we consider pairs.

Lemma 15.5 *Let $\mathcal{A}$, $\mathcal{B}$ be vector spaces over the same field F (everything is countable). Suppose the tuples $\bar{a}$ in $\mathcal{A}$ and $\bar{b}$ in $\mathcal{B}$ satisfy the same open formulas.*
(a) If F is finite, then

$$(\mathcal{A}, \bar{a}) \leq_1 (\mathcal{B}, \bar{b}) \iff dim(\mathcal{B}) \leq dim(\mathcal{A}) ,$$

and for $\beta \geq 2$,

$$(\mathcal{A}, \bar{a}) \leq_\beta (\mathcal{B}, \bar{b}) \iff dim(\mathcal{A}) = dim(\mathcal{B}) .$$

(b) If F is infinite, then necessarily $(\mathcal{A}, \bar{a}) \leq_1 (\mathcal{B}, \bar{b})$. In addition,

$$(\mathcal{A}, \bar{a}) \leq_2 (\mathcal{B}, \bar{b}) \iff dim(\mathcal{B}) \leq dim(\mathcal{A}) ,$$

and for $\beta \geq 3$,

$$(\mathcal{A}, \bar{a}) \leq_\beta (\mathcal{B}, \bar{b}) \iff dim(\mathcal{A}) = dim(\mathcal{B}) .$$

Each part of Lemma 15.5 should be clear after just a little thought.

The next result asserts the existence of α-friendly families of vector spaces.

Proposition 15.6 *Let F be a computable field. For any computable ordinal α, and any computable sequence of possible dimensions d_i, finite or countably infinite, there is an α-friendly family $K = \{\mathcal{A}_0, \mathcal{A}_1, \ldots\}$ of vector spaces over F, such that $\mathcal{A}_n$ has dimension d_n.*

Proof: For each n, let $\mathcal{A}_n$ be a vector space over F, with basis $\bar{b}_n$, of size d_n, such that $(\mathcal{A}_n, \bar{b}_n)$ is computable, uniformly in n. Given $\bar{a}$ in $\mathcal{A}_n$, we can effectively express the elements in terms of the basis $\bar{b}_n$; and as above, we can determine whether $\bar{a}$ in $\mathcal{A}_m$ and $\bar{b}$ in $\mathcal{A}_n$ satisfy the same open formulas. Then by Lemma 15.5, K is α-friendly.

15.3.3 Linear orderings

The lemmas below are useful in calculating the back-and-forth relations for linear orderings.

Lemma 15.7 . *Suppose $\mathcal{A}$ and $\mathcal{B}$ are linear orderings. Let $\bar{a} = (a_0, \ldots, a_{n-1})$, $\bar{b} = (b_0, \ldots, b_{n-1})$ be increasing tuples from $\mathcal{A}$, $\mathcal{B}$, respectively, and let $\mathcal{A}_i$, $\mathcal{B}_i$ be intervals such that*

$$\mathcal{A} = \mathcal{A}_0 + \{a_0\} + \mathcal{A}_1 + \ldots + \mathcal{A}_{n-1} + \{a_{n-1}\} + \mathcal{A}_n ,$$

$$\mathcal{B} = \mathcal{B}_0 + \{b_0\} + \mathcal{B}_1 + \ldots + \mathcal{B}_{n-1} + \{b_{n-1}\} + \mathcal{B}_n .$$

Then $(\mathcal{A}, \bar{a}) \leq_\beta (\mathcal{B}, \bar{b})$ if and only if for all $0 \leq i \leq n$, $\mathcal{A}_i \leq_\beta \mathcal{B}_i$.

Proof: This is easily proved by induction on β.

Lemma 15.8 *Suppose $\mathcal{A}$ and $\mathcal{B}$ are linear orderings. Then $\mathcal{A} \leq_1 \mathcal{B}$ if and only if $\mathcal{A}$ is infinite or at least as large as $\mathcal{B}$. For $\beta > 1$, $\mathcal{A} \leq_\beta \mathcal{B}$ if and only if for any $1 \leq \gamma < \beta$ and any partition of $\mathcal{B}$ into intervals $\mathcal{B}_0, \ldots, \mathcal{B}_n$, with endpoints in $\mathcal{B}$, there is a corresponding partition of $\mathcal{A}$ into intervals $\mathcal{A}_0, \ldots, \mathcal{A}_n$, with endpoints in $\mathcal{A}$, such that $\mathcal{B}_i \leq_\gamma \mathcal{A}_i$.*

Proof: This follows from Lemma 15.7.

We give a couple of simple examples.

Example 1: $\omega \leq_1 5$ and $5 \not\leq_1 \omega$.

Example 2: $\omega \leq_2 \omega^* + \omega$ and $\omega^* + \omega \not\leq_2 \omega$.

Proof for Example 2: For any partition of $\omega^* + \omega$, there is a partition of ω that matches except that the first interval in ω is finite, while that in $\omega^* + \omega$ has type ω^*. Since $\omega^* \leq_1 n$, for all finite n, we have $\omega \leq_2 \omega^* + \omega$. If we partition ω and $\omega^* + \omega$ each into two subintervals, the second intervals match, but the first interval in ω is finite, while that in $\omega^* + \omega$ is infinite. Since $n \not\leq_1 \omega^*$, for finite n, we have $\omega^* + \omega \not\leq_2 \omega$.

The next result gives the back-and-forth relations for well orderings of type ω^α, for various α, as calculated in [3]

Lemma 15.9 *Let α, β, γ be countable ordinals. Then*
  (a) for $\gamma \geq 1$, $\omega^\beta \leq_{2\gamma} \omega^\alpha \Leftrightarrow \alpha, \beta \geq \gamma$ or $\alpha = \beta$,
  (b) $\omega^\beta \leq_{2\gamma+1} \omega^\alpha \Leftrightarrow (\alpha \geq \gamma \ \& \ \beta > \gamma)$ or $\alpha = \beta$.

15.3. EXAMPLES

Proof: The $\Leftarrow$ portions of (a) and (b) are proved by a tedious induction on γ. We consider the case where $\gamma = 1$.

(a) Let $\alpha, \beta \geq 1$. For any partition of ω^α into subintervals α_i, there is a corresponding partition of ω^β into subintervals β_i such that if α_i has type n, for $n \in \omega$, then β_i has the same type. For all i, we have $\alpha_i \leq_1 \beta_i$. Therefore, by Lemma 15.8, $\omega^\beta \leq_2 \omega^\alpha$.

(b) Suppose $\alpha \geq 1$ and $\beta > 1$. For any partition of ω^α, we may add points so that each subinterval α_i has type n, for some $n \in \omega$, or ω^ξ, for some $\xi \geq 1$. Then there is a corresponding partition of ω^β into subintervals β_i such that if α_i has type n, then β_i has the same type and if α_i has type ω^ξ for some $\xi \geq 1$, then β_i has type ω^ζ for some $\zeta \geq 1$. Then $\alpha_i \leq_2 \beta_i$, for all i. Therefore, by Lemma 15.8, $\omega^\beta \leq_3 \omega^\alpha$.

We turn to the $\Rightarrow$ portions of (a) and (b). First, note that if $\beta < \alpha$, then there is a $\Sigma_{2\beta+2}$ sentence true in ω^α and not true in ω^β. If $\beta > \alpha$, then there is a $\Pi_{2\alpha+1}$ sentence true in ω^α and not true in ω^β. Supposing that $\omega^\beta \leq_{2\gamma} \omega^\alpha$, if $\beta \leq \gamma$, then $\alpha \leq \beta$, and if $\alpha < \gamma$, then $\beta \leq \alpha$. Therefore, either $\alpha, \beta \geq \gamma$ or else $\alpha = \beta$. Similarly, supposing that $\omega^\beta \leq_{2\gamma+1} \omega^\alpha$, if $\beta \leq \gamma$, then $\beta \geq \alpha$, and if $\alpha < \gamma$, then $\beta \leq \alpha$. Therefore, either $\alpha \geq \gamma$ and $\beta > \gamma$, or else $\alpha = \beta$.

The next result gives precise calculations of the back-and-forth relations for arbitrary countable well orderings.

Lemma 15.10 *Let δ be either a limit ordinal or 0, and let α and β be computable ordinals, both divisible by ω^ξ. Let*

$$\alpha = \omega^\xi \cdot \alpha_\xi + \rho_\xi ,$$

$$\beta = \omega^\xi \cdot \beta_\xi + \sigma_\xi ,$$

where $\rho_\xi, \sigma_\xi < \omega^\xi$, and let m_ξ, n_ξ be the coefficients of ω^ξ in the Cantor normal form expressions for α, β, respectively. Then

(a) $\alpha \leq_{\delta+2n+1} \beta$ *if and only if one of the following holds:*
  (i) $\alpha = \beta$,
  (ii) $\rho_{\delta+n} = \sigma_{\delta+n}$, $\alpha_{\delta+n+1} \geq 1$, and $\beta_{\delta+n+1} \geq 1$,
  (iii) $\rho_{\delta+n} = \sigma_{\delta+n}$, $\alpha_{\delta+n+1} = \beta_{\delta+n+1} = 0$, and $m_{\delta+n} \geq n_{\delta+n} > 0$.
(b) $\alpha \leq_{\delta+2n+2} \beta$ *if and only if one of the following holds:*
  (i) $\alpha = \beta$,
  (ii) $\rho_{\delta+n} = \sigma_{\delta+n}$, $\alpha_{\delta+n+1} \geq 1$, $\beta_{\delta+n+1} \geq 1$, and $m_{\delta+n} \geq n_{\delta+n}$,
(c) *for δ a limit ordinal, $\alpha \leq_\delta \beta$ if and only if one of the following holds:*
  (i) $\alpha = \beta$,
  (ii) $\rho_\delta = \sigma_\delta$, $\alpha_\delta \geq 1$, and $\beta_\delta \geq 1$.

Proof: The "if" statements are proved using Lemmas 15.8 and 15.9. The "only if" statements are proved by considering particular sentences.

The next result gives the existence of α-friendly well orderings and families of well orderings.

Proposition 15.11 *Let α be a computable ordinal.*

(a) For any computable ordinal β, there is an α-friendly ordering $\mathcal{A}$ of order type β.

(b) For any a pair of computable ordinals β_1, β_2, there is an α-friendly pair of orderings $\{\mathcal{A}_1, \mathcal{A}_2\}$ such that $\mathcal{A}_i$ has type β_i.

(c) Let $(\beta_n)_{n \in \omega}$ be a sequence of computable ordinals, with a computable sequence of corresponding notations $(b_n)_{n \in \omega}$, all on a single path in O. Then there is an α-friendly family of orderings $\{\mathcal{A}_0, \mathcal{A}_1, \ldots\}$ such that $\mathcal{A}_n$ has type β_n.

Proof: (a) By Theorem 4.19, if $|b| = \beta$, then we have a computable ordering $\mathcal{A}$ of type β with the feature that for any interval with endpoints in $\mathcal{A}$, we can effectively calculate the Cantor normal form for an interval, using notations $\leq_O b$ for the powers of ω. Then it follows from Lemma 15.10 that $\mathcal{A}$ is α-friendly.

(b) Take $b^* \in O$ such that β_1, β_2 have notations $b_i <_O b^*$. Theorem 4.19 yields computable orderings $\mathcal{A}_1$, $\mathcal{A}_2$ in which we can calculate the Cantor normal form for an interval, using notations $d <_O b^*$ for the powers of ω. It follows from Lemma 15.10 that this pair is α-friendly.

(c) Take $b^* \in O$ such that $b_n <_O b^*$. We may let $b^* = 3 \cdot 5^e$, where e is an index for the sequence $(b_n)_{n \in \omega}$. Theorem 4.19 yields uniformly computable orderings $\mathcal{A}_i$ (where $\mathcal{A}_i$ has type β_i) in which we can calculate the Cantor normal form for an interval, using notations $<_O b^*$, for the powers of ω. It follows from Lemma 15.10 that this family is α-friendly.

Remark: The fact that we can effectively calculate Cantor normal form in these well orderings has importance beyond α-friendliness.

15.3.4 Boolean algebras

Let $\mathcal{A}$ be a Boolean algebra. Any tuple $\overline{a}$ in $\mathcal{A}$ generates a sub-algebra. In this subalgebra, the elements obtained by choosing a or the complement a' for each $a \in \overline{a}$ and taking the meet are either atoms or 0. These form a "partition" of $\mathcal{A}$. For $a \in \mathcal{A}$, we may identify a with the Boolean algebra consisting of the elements $a \wedge y$, for $y \in \mathcal{A}$ (a plays the role of 1).

The next result is the analogue of Lemma 15.7.

Lemma 15.12 *Suppose that $\mathcal{A}$ and $\mathcal{B}$ are Boolean algebras. Let $\overline{a}$, $\overline{b}$ be tuples of the same length from $\mathcal{A}$, $\mathcal{B}$, respectively. Let $\mathcal{A}_i$, for $i = 1, \ldots, n$ be the elements of the partition of $\mathcal{A}$ generated by $\overline{a}$, and let $\mathcal{B}_i$, $i = 0, \ldots, n$ be the corresponding elements of the partition of $\mathcal{B}$ generated by $\overline{b}$. Then $(\mathcal{A}, \overline{a}) \leq_\beta (\mathcal{B}, \overline{b})$ if and only if $\mathcal{A}_i \leq_\beta \mathcal{B}_i$, for $i = 0, \ldots, n$.*

The next result is the analogue of Lemma 15.8.

Lemma 15.13 *Suppose that $\mathcal{A}$ and $\mathcal{B}$ are Boolean algebras. Then $\mathcal{A} \leq_1 \mathcal{B}$ if and only if $\mathcal{A}$ is infinite or can be split into at least as many disjoint parts as $\mathcal{B}$ (if $\mathcal{A}$ is the join of n atoms, then $\mathcal{B}$ is the join of m atoms, for some $m \leq n$).*

For $\beta > 1$, $\mathcal{A} \leq_\beta \mathcal{B}$ if and only if for any $1 \leq \gamma < \beta$ and any finite partition of $\mathcal{B}$ into elements $\mathcal{B}_1, \ldots, \mathcal{B}_n$, there is a corresponding partition of $\mathcal{A}$, with atoms $\mathcal{A}_1, \ldots, \mathcal{A}_n$ such that $\mathcal{B}_i \leq_\gamma \mathcal{A}_i$.

We gave a complete description of the back-and-forth relations for well orderings. We can do the same for the superatomic Boolean algebras. The calculations are from [2]. Superatomic Boolean algebras are represented by the interval algebras $I(\alpha)$, where α is an ordinal (see the Appendix). Moreover, if $\omega^\beta \cdot n$ is the leading term in the expression for α in Cantor normal form, then $I(\alpha) \cong I(\omega^\beta \cdot n)$. Any element a of $I(\alpha)$ is a finite join of elements corresponding to intervals in α, say of order types $\alpha_1, \ldots, \alpha_n$. The element a, thought of as a Boolean algebra in itself, is isomorphic to $I(\omega^\gamma \cdot m)$, where $\omega^\gamma \cdot m$ is the leading term in the commutative sum of the ordinals α_i.

Proposition 15.14 *Let δ be either a limit ordinal or 0, let α, β be arbitrary ordinals, and let m, n, $k < \omega$. Then*
(a) $I(\omega^\alpha \cdot m) \leq_{\delta+2k+1} I(\omega^\beta \cdot n)$ if and only if one of the following holds:
(i) $\alpha = \beta$ and $m = n$,
(ii) $\alpha = \beta = \delta + k$ and $m \geq n$,
(iii) $\alpha \geq \delta + k + 1$ and $\beta \geq \delta + k$,
(b) assuming that δ and k are not both 0, $I(\omega^\alpha \cdot m) \leq_{\delta+2k} I(\omega^\beta \cdot n)$ if and only if one of the following holds:
(i) $\alpha = \beta < \delta + k$ and $m = n$,
(ii) $\alpha, \beta \geq \delta + k$.

Proof: This is a tedious induction on δ and k.

Proposition 15.15 *Let α and β be computable ordinals. Then there is an α-friendly Boolean algebra $\mathcal{A} \cong I(\beta)$.*

Proof: As in the proof of Proposition 15.11, we use Theorem 4.19. Let $\mathcal{B}$ be a computable ordering of type β in which we can effectively calculate the Cantor normal form for intervals. Let $\mathcal{A} = I(\mathcal{B})$. Now, take a finite subalgebra of $\mathcal{A}$, with atoms $\mathcal{A}_1, \ldots, \mathcal{A}_n$. For each i, $\mathcal{A}_i$ is a finite join of disjoint intervals in $\mathcal{B}$. We have $\mathcal{A}_i \cong I(\omega^{\beta_i} \cdot m_i)$, where $\omega^{\beta_i} \cdot m_i$ is the leading term in the expression formed by taking the commutative sum of the order types of the disjoint intervals making up $\mathcal{A}_i$. We can effectively calculate these $\omega^{\beta_i} \cdot m_i$. Therefore, using Proposition 15.14, we see that $\mathcal{A}$ is α-friendly.

Remark: In addition to being α-friendly, the Boolean algebras above have the useful feature that for any element a, we can effectively determine the invariants β and n, where a is the join of n β-atoms.

15.4 Ranks

In Chapter 6, we introduced three notions of rank, which we denoted by r, R, and SR. In Chapter 6, we calculated ranks only for very simple examples. At the end of Chapter 8, we gave some general results on ranks. In particular, we saw that if $\mathcal{A}$ is hyperarithmetical, then

$$r(\mathcal{A}) \leq \omega_1^{CK},$$

while

$$SR(\mathcal{A}) \leq R(\mathcal{A}) \leq \omega_1^{CK} + 1.$$

Now, we are prepared to calculate the ranks for some more interesting examples.

The next result gives the ranks of ω^α, for arbitrary countable ordinals α.

Proposition 15.16 *If α is a countable ordinal, then $r(\omega^\alpha) = 2\alpha$ — if α is a limit ordinal, then this is just α. If α is a successor ordinal, then $R(\omega^\alpha) = 2\alpha + 1$, and if α is a limit ordinal, then $R(\omega^\alpha) = \alpha$.*

Proof: We have already shown this for $\alpha = 1$. By Lemma 15.9, if $\delta, \gamma < \omega^{\beta+1}$, then

$$\delta \leq_{2\beta+2} \gamma \;\Rightarrow\; \delta = \gamma.$$

Also by Lemma 15.9, we have $\omega^\beta \cdot 2 \leq_{2\beta+1} \omega^\beta$. It follows that if $\alpha = \beta + 1$, then $r(\omega^\alpha) = 2\beta + 2 = 2\alpha$. Since $r(\omega^\beta \cdot 2) = 2\alpha$, $R(\omega^\alpha) = 2\alpha + 1$. If α is a limit ordinal, then $r(\omega^\alpha) = \alpha$, and $R(\omega^\alpha) = \alpha$.

Recall that ω_1^{CK} is the first non-computable ordinal. By Theorem 8.10, there is no hyperarithmetical ordering of type ω_1^{CK}. We can calculate the ranks.

Proposition 15.17 $r(\omega_1^{CK}) = R(\omega_1^{CK}) = SR(\omega_1^{CK}) = \omega_1^{CK}$.

Proof: We could use Proposition 15.16, or we could argue directly, as follows. In Chapter 6, we saw that each element of ω_1^{CK} has a computable infinitary defining formula. This is Π_α for some $\alpha < \omega_1^{CK}$. By Lemma 15.9, for any computable ordinal α, there exist distinct computable ordinals β and γ such that $\beta \leq_\alpha \gamma$, so the defining formulas are not all of complexity α for any one computable ordinal α. From this, it follows that $r(\omega_1^{CK}) = \omega_1^{CK}$ and $R(\omega_1^{CK}) = \omega_1^{CK}$. Since the value of R is a limit ordinal, SR has the same value.

Exercise: Note that $\omega^{\omega_1^{CK}} = \omega_1^{CK}$. Apply Proposition 15.16 to show that $r(\omega_1^{CK}) = R(\omega_1^{CK}) = \omega_1^{CK}$. As above, conclude that $SR(\omega_1^{CK}) = \omega_1^{CK}$.

By Harrison's Theorem (Theorem 8.11), there is a computable ordering of type $\omega_1^{CK}(1 + \eta)$. Below, we calculate the ranks.

Proposition 15.18 *(a) $r(\omega_1^{CK}(1 + \eta)) = \omega_1^{CK}$,*
(b) $R(\omega_1^{CK}(1 + \eta)) = SR(\omega_1^{CK}(1 + \eta)) = \omega_1^{CK} + 1$.

Proof: By Corollary 8.23, $r(\omega_1^{CK}(1 + \eta)) \leq \omega_1^{CK}$. Just in this proof, we call the elements of the initial copy of ω_1^{CK} *standard*, and we call the other elements *non-standard*. Each standard element is definable by some formula of $L_{\omega_1,\omega}$—in fact it is definable by a computable infinitary formula. We can argue

15.4. RANKS

as we did in Proposition 15.17 that the definitions are not all of complexity α for any one computable ordinal α. It follows that $r(\omega_1^{CK}(1+\eta)) = \omega_1^{CK}$. By Proposition 6.8, $R(\omega_1^{CK}(1+\eta))$ is either ω_1^{CK} or $\omega_1^{CK}+1$. Then $SR(\omega_1^{CK}(1+\eta))$ is the same.

To show that $R(\omega_1^{CK}(1+\eta)) = \omega_1^{CK}+1$, it is enough to show that for some element b, $r(b) = \omega_1^{CK}$. Let b be a non-standard element, say it is first in its copy of ω_1^{CK}. Then the interval to the left of b has type $\omega_1^{CK}(1+\eta)$. We obtain the fact that $r(b) = \omega_1^{CK}$ by showing that for any computable ordinal α, there is a standard element a such that all $\Pi_{2\alpha}$ formulas true of b are also true of a. Take a such that the interval to the left of a has type ω^α. The intervals to the right of b and a are isomorphic. Then the lemma below will complete the proof, since it tells us that the $\Pi_{2\alpha}$ formulas true of b are true of a.

Lemma 15.19 *For each computable ordinal α, $\omega_1^{CK}(1+\eta) \leq_{2\alpha} \omega^\alpha$.*

Proof: We proceed by induction on $\alpha \geq 1$.

Case 1: Let $\alpha = 1$.

For any finite partition of ω, there is a partition of $\omega_1^{CK}(1+\eta)$ such that corresponding subintervals are isomorphic, except that the last interval in ω has type ω, while the last interval in $\omega_1^{CK}(1+\eta)$ has type $\omega_1^{CK}(1+\eta)$. Since ω is infinite, $\omega \leq_1 \omega_1^{CK}(1+\eta)$, by Lemma 15.5. It follows that

$$\omega_1^{CK}(1+\eta) \leq_2 \omega .$$

Case 2: Let α be a limit ordinal.

Suppose that the statement holds for $\beta < \alpha$. Then for $\beta < \alpha$, we have $\omega_1^{CK}(1+\eta) \leq_{2\beta} \omega^\beta$, by H.I., and $\omega^\beta \leq_{2\beta} \omega^\alpha$, by Lemma 15.9. Then

$$\omega_1^{CK}(1+\eta) \leq_{2\alpha} \omega^\alpha .$$

Case 3: Let $\alpha = \beta + 1$.

Suppose that the statement holds for β. For any partition of ω^α, there is a partition of $\omega_1^{CK}(1+\eta)$ such that corresponding intervals are isomorphic, except that the last interval in ω^α has type ω^α, while the last interval in $\omega_1^{CK}(1+\eta)$ has type $\omega_1^{CK}(1+\eta)$. Therefore, it is enough to show that $\omega^\alpha \leq_{2\beta+1} \omega_1^{CK}(1+\eta)$. Say

$$\omega_1^{CK}(1+\eta) = \mathcal{A}_0 + \{a_0\} + \mathcal{A}_1 + \ldots + \{a_{n-1}\} + \mathcal{A}_n .$$

We may suppose that for each i, $\mathcal{A}_i$ is finite, or of type ω^γ, for some computable ordinal γ, or of $\omega_1^{CK}(1+\eta)$.

We can express ω^α in the form

$$\mathcal{B}_0 + \{b_0\} + \mathcal{B}_1 + \ldots + \{b_{n-1}\} + \mathcal{B}_n ,$$

where
  (i) if $\mathcal{A}_i$ is finite or has type ω^γ for $\gamma \leq \beta$, then $\mathcal{B}_i \cong \mathcal{A}_i$,
  (ii) if $\mathcal{A}_i$ has type ω^γ for $\gamma > \beta$, then $\mathcal{B}_i$ has type ω^β, and
  (iii) if $\mathcal{A}_i$ has type $\omega_1^{CK}(1+\eta)$, then $\mathcal{B}_i$ has type ω^β, or ω^α (in the case where $i = n$).

By Lemma 15.9, $\mathcal{A}_i \leq_{2\beta} \mathcal{B}_i$ if $\mathcal{A}_i$ has type ω^γ for some $\gamma > \beta$ and $\mathcal{B}_i$ has type ω^β. By H.I., $\mathcal{A}_i \leq_{2\beta} \mathcal{B}_i$ if $\mathcal{A}_i$ has type $\omega_1^{CK}(1+\eta)$ and $\mathcal{B}_i$ has type ω^β. Again by Lemma 15.9, $\omega^\beta \leq_{2\beta} \omega^\alpha$, so $\mathcal{A}_i \leq_{2\beta} \mathcal{B}_i$ if $\mathcal{A}_i$ has type $\omega_1^{CK}(1+\eta)$ and $\mathcal{B}_i$ has type ω^α. Then $\omega^\alpha \leq_{2\beta+1} \omega_1^{CK}(1+\eta)$.

We may consider the Boolean algebras obtained as interval algebras from the orderings above—$I(\omega^\alpha)$, where α is a countable ordinal, and $I(\omega_1^{CK}(1+\eta))$. Imitating what we have just done, we can calculate the ranks..

Exercise: Prove the following.
(a) $r(I(\omega_1^{CK}(1+\eta))) = \omega_1^{CK}$
(b) $R(I(\omega_1^{CK}(1+\eta))) = SR(I(\omega_1^{CK}(1+\eta))) = \omega_1^{CK} + 1$.

Exercise: For arbitrary countable ordinals α, find $r(I(\omega^\alpha))$ and $R(I(\omega^\alpha))$.

15.5 Stronger back-and-forth relations

There are times when it is convenient not to use the standard relations $\leq_\beta$. Using stronger relations may simplify our calculations. We illustrate this below.

Example 1: Let $\mathcal{A}$ and $\mathcal{B}$ be orderings of types $\omega^\alpha + \eta$ and $\omega^\beta + \eta$, respectively, with an added predicate for the dense part. This has the effect of making the back-and-forth relations stronger. Suppose $\bar{a}$ is in $\mathcal{A}$ and $\bar{b}$ is in $\mathcal{B}$. Let $\bar{a} = \bar{a}_1, \bar{a}_2$, where $\bar{a}_1$ is outside the dense part, $\bar{a}_2$ inside, and let $\bar{b}_1, \bar{b}_2$ be the corresponding parts of $\bar{b}$—i.e., the kth term of $\bar{a}$ is in $\bar{a}_1$ if and only if the kth term of $\bar{b}$ is in $\bar{b}_1$. Then for all $\gamma \geq 1$, we have

$$(\mathcal{A}, \bar{a}) \leq_\gamma (\mathcal{B}, \bar{b})$$

if and only if $\bar{b}_1$ is outside the dense part, $\bar{b}_2$ is inside the dense part, $\bar{a}_2$ and $\bar{b}_2$ are ordered in the same way, and

$$(\omega^\alpha, \bar{a}_1) \leq_\gamma (\omega^\beta, \bar{b}_1) .$$

From this, it is easy to see that there are copies of $\omega^\alpha + \eta$ and $\omega^\beta + \eta$, with the extra predicate, forming an α-friendly family with respect to the stronger back-and-forth relations.

Example 2: Let J and K be families of linear orderings, and consider the shuffle sums $\sigma(J)$, $\sigma(K)$ (see the Appendix). We add an equivalence relation,

15.6. ABSTRACT BACK-AND-FORTH RELATIONS

where an equivalence class is an interval representing a copy of some $\mathcal{A}$ in J, or $\mathcal{B}$ in K. Let
$$\bar{a} = \bar{a}_1 \ldots \bar{a}_n \text{ and } \bar{b} = \bar{b}_1 \ldots \bar{b}_n,$$
where $\bar{a}_i$ is the part of $\bar{a}$ in a single equivalence class $\mathcal{A}_i$ and $\bar{b}_i$ is the corresponding part of $\bar{b}$. For $\gamma \geq 1$,
$$(\sigma(J), \bar{a}) \leq_\gamma (\sigma(K), \bar{b})$$
if and only if the following conditions hold:
 (a) $\bar{b}_i$ is the part of $\bar{b}$ in a single equivalence class $\mathcal{B}_i$,
 (b) $(\mathcal{A}_i, \bar{a}_i) \leq_\gamma (\mathcal{B}_i, \bar{b}_i)$, and
 (c) for all $\mathcal{B} \in K$ and all $\delta < \gamma$, there exists $\mathcal{A} \in J$ such that $\mathcal{B} \leq_\delta \mathcal{A}$.

As a special case, consider $\sigma(\mathcal{A}, \mathcal{B})$ and $\sigma(\mathcal{A})$, with the added equivalence relation. We have $\sigma(\mathcal{A}, \mathcal{B}) \leq_\gamma \sigma(\mathcal{A})$ if and only if for all $\delta < \gamma$, $\mathcal{A} \leq_\delta \mathcal{B}$. If $\mathcal{A}$ and $\mathcal{B}$ form an α-friendly family, then there are copies of $\sigma(\mathcal{A}, \mathcal{B})$ and $\sigma(\mathcal{A})$ that form an α-friendly family with respect to the stronger back-and-forth relations.

15.6 Abstract back-and-forth relations

In an α-system, there is an abstract enumeration function E, and there are abstract relations $\leq_\beta$. Often we have a further relation $\leq_\alpha$ preserved along paths in the tree, and implying $\leq_\beta$ for $\beta < \alpha$. (Typically, the relation $\leq_\alpha$ represents extension.)

Proposition 15.20 *Let $(L, U, \hat{\ell}, E, P, (\leq_\beta)_{\beta < \alpha})$ be an α-system, and suppose that all elements of L occur as terms in the sequences in P (we could always restrict L to make this true). Let $\leq_\alpha$ be a further transitive, reflexive relation on L, preserved along paths in P, such that $\leq_\alpha$ implies $\Rightarrow \leq_\beta$, for all $\beta < \alpha$. Then E and the relations $\leq_\beta$ satisfy the following:*
 (i) $\ell \leq_0 \ell'$ implies $E(\ell) \subseteq E(\ell')$,
 (ii) for any $\ell \leq_\beta \ell' \leq_\alpha \ell''$ and any $\gamma < \beta$, there exists ℓ^* such that $\ell \leq_\alpha \ell^*$ and $\ell'' \leq_\gamma \ell^*$.

Proof: Condition (3) yields (i). For Property (ii), take $\sigma u \in P$ where ℓ is last in σ. Since $\ell' \leq_\alpha \ell''$ implies $\ell' \leq_\beta \ell''$, and by Condition 1, $\leq_\beta$ is transitive, we have $\ell \leq_\beta \ell''$. By Condition (4), there exists ℓ^* such that $\sigma u \ell^* \in P''$, which yields $\ell \leq_\alpha \ell^*$, and $\ell'' \leq \ell^*$.

Given an enumeration function E on L, and a transitive, reflexive relation $\leq_\alpha$ on L such that
$$\ell \leq_\alpha \ell' \;\Rightarrow\; E(\ell) \subseteq E(\ell'),$$
we derive a family of relations $(\leq_\beta)_{\beta < \alpha}$ as follows:
 (a) Let $\ell \leq_0 \ell'$ if and only if $E(\ell) \subseteq E(\ell')$.
 (b) For $0 < \beta < \alpha$, let $\ell \leq_\beta \ell'$ if and only if for all ℓ'' such that $\ell' \leq_\alpha \ell''$ and all $\gamma < \beta$, there exists ℓ^* such that $\ell \leq_\alpha \ell^*$ and $\ell'' \leq_\gamma \ell^*$.

We can show, by induction on β, that $\leq_\beta$ is transitive and reflexive, and $\leq_\beta$ implies $\leq_\gamma$, for $\gamma < \beta \leq \alpha$. The relations may not be distinct—$\leq_\alpha$ may be the same as $\leq_0$.

It is not difficult to show that the derived relations $\leq_\beta$ satisfy (i) and (ii) of Proposition 15.20. The next result says that they are the weakest relations with these properties.

Proposition 15.21 *Let E be an enumeration function on L, and let $\leq_\alpha$ be a transitive, reflexive relation on L such that $\ell \leq_\alpha \ell'$ implies $E(\ell) \supseteq E(\ell')$. Let $(\leq_\beta)_{\beta < \alpha}$ be the derived relations, defined as above, and let $(\leq'_\beta)_{\beta<\alpha}$ be another family of reflexive, transitive relations on L satisfying (i) and (ii) with the given E and $\leq_\alpha$. Then $\ell \leq_\beta \ell'$ implies $\ell \leq'_\beta \ell'$.*

Proof: This is an easy induction on β.

In a number of applications of the metatheorems from Chapters 13 and 14, an element of L represents a finite partial isomorphism p, from some $\mathcal{B}$, under construction, with computable universe B, to a given structure $\mathcal{A}$. The aim of the construction is to form a sequence of partial isomorphisms $(p_n)_{n \in \omega}$ such that the union is a total isomorphism, with $\cup_n E(p_n) = D(\mathcal{B})$, so that various high-level requirements are satisfied, and $D(\mathcal{B})$ is computable.

Let $\mathcal{F}$ be the set of finite partial one-one functions from B to $\mathcal{A}$. Let $\leq_\alpha$ be the relation of extension on $\mathcal{F}$. There is a natural enumeration function E on $\mathcal{F}$, defined in terms of the standard enumeration function E on tuples from $\mathcal{A}$, where for $p \in \mathcal{F}$ mapping $\bar{b}$ to $\bar{a}$,

$$E(p) = \{\varphi(\bar{b}) : \varphi(\bar{x}) \in E(\bar{a})\} \ .$$

In this case, the weakest relations $\leq_\beta$ on $\mathcal{F}$ that, with this E and $\leq_\alpha$, satisfy (i) and (ii) turn out to match the standard back-and-forth relations $\leq_\beta$ on tuples from $\mathcal{A}$ in the following way. If p maps $\bar{b}$ to $\bar{a}$ and q maps $\bar{b}' \supseteq \bar{b}$ to $\bar{a}'$, then $p \leq_\beta q$ if and only if $\bar{a} \leq_\beta \bar{b}$.

15.7 Open problems

It would be useful to calculate the standard back-and-forth relations for further classes of structures, and show the existence of further α-friendly families.

Open problem: Determine the standard back-and-forth relations for reduced Abelian p-groups (in the way that we have done those for well orderings and superatomic Boolean algebras). There are partial results in this direction by Barker [21].

Open problem: Investigate the closure properties for the class of linear orderings with α-friendly copies. Is the class closed under finite sum, product, computable sum etc.? (Ash asked about this in 1987.)

Chapter 16

Theorems of Barker and Davey

In this chapter, we use Theorem 14.1 to lift results from Chapter 11 to higher levels. We let $\mathcal{A}$ be a computable structure and consider the problems below. In the first two problems, R is a further relation on $\mathcal{A}$. In the third problem, R and S are a pair of further relations on $\mathcal{A}$, of the same arity.

Problem 1 *When is there a computable copy of $\mathcal{A}$ in which the image of R is not Σ^0_α?*

Problem 2 *When is there a computable copy of $\mathcal{A}$ in which the image of R is not Δ^0_α?*

Problem 3 *When is there a computable copy of $\mathcal{A}$ in which the images of R and S are not separated by any Δ^0_α relation?*

Note that Problem 2 is the special case of Problem 3, where $S = \neg R$. In Chapter 10, we considered problems analogous to these, with arbitrary copies of $\mathcal{A}$ instead of just computable copies.

16.1 Barker's Theorems

We give results of Barker [20] on Problems 1 and 2. There is some terminology associated with these problems. A relation R is *intrinsically Σ^0_α* on $\mathcal{A}$ if in every computable copy of $\mathcal{A}$, the image of R is Σ^0_α. Similarly, R is *intrinsically Δ^0_α* on $\mathcal{A}$ if in every computable copy of $\mathcal{A}$, the image of R is Δ^0_α.

We begin with Problem 1. In Chapter 10, we showed that R is *relatively* intrinsically Σ^0_α on $\mathcal{A}$ if and only if it is formally Σ^0_α on $\mathcal{A}$ (i.e., R is defined by a computable Σ_α formula). Barker showed that under some effectiveness assumptions, R is intrinsically Σ^0_α on $\mathcal{A}$ if and only if it is formally Σ^0_α on $\mathcal{A}$.

The next two results, taken together, say that, under certain effectiveness conditions, if R is not definable in $\mathcal{A}$ by a computable Σ_α formula, then it is not intrinsically Σ_α^0 on $\mathcal{A}$. In the hypotheses, we refer to the standard enumeration function and the standard back-and-forth relations, and the notion of α-friendliness that we use is based on these. However, we could substitute a different enumeration function, or we could use stronger back-and-forth relations, with the corresponding notion of α-friendliness, as described at the end of Chapter 15.

We define a notion of "freeness". This is not the first such notion that we have defined, and it will not be the last. In Chapter 11, we defined notions for use in connection with intrinsically n-c.e. and intrinsically β-c.e. relations. The following notion is appropriate for the setting of Problem 1. Let $\bar{c}$ and $\bar{a}$ be tuples in $\mathcal{A}$, where $length(\bar{a})$ is equal to the arity of R. We say that $\bar{a}$ is α-*free over* $\bar{c}$ if

(i) $\bar{a} \in R$, and
(ii) for all $\bar{a}_1$ and all $\beta < \alpha$, there exist $\bar{a}' \notin R$ and $\bar{a}'_1$ such that

$$\bar{c}, \bar{a}, \bar{a}_1 \leq_\beta \bar{c}, \bar{a}', \bar{a}'_1 \ .$$

The result below says that if $\bar{c}$ is a tuple in $\mathcal{A}$ over which no tuple $\bar{a}$ is α-free, then, under some effectiveness conditions, R is defined by a computable Σ_α formula $\varphi(\bar{c}, \bar{x})$, with parameters $\bar{c}$.

Proposition 16.1 . *Let $\mathcal{A}$ be an α-friendly structure whose finitary Σ_1 diagram is computable, and let R be a further computable relation on $\mathcal{A}$. Suppose $\bar{c}$ is a tuple over which no $\bar{a}$ is α-free (in the current sense). Finally, suppose that for each tuple $\bar{a} \in R$, we can find $\bar{a}_1$ and $\beta < \alpha$ witnessing the fact that $\bar{a}$ is not α-free over $\bar{c}$. Then there is a computable Σ_α formula $\varphi(\bar{c}, \bar{x})$ defining R.*

Proof: Given $\bar{a} \in R$, we determine $\bar{a}_1$ and $\beta < \alpha$ witnessing the fact that $\bar{a}$ is not α-free over $\bar{c}$. By Theorem 15.2, we can find a computable Π_β formula $\varphi^\beta_{\bar{c},\bar{a},\bar{a}_1}(\bar{c}, \bar{x}, \bar{u})$ saying (in $\mathcal{A}$) that $\bar{c}, \bar{a}, \bar{a}_1 \leq_\beta \bar{c}, \bar{x}, \bar{u}$. The formula

$$\psi_{\bar{a}}(\bar{c}, \bar{x}) = \exists \bar{u}\, \varphi^\beta_{\bar{c},\bar{a},\bar{a}_1}(\bar{c}, \bar{x}, \bar{u})$$

is true of $\bar{a}$ and is not true of any tuple outside R. Therefore, R is defined by the formula

$$\bigvee_{\bar{a} \in R} \psi_{\bar{a}}(\bar{c}, x) \ ,$$

which is computable Σ_α.

Remarks: Suppose no $\bar{a}$ is α-free over $\bar{c}$. Above, we assumed that there is an effective procedure for finding witnesses. For $\alpha = 1$, this is implied by the assumption that we can determine whether a given finitary Σ_1 formula $\varphi(\bar{c}, \bar{x})$ is satisfied in $\neg R$. For $\alpha > 1$, the assumption that we can find witnesses can be

16.1. BARKER'S THEOREMS

dropped. If $\mathcal{A}$ is α-friendly and R is computable, then using Δ_2^0, we can find a tuple $\bar{a}_1$ and an ordinal $\beta < \alpha$ such that there do not exist $\bar{a}'$, $\bar{a}_1'$ with

$$\bar{c}, \bar{a}, \bar{a}_1 \leq_\beta \bar{c}, \bar{a}', \bar{a}_1' .$$

If, in addition, the finitary Σ_1 diagram of $\mathcal{A}$ is computable, then we have the formulas $\varphi_{\bar{c},\bar{a},\bar{a}_1}^\beta(\bar{c}, \bar{x}, \bar{u})$, as above. Then R is defined by the Σ_2^0 disjunction of the computable Σ_α formulas

$$\exists \bar{u} \, \varphi_{\bar{c},\bar{a},\bar{a}_1}^\beta(\bar{c}, \bar{x}, \bar{u}) .$$

By Proposition 7.14, we can replace this by a computable Σ_α formula.

The next result says that if R is not definable by a computable Σ_α formula, then, under some effectiveness conditions, R is not intrinsically Σ_α^0 on $\mathcal{A}$.

Theorem 16.2 *Let $\mathcal{A}$ be an α-friendly structure, and let R be a further computable relation on $\mathcal{A}$. Suppose that for each tuple $\bar{c}$ in $\mathcal{A}$, we can find a tuple $\bar{a}$ that is α-free over $\bar{c}$ (in the current sense). Then $\mathcal{A}$ has a computable copy in which the image of R is not Σ_α^0.*

Proof: For simplicity, we suppose that R is unary. Let B be an infinite computable set of constants, for the universe of $\mathcal{B}$. We shall determine a $1-1$ function F from $\mathcal{A}$ onto B, and let $\mathcal{B}$ be the structure induced by F on B. In addition to putting elements into the domain and range of F, we satisfy the following requirements, in what, viewed from level α, is a finite-injury priority construction.

$$R_e : W_e^{\Delta_\alpha^0} \neq F(R)$$

Let $\mathcal{F}$ be the set of finite partial $1-1$ functions p from B to $\mathcal{A}$. These are possible parts of F^{-1}. The strategy for R_e is as follows. Given $p \in \mathcal{F}$ representing work on earlier requirements, we let $p' \supseteq p$ map a new constant b to the element a effectively chosen to be α-free over $ran(p)$. If b ever appears in $W_e^{\Delta_\alpha^0}$, then we replace p' by some $p'' \supseteq p$, where $p''(b) \notin R$.

To apply the metatheorem (Theorem 14.1), we define an α-system

$$(L, U, \hat{\ell}, E, P, (\leq_\beta)_{\beta < \alpha})$$

and a Δ_α^0 instruction function q. Let U be the set of finite sets of statements of the form "$b \in W_e^{\Delta_\alpha^0}$", true or false. Let L be the set of finite sequences

$$\ell = p_0 b_0 p_1 b_1 ... b_{r-1} p_{r-1}$$

satisfying the following conditions:
  (a) $p_0 = \emptyset$,
  (b) $p_n \subseteq p_{n+1}$,

(c) $dom(p_n)$, $ran(p_n)$ include the first n elements of $\mathcal{B}$, $\mathcal{A}$, respectively,
(d) $b_n \in dom(p_{n+1})$.

Let $\hat{\ell} = p_0$, where $p_0 = \emptyset$.

In Chapter 15, we defined the standard enumeration function E and the standard back-and-forth relations $\leq_\beta$ on tuples from a structure $\mathcal{A}$, and we indicated how to extend the definitions to the set $\mathcal{F}$ described above. Now, we extend the definitions further to L. For $\ell \in L$ ending in p, we let

$$E(\ell) = E(p) .$$

For ℓ and ℓ' in L, where ℓ ends in p and ℓ' ends in p', we let

$$\ell \leq_\beta \ell' \Leftrightarrow p \leq_\beta p' .$$

The tree P is defined so that a finite-injury priority construction is carried out along the paths, with action on the requirements R_n in a given $\ell \in L$ based on information from the current $u \in U$. For $u \in U$, let $L(u)$ be the set of elements of L of the form

$$p_0 b_0 p_1 b_1 \ldots b_{r-1} p_r ,$$

where for each $n < r$, b_n and p_{n+1} "take care of requirement R_n according to the information given by u". By this, we mean that if the statement "$b_n \in W_n^{\Delta_n^0}$" is in u, then $p_{n+1}(b_n) \notin R$; and otherwise, $p_{n+1}(b_n)$ is the element of R effectively chosen to be α-free over $ran(p_n)$.

Now, we let P be the set of finite sequences

$$\sigma = \hat{\ell} u_1 \ell_1 u_2 \ell_2 \ldots$$

satisfying the following conditions:
  (a) $u_k \in U$,
  (b) $\ell_k \in L(u_k)$,
  (c) $u_k \subseteq u_{k+1}$,
  (d) $\ell_k \leq_0 \ell_{k+1}$,
  (e) if $\ell_k = p_0 b_0 \ldots b_{r-1} p_r$, then either
    (i) $\ell_k \in L(u_{k+1})$ and ℓ_{k+1} has the form $p_0 b_0 p_1 \ldots p_r b_r p_{r+1}$, or else
    (ii) $\ell_k \notin L(u_{k+1})$ and ℓ_{k+1} has the form $p_0 b_0 p_1 \ldots p_n b_n p_{n+1}^*$, where n is first such that p_{n+1} and b_n do not take care of R_n according to u_{k+1}.

Remark: Let $\pi = \hat{\ell} u_1 \ell_1 u_2 \ell_2 \ldots$ be a path through P. There is a sequence of numbers $k(n)$ such that $\ell_{k(n)}$ has the form

$$p_0 b_0 \ldots p_n b_n p_{n+1}$$

and for all $k > k(n)$, $\ell_k \supseteq \ell_{k(n)}$. Then we have a $1-1$ function

$$\cup_n p_{k(n)}$$

16.1. BARKER'S THEOREMS

from B onto $\mathcal{A}$. The inverse function F, mapping $\mathcal{A}$ $1-1$ onto B, induces a structure $\mathcal{B}$ on B, such that $D(\mathcal{B}) = E(\pi)$. For this F, the requirements are all satisfied, according to the information in $\cup_n u_n$.

We have defined all of the ingredients of the α-system. Before verifying the conditions, we define the instruction function q. For $\sigma \in P$ of length $2n+1$, we let $q(\sigma)$ be the set of true statements "$b \in W_e^{\Delta_\alpha^0}$" that are enumerated in n steps using the Δ_α^0 oracle.

The first three conditions for α-systems clearly hold. Toward Condition (4), let $\sigma u \in P$, where the last term of σ is ℓ^0, and

$$\ell^0 \leq_{\beta_0} \ldots \leq_{\beta_{k-1}} \ell^k ,$$

and $\alpha > \beta_0 > \ldots > \beta_{k-1} > \beta_k$. We need ℓ^* such that $\sigma u \ell^* \in P$ and $\ell^i \leq_{\beta_i} \ell^*$.

Claim: Suppose ℓ^i ends in q_i. Then there exists $p \supseteq q_0$ such that $q_i \leq_{\beta_i} p$ for $0 < i \leq k$.

Proof of Claim: Let $q_k* = q_k$, and for $0 < i \leq k$, let $q_{i-1}^* \supseteq q_{i-1}$, where $q_i^* \leq_{\beta_i} q_{i-1}^*$. Then the desired p is q_0^*.

Let $\ell^0 = p_0 b_0 \ldots b_{r-1} p_r$, so $p_r = q_0$. We consider two cases.

Case 1: Suppose $\ell^0 \notin L(u)$.

Take the first $n < r$ such that b_n and p_{n+1} fail to satisfy R_n according to the information in u. The statement "$b_n \in W_n^{\Delta_\alpha^0}$" is in u, and $p_{n+1}(b_n) \in R$ is the element a effectively chosen to be α-free over $ran(p_n)$. For p as in the claim above, there exist $a' \notin R$ and $p'_{n+1} \supseteq p_n$ such that $p'_{n+1}(b_n) = a'$ and $p \leq_{\beta_0} p'_{n+1}$. Then $p_0 b_0 \ldots p_n b_n p'_{n+1}$ is the required ℓ^*.

Case 2: Suppose $\ell^0 \in L(u)$.

We have $p_{r+1} \supseteq q_0 = p_r$, where p_{r+1} is as in the claim. Let a be the element effectively chosen to be α-free over $ran(p_r)$. We may suppose that $a \in ran(p_{r+1})$ (extending, if necessary). Say $p_{r+1}(b_r) = a$. If "$b_r \in W_r^{\Delta_\alpha^0}$" is not in u, then the required ℓ^* is $p_0 b_0 \ldots b_{r-1} p_r b_r p_{r+1}$. Suppose "$b_r \in W_r^{\Delta_\alpha^0}$" is in u. There exist $a' \notin R$ and $p'_{r+1} \supseteq p_r$ such that $p_{r+1}(b_r) = a'$ and $p_{r+1} \leq_{\beta_0} p'_{r+1}$. Then the required ℓ^* is $p_0 b_0 \ldots b_{r-1} p_r b_r p'_{r+1}$.

We are in a position to apply Theorem 14.1. We obtain a Δ_α^0 run π of (P, q) such that $E(\pi)$ is c.e. By the remark above, π yields an isomorphism F from $\mathcal{A}$ onto a structure $\mathcal{B}$ such that $E(\pi) = D(\mathcal{B})$. Since $E(\pi)$ is c.e., $\mathcal{B}$ is computable. The requirements are all satisfied using the true information given by the instruction function q, so $F(R)$ is not Σ_α^0. This completes the proof of Theorem 16.2.

Next, we turn to Problem 2. If R and $\neg R$ are both definable in $\mathcal{A}$ by computable Σ_α formulas, then in any computable copy of $\mathcal{A}$, the image of R is Δ^0_α. The next two results say that under suitable effectiveness conditions, the converse holds.

We define a new notion of "freeness", appropriate for the new setting. For a tuple $\bar{a}$ whose length matches the arity of R, we now say that $\bar{a}$ is α-*free over* $\bar{c}$ if for all $\bar{a}_1$ and all $\beta < \alpha$, there exist $\bar{a}'$ and $\bar{a}'_1$ such that
 (i) $\bar{a} \in R \Leftrightarrow \bar{a}' \notin R$, and
 (ii) $\bar{c}, \bar{a}, \bar{a}_1 \leq_\beta \bar{c}, \bar{a}', \bar{a}'_1$.

The result below says that if there is some $\bar{c}$ in $\mathcal{A}$ over which no $\bar{a}$ is α-free, in the current sense, then, under some effectiveness conditions, there are computable Σ_α formulas, with parameters $\bar{c}$, defining both R and $\neg R$.

Proposition 16.3 *Let $\mathcal{A}$ be α-friendly, with computable finitary Σ_1 diagram, and let R be a further computable relation on $\mathcal{A}$. Let $\bar{c}$ be a tuple from $\mathcal{A}$ such that no $\bar{a}$ is α-free over $\bar{c}$ (in the current sense). Suppose, moreover, that for each tuple $\bar{a}$ of the appropriate length, we can find $\bar{a}_1$ and $\beta < \alpha$ witnessing the fact that $\bar{a}$ is not α-free over $\bar{c}$. Then R and $\neg R$ are each definable in $\mathcal{A}$ by a computable Σ_α formula with parameters $\bar{c}$.*

Proof: If $\bar{a} \in R$, then $\bar{a}$ is not α-free over $\bar{c}$ either in the current sense or in the sense of the definition preceding Proposition 16.1. Moreover, we can find the witnesses. Then by Propositon 16.1, R is definable in $\mathcal{A}$ by a computable Σ_α formula with parameters $\bar{c}$. A similar application of Proposition 16.1, with $\neg R$ playing the role of R, yields the fact that $\neg R$ is definable by a computable Σ_α formula with parameters $\bar{c}$.

Note: The remarks after Proposition 16.1 apply here as well.

The next result says that if for each $\bar{c}$ in $\mathcal{A}$, some $\bar{a}$, in R or $\neg R$, is α-free over $\bar{c}$, then, under some effectiveness conditions, R is not intrinsically Δ^0_α on $\mathcal{A}$.

Theorem 16.4 *Let $\mathcal{A}$ be an α-friendly structure and let R be a further computable relation on $\mathcal{A}$. Suppose that for any $\bar{c}$, we can find a tuple $\bar{a}$ that is α-free over $\bar{c}$ (in the current sense). Then $\mathcal{A}$ has a computable copy in which the image of R is not Δ^0_α.*

Proof: Let $\bar{c}_n$ consist of the first n elements of $\mathcal{A}$, and let $\bar{a}_n$ be the tuple effectively chosen to be α-free over $\bar{c}$. If $\bar{a}_n \in R$ for infinitely many n, then for any $\bar{c}$, we can find $\bar{a} \in R$ α-free over $\bar{c}$ not only in the current sense but also in the sense of the definition preceding Proposition 16.1. Then by Theorem 16.2, there is an isomorphism F from $\mathcal{A}$ onto a computable copy $\mathcal{B}$ such that $F(R)$ is not Σ^0_α. On the other hand, if $\bar{a}_n \in R$ for only finitely many n, then for any $\bar{c}$, we can find $\bar{a} \notin R$ such that $\bar{a}$ is α-free over $\bar{c}$ in the current sense, and, letting $\neg R$ play the role of R, $\bar{a}$ is also α-free over $\bar{c}$ in the old sense. We are in a position to apply Theorem 16.2, and we obtain an isomorphism F from $\mathcal{A}$ onto a computable $\mathcal{B}$ such that $F(R)$ is not Δ^0_α.

16.2 Davey's Theorems

We now turn to Problem 3, and give results of Davey [33]. Let $\mathcal{A}$ be a computable structure, and let R and S be a pair of further relations on $\mathcal{A}$, of the same arity. Suppose that there are computable Σ_α formulas $\varphi(\bar{c}, \bar{x})$ and $\psi(\bar{c}, \bar{x})$ such that

   (i) every tuple $\bar{a}$ (of the appropriate length) satisfies $\varphi(\bar{c}, \bar{x}) \vee \psi(\bar{c}, \bar{x})$,
   (ii) $\varphi(\bar{c}, \bar{x})$ is satisfied by all $\bar{a} \in R$ and no $\bar{a} \in S$, and
   (iii) $\psi(\bar{c}, \bar{x})$ is satisfied by all $\bar{a} \in S$ and no $\bar{a} \in R$.

Then by the easy direction of Theorem 10.12, in any computable copy of $\mathcal{A}$, the images of R and S have a Δ^0_α separator. The next two results, together, say that under some effectiveness conditions, the converse holds. In Theorem 11.11, we considered the case where $\alpha = 1$.

We define another notion of "freeness". We say that $\bar{a}$ is α-free over $\bar{c}$ if for all $\bar{a}_1$ and all $\beta < \alpha$, there exist $\bar{a}' \in R$, $\bar{a}'' \in S$, with associated tuples $\bar{a}'_1$, and $\bar{a}''_1$, such that

$$\bar{c}, \bar{a}, \bar{a}_1 \leq_\beta \bar{c}, \bar{a}', \bar{a}'_1 \quad \text{and} \quad \bar{c}, \bar{a}, \bar{a}_1 \leq_\beta \bar{c}, \bar{a}'', \bar{a}''_1 \ .$$

The result below says that if there is some $\bar{c}$ such that no $\bar{a}$ is α-free over $\bar{c}$, in this sense, then, under some effectiveness conditions, there exist computable Σ_α formulas $\varphi(\bar{c}, \bar{x})$ and $\psi(\bar{c}, \bar{x})$ such that

   (i) every tuple $\bar{a}$ (having the arity of R and S) satisfies $\varphi(\bar{c}, \bar{x})$ or $\psi(\bar{c}, \bar{x})$,
   (ii) $\varphi(\bar{c}, \bar{x})$ is satisfied by all $\bar{a} \in R$ and no $\bar{a} \in S$, and
   (ii) $\psi(\bar{c}, \bar{x})$ is satisfied by all $\bar{a} \in S$ and no $\bar{a} \in R$.

Proposition 16.5 *Let $\mathcal{A}$ be an α-friendly structure with a computable finitary Σ_1 diagram, and let R and S be further computable relations on $\mathcal{A}$. Suppose for some $\bar{c}$ in $\mathcal{A}$, no $\bar{a}$ is α-free over $\bar{c}$ (in the current sense). Suppose, moreover, that for each $\bar{a}$, we can find $\bar{a}_1$ and $\beta < \alpha$ witnessing the fact that $\bar{a}$ is not α-free over $\bar{c}$, and we can determine whether there exist $\bar{a}' \in R \cup S$ and $\bar{a}'_1$ such that*

$$\bar{c}, \bar{a}, \bar{a}_1 \leq_\beta \bar{c}, \bar{a}', \bar{a}'_1 \ .$$

Then there exist computable Σ_α formulas $\varphi(\bar{c}, \bar{x})$ and $\psi(\bar{c}, \bar{x})$ such that

   (a) every tuple $\bar{a}$ (of the appropriate length) satisfies $\varphi(\bar{c}, (\bar{x}) \vee \psi(\bar{c}, \bar{x})$,
   (b) $\varphi(\bar{c}, \bar{x})$ is satisfied by all $\bar{a} \in R$ and no $\bar{a} \in S$, and
   (c) $\psi(\bar{c}, \bar{x})$ is satisfied by all $\bar{a} \in S$ and no $\bar{a} \in R$.

Proof: Given $\bar{a}$, let $\bar{a}_1$ and $\beta < \alpha$ witness the fact that $\bar{a}$ is not α-free over $\bar{c}$. By Theorem 15.2, we can find a computable Π_β formula $\varphi^\beta_{\bar{c}, \bar{a}, \bar{a}_1}(\bar{c}, \bar{x}, \bar{u})$ saying that $\bar{c}, \bar{a}, \bar{a}_1 \leq_\beta \bar{c}, \bar{x}, \bar{u}$. Let $\theta_{\bar{a}}(\bar{c}, \bar{x}) = \exists \bar{u}\, \varphi^\beta_{\bar{c}, \bar{a}, \bar{a}_1}(\bar{c}, \bar{x}, \bar{u})$. Let $\varphi(\bar{c}, \bar{x})$ be the disjunction of the formulas $\theta_{\bar{a}}(\bar{c}, \bar{x})$ that are not satisfied by any $\bar{a}' \in S$— this is the disjunction of the c.e. set of formulas $\theta_{\bar{a}}(\bar{c}, \bar{x})$ satisfied in R and the computable set of formulas $\theta_{\bar{a}}(\bar{c}, \bar{x})$ that are not satisfied in $R \cup S$. Similarly, let $\psi(\bar{c}, \bar{x})$ be the disjunction of the formulas $\theta_{\bar{a}}(\bar{c}, \bar{x})$ not satisfied by any $\bar{a}'' \in R$.

Remarks: Above, we assumed that given $\bar{a}$, we can effectively find witnesses for the fact that $\bar{a}$ is not α-free over $\bar{c}$. In the case where $\alpha = 1$, this assumption is implied by the conditions of Theorem 11.11 (in particular, the fact that we can determine whether a given finitary existential formula $\varphi(\bar{c}, \bar{x})$ is satisfied by a tuple from R, or by one from S). In the case where $\alpha > 1$, the assumption can be dropped, in the presence of our other assumptions.

If $\mathcal{A}$ is α-friendly, and R and S are computable, then using Δ_2^0, we can decide whether there exist $\bar{a}' \in R$ and $\bar{a}_1'$ such that

$$\bar{c}, \bar{a}, \bar{a}_1 \leq_\beta \bar{c}, \bar{a}', \bar{a}_1',$$

or $\bar{a}'' \in S$ and $\bar{a}_1''$ such that

$$\bar{c}, \bar{a}, \bar{a}_1 \leq_\beta \bar{c}, \bar{a}'', \bar{a}_1'',$$

and we can find the desired witnesses $\bar{a}_1$ and β. If, in addition, the finitary Σ_1 diagram of $\mathcal{A}$ is computable, we have the formulas $\varphi^\beta_{\bar{c},\bar{a},\bar{a}_1}(\bar{c}, \bar{x}, \bar{u})$, as above, and from these, we obtain formulas $\varphi(\bar{c}, \bar{x})$ and $\psi(\bar{c}, \bar{x})$ with the properties in the statement of Proposition 16.5, except that the disjunctions are Σ_2^0 instead of c.e. Using Proposition 7.12, we can replace the Σ_2^0 disjunctions by c.e. disjunctions.

The next result says that if for each tuple $\bar{c}$ in $\mathcal{A}$, there exists a tuple $\bar{a}$ that is α-free over $\bar{c}$, then, under some effectiveness conditions, there is a computable copy of $\mathcal{A}$ in which the images of R and S are not Δ_α^0 separable.

Theorem 16.6 *Let $\mathcal{A}$ be a computable structure and let R and S be further relations on $\mathcal{A}$. Suppose $\mathcal{A}$ is α-friendly. Suppose that for any tuple $\bar{c}$ in $\mathcal{A}$, we can find another tuple $\bar{a}$ that is α-free over $\bar{c}$ (in the current sense). Then there is an isomorphism F from $\mathcal{A}$ onto a computable $\mathcal{B}$ such that $F(R)$ and $F(S)$ are not Δ_α^0 separable.*

Proof: For simplicity, we suppose that R and S are unary relations. As usual, let B be an infinite computable set of constants, and let $\mathcal{F}$ be the set of finite partial $1-1$ functions from B to $\mathcal{A}$. The construction involves carrying out a finite injury priority construction at level α. In addition to putting elements into the domain and range of F, we satisfy the following requirements.

$$R_{(e,e')}\colon\ B \subseteq W_e^{\Delta_\alpha^0} \cup W_{e'}^{\Delta_\alpha^0}\ \Rightarrow\ (W_e^{\Delta_\alpha^0} \cap F(S) \neq \emptyset \vee W_{e'}^{\Delta_\alpha^0} \cap F(R) \neq \emptyset).$$

The strategy for a single requirement $R_{(e,e')}$ is as follows. Given $p \in \mathcal{F}$ taking care of earlier requirements, let a be α-free over $ran(p)$, and let $p' \supseteq p$ map a new constant b to a. If b ever appears in $W_e^{\Delta_\alpha^0} \cup W_{e'}^{\Delta_\alpha^0}$, then we replace p' by $p'' \supseteq p$ such that either

$$p''(b) \in S\ \&\ b \in W_e^{\Delta_\alpha^0},\quad \text{or else}\quad p''(b) \in R\ \&\ b \in W_{e'}^{\Delta_\alpha^0}.$$

16.2. DAVEY'S THEOREMS

We define an α-system. Let U be the set of finite sets of statements of the form "$b \in W_e^{\Delta_\alpha^0}$", not necessarily true. Let L consist of the finite sequences

$$\ell = p_0 b_0 p_1 b_1 \ldots b_{r-1} p_r$$

satisfying the following conditions:
  (a) $p_0 = \emptyset$,
  (b) $p_{n+1} \supseteq p_n$,
  (c) $dom(p_n)$, $ran(p_n)$ include the first n constants from B, $\mathcal{A}$, respectively, and
  (c) $b_n \in dom(p_{n+1})$.

Let $\hat{\ell} = p_0$, where $p_0 = \emptyset$. We define E and $\leq_\beta$ in terms of the standard enumeration function and the standard back-and-forth relations on $\mathcal{F}$. For $\ell \in L$ ending in p, let $E(\ell) = E(p)$. For ℓ and ℓ' in L, where ℓ ends in p and ℓ' ends in q, let

$$\ell \leq_\beta \ell' \Leftrightarrow p \leq_\beta q \ .$$

As in Theorem 16.2, the tree P will have the feature that a finite-injury priority construction is carried out along the paths. For $u \in U$, let $L(u)$ be the set of $\ell = p_0 b_0 p_1 b_1 \ldots b_{r-1} p_r$ in L such that for each $n < r$, p_{n+1} and b_n take care of R_n according to u, where this means that if $n = \langle e, e' \rangle$, then one of the following holds:

  (i) "$b_n \in W_e^{\Delta_\alpha^0}$" is in u and $p_{n+1}(b_n) \in S$,
  (ii) "$b_n \in W_{e'}^{\Delta_\alpha^0}$" is in u and $p_{n+1}(b_n) \in R$, or
  (iii) u does not contain either of "$b_n \in W_e^{\Delta_\alpha^0}$", "$b_n \in W_{e'}^{\Delta_\alpha^0}$", and $p_{n+1}(b_n)$ is the element a effectively chosen to be α-free over $ran(p_n)$.

Now, we let P be the set of finite sequences

$$\hat{\ell} u_1 \ell_1 u_2 \ell_2 \ldots$$

such that the following conditions hold:
  (a) $u_n \in U$,
  (b) $\ell_n \in L(u_n)$,
  (c) $u_n \subseteq u_{n+1}$,
  (d) $p_r \leq_0 q$, and
  (d) if $\ell_n = p_0 b_0 \ldots b_{r-1} p_r$, then either
    (i) $\ell_n \in L(u_{n+1})$, in which case, ℓ_{n+1} has the form $p_0 b_0 \ldots b_{r-1} p_r b_r p_{r+1}$, or
    (ii) $\ell_n \notin L(u_{n+1})$, in which case, ℓ_{n+1} has the form $p_0 b_0 \ldots p_k b_k p'_{k+1}$, for the first $k < r$ such that p_{k+1} does not satisfy R_k according to u_{k+1}.

Remark: For any path $\pi = \hat{\ell} u_1 \ell_1 u_2 \ell_2 \ldots$ through P, there is a sequence of numbers $k(n)$ such that $\ell_{k(n)}$ has the form $p_0 b_0 \ldots b_{n-1} p_n$, and for all $k > k(n)$, $\ell_k \supseteq \ell_{k(n)}$. Then $\cup_n p_n$ is a one-one function from B onto $\mathcal{A}$. The inverse F of this function induces a structure $\mathcal{B}$ on B such that $E(\pi) = D(\mathcal{B})$. Moreover, this F satisfies all of the requirements, according to the information in $\cup_n u_n$.

We have defined all of the ingredients of the α-system. Before checking the conditions, we define the Δ_α^0 instruction function q. If σ is a sequence in P of length $2n+1$, then $q(\sigma)$ is the set of true statements "$b \in W_e^{\Delta_\alpha^0}$" that are enumerated by step n using the oracle Δ_α^0.

As usual, the first three conditions clearly hold. Toward Condition (4), suppose that $\sigma u \in P$, where σ ends in ℓ^0, and

$$\ell^0 \leq_{\beta_0} \ell^1 \leq_{\beta_1} \ldots \leq_{\beta_{k-1}} \ell^k ,$$

for $\alpha > \beta_0 > \ldots > \beta_k$. We need ℓ^* such that $\sigma u \ell^* \in P$ and $\ell^i \leq_{\beta_i} \ell^*$. Say $\ell^0 = p_0 b_0 \ldots b_{r-1} p_r$. We have $p_{r+1} \supseteq p_r$ such that for $1 \leq i \leq k$, if ℓ^i ends in p_i, then $p_i \leq_{\beta_i} p_{r+1}$. There are two cases to consider.

Case 1: Suppose that $\ell^0 \notin L(u)$.

Let n be first such that p_{n+1} does not take care of R_n, where $n = \langle e, e' \rangle$. This means that u contains one of the statements "$b_n \in W_e^{\Delta_\alpha^0}$" or "$b_n \in W_{e'}^{\Delta_\alpha^0}$", and $p_{n+1}(b_n)$ is equal to the element a effectively chosen to be α-free over $ran(p_n)$. Take $p'_{n+1} \supseteq p_n$ such that $p_{r+1} \leq_{\beta_0} p'_{n+1}$ and either $p'_{n+1}(b_n) \in S$, while "$b_n \in W_e^{\Delta_\alpha^0}$" is in u, or else $p'_{n+1}(b_n) \in R$, while "$b_n \in W_{e'}^{\Delta_\alpha^0}$" is in u.

Case 2: Suppose that $\ell^0 \in L(u)$.

Let a be α-free over $ran(p_r)$. We may suppose that $a \in ran(q)$, say $q(b_r) = a$. If u does not contain either of the statements "$b_r \in W_e^{\Delta_\alpha^0}$", "$b_r \in W_{e'}^{\Delta_\alpha^0}$", then the required ℓ^* is $p_0 b_0 \ldots b_{r-1} p_r b_r q$. If u contains one of the statements, then we take $q' \supseteq p_r$ such that $q \leq_{\beta_0} q'$ and either $q'(b) \in S$, while "$b \in W_e^{\Delta_\alpha^0}$" is in u, or else $q'(b) \in R$, while "$b \in W_{e'}^{\Delta_\alpha^0}$" is in u. Then the required ℓ^* is $p_0 b_0 \ldots b_{r-1} p_r b q'$.

We are in a position to apply Theorem 14.1. We obtain a run π of (P, q) such that $E(\pi)$ is c.e. By the remark above, this yields an isomorphism F from $\mathcal{A}$ onto a structure $\mathcal{B}$ such that $D(\mathcal{B}) = E(\pi)$. Then $\mathcal{B}$ is computable, and the requirements are all satisfied according to the true information given by the instruction function q. Therefore, $F(R)$ and $F(S)$ have no Δ_α^0 separator.

Chapter 17

Δ^0_α stability and categoricity

In this chapter, we lift results from Chapter 12 to higher levels in the hyperarithmetical hierarchy. We consider the following problems.

Problem 1 *For a computable structure $\mathcal{A}$, when is there a computable copy $\mathcal{B}$ with an isomorphism F from $\mathcal{A}$ onto $\mathcal{B}$ such that F is not Δ^0_α?*

Problem 2 *For a computable structure $\mathcal{A}$, when is there a computable copy $\mathcal{B}$ such that no isomorphism from $\mathcal{A}$ onto $\mathcal{B}$ is Δ^0_α?*

Problem 1 was the subject of [1] and [3], and Problem 2 was the subject of [2]. In Chapter 12, we discussed results of Goncharov on these problems, for the case where $\alpha = 1$. We defined the notions of *computable categoricity* and *computable stability*, which we now generalize. A computable structure $\mathcal{A}$ is Δ^0_α *categorical* if for every computable $\mathcal{B} \cong \mathcal{A}$, there is a Δ^0_α isomorphism from $\mathcal{A}$ onto $\mathcal{B}$, and $\mathcal{A}$ is Δ^0_α *stable* if for every computable $\mathcal{B} \cong \mathcal{A}$, every isomorphism from $\mathcal{A}$ onto $\mathcal{B}$ is Δ^0_α.

To prove general theorems on Δ^0_α categoricity and Δ^0_α stability, we use the metatheorem on α-systems, from Chapter 14. In fact, these were the first applications of the metatheorem. Using the general theorems on categoricity and stability, together with some calculations of back-and-forth relations, which we discussed in Chapter 15, we can characterize the Δ^0_α stable well orderings, and the Δ^0_α categorical superatomic Boolean algebras, for an arbitrary computable ordinal α. The general results and the applications are all in [1], [2], and [3]. Throughout the chapter, we refer to the standard back-and-forth relations $\leq_\beta$ and the standard enumeration function E.

17.1 Relations between notions

Recall that $\mathcal{A}$ is *almost rigid* if it becomes rigid after naming finitely many elements. In Chapter 12, we showed that $\mathcal{A}$ is computably stable if and only if it is computably categorical and almost rigid. This result extends.

Proposition 17.1 *Let $\mathcal{A}$ be a computable structure. Then $\mathcal{A}$ is Δ^0_α stable if and only if it is Δ^0_α categorical and almost rigid.*

The proof for Proposition 17.1 is essentially the same as that for Proposition 12.1.

17.2 Δ^0_α stability

Recall from Chapter 10 that a formally Σ^0_α defining family for $\mathcal{A}$ is a c.e. defining family made up of computable Σ_α formulas, with a fixed finite tuple of parameters from $\mathcal{A}$. We saw that a Σ^0_α defining family consisting of computable Σ_α formulas can be converted into a c.e. defining family, also consisting of computable Σ_α formulas. By Theorem 10.9, for any computable structure $\mathcal{A}$, the following are equivalent:
(1) $\mathcal{A}$ is relatively Δ^0_α stable,
(2) $\mathcal{A}$ is has a formally Σ^0_α defining family.

Obviously, if a computable structure $\mathcal{A}$ is relatively Δ^0_α stable, then it is Δ^0_α stable. The next two results, taken together, say that under certain effectiveness conditions on $\mathcal{A}$, the converse holds. We must define an appropriate notion of freeness. For a tuple $\bar{c}$ and an element a in $\mathcal{A}$, we say that a is α-*free over* $\bar{c}$ if for any $\bar{a}_1$ and any $\beta < \alpha$, there exist $a' \neq a$ and $\bar{a}'_1$ such that

$$\bar{c}, a, \bar{a}_1 \leq_\beta \bar{c}, a', \bar{a}'_1 \ .$$

The result below says that if $\mathcal{A}$ has a tuple of elements over which no further element is α-free, then under some effectiveness conditions, $\mathcal{A}$ has a formally Σ^0_α defining family.

Proposition 17.2 *Let $\mathcal{A}$ be an α-friendly structure with computable existential diagram. Suppose for some tuple $\bar{c}$, no element is α-free over $\bar{c}$. Then $\mathcal{A}$ has a formally Σ^0_α defining family with parameters $\bar{c}$.*

Proof: For each a in $\mathcal{A}$, there exist $\bar{a}_1$ and $\beta < \alpha$ such that for all a' and $\bar{a}'_1$,

$$\bar{c}, a, \bar{a}_1 \leq_\beta \bar{c}, a', \bar{a}'_1 \Rightarrow a = a' \ .$$

We refer to $\bar{a}_1$ and β as *witnesses* corresponding to a.

Claim: (a) If $\alpha = 1$, then there is an effective procedure for finding witnesses. (In this case, $\beta = 0$, so we just need to find the witnessing *tuples*.)
(b) If $\alpha > 1$, then there is a Δ^0_2 procedure for finding witnesses.

Proof of Claim: (a) Let E be the standard enumeration function on tuples from $\mathcal{A}$. Given a, we search for $\bar{a}_1$ such that if $\delta(\bar{c}, a, \bar{a}_1)$ is the conjunction of the sentences of $E(\bar{c}, a, \bar{a}_1)$, then

$$\mathcal{A} \models \neg\, \exists x\, \exists \bar{u}\, (\, \delta(\bar{c}, x, \bar{u})\, \&\, x \neq a\,) \ .$$

17.2. Δ_α^0 STABILITY

(b) Given a, we can apply an effective procedure to any $\bar{a}_1$ and any $\beta < \alpha$ to search for $a' \neq a$ and $\bar{a}'_1$ such that

$$\bar{c}, a, \bar{a}_1 \leq_\beta \bar{c}, a', \bar{a}'_1 .$$

Using Δ_2^0, we can decide, for any given $\bar{a}_1$ and β, whether this search halts, so by testing systematically, we can find $\bar{a}_1$ and β for which it does not. This proves the claim.

Given a, and having located appropriate witnesses $\bar{a}_1$ and $\beta < \alpha$, we can find a computable Π_β formula $\varphi^\beta_{a,\bar{a}_1}(\bar{c}, x, \bar{u})$ saying that $\bar{c}, a, \bar{a}_1 \leq_\beta \bar{c}, x, \bar{u}$. (This is by Theorem 15.2.) Then a is defined by the computable Σ_α formula

$$\psi_a(\bar{c}, x) = \exists \bar{u}\, \varphi^\beta_{a,\bar{a}_1}(\bar{c}, x, \bar{u}) .$$

Then the set

$$\{\psi_a(\bar{c}, x) : a \in \mathcal{A}\}$$

is the desired formally Σ_α^0 defining family.

The next result says that if for each tuple $\bar{c}$ in $\mathcal{A}$, some element is α-free over $\bar{c}$, then, under certain effectiveness conditions, $\mathcal{A}$ is not Δ_α^0 stable.

Theorem 17.3 *Suppose that $\mathcal{A}$ is α-friendly, and for each tuple $\bar{c}$, in $\mathcal{A}$, we can find an element a that is α-free over $\bar{c}$. Then there is an isomorphism F from $\mathcal{A}$ onto a computable copy $\mathcal{B}$ such that F is not Δ_α^0.*

Proof: Let B be an infinite computable set of constants, for the universe of $\mathcal{B}$. We shall construct a $1-1$ function F from $\mathcal{A}$ onto B and let $\mathcal{B}$ be the structure induced by F. In addition to putting elements into the domain and range of F, we satisfy the following requirements:

$$R_e : \varphi_e^{\Delta_\alpha^0} \neq F$$

Viewed at level $\alpha + 1$, the strategy for R_e is as follows: Given a finite part p of F^{-1} taking care of earlier requirements, where p maps $\bar{d}$ to $\bar{c}$, we find an element a that is α-free over $\bar{c}$. If $\varphi_e^{\Delta_\alpha^0}(a) \uparrow$, then the requirement is satisfied trivially. If $\varphi_e^{\Delta_\alpha^0}(a) \downarrow = b$, then we satisfy the requirement by taking $q \supseteq p$, where q maps b to some $a' \neq a$.

There is no priority construction on the top level. As a consequence, the $(\alpha + 1)$-system

$$(L, U, \hat{\ell}, P, E, (\leq_\beta)_{\beta \leq \alpha})$$

and the $\Delta_{\alpha+1}^0$ instruction function q that we describe below are relatively simple. Let L be the set of finite partial $1-1$ functions p from B to $\mathcal{A}$. Let U consist of all statements of the forms $\varphi_e^{\Delta_\alpha^0}(a) = b$ and $\varphi_e^{\Delta_\alpha^0}(a) \uparrow$ (true or not). Let $\hat{\ell} = \emptyset$.

Let E be the standard enumeration function on L. For $\beta < \alpha$, let $\leq_\beta$ be the standard back-and-forth relation on L, and let $\leq_\alpha$ be $\subseteq$.

We are ready to define the tree P. This consists of the finite alternating sequences
$$p_0 u_1 p_1 u_2 p_2 \ldots ,$$
such that $u_n \in U$, $p_n \in L$, and the following conditions are satisfied:

(a) $p_0 = \emptyset \ (= \hat{\ell})$,
(b) $p_n \subseteq p_{n+1}$,
(c) $dom(p_n)$, $ran(p_n)$ include the first n elements of B, $\mathcal{A}$, respectively,
(d) if a is the element effectively chosen to be α-free over $ran(p_n)$, then u_{n+1} has form either $\varphi_n^{\Delta_\alpha^0}(a) \uparrow$, or $\varphi_n^{\Delta_\alpha^0}(a) = b$, and if it is $\varphi_n^{\Delta_\alpha^0}(a) = b$, then p_{n+1} takes b to some $a' \neq a$ (so p_{n+1} takes care of requirement R_e according to the information in u_{n+1}).

Remarks: For any path $\pi = p_0 u_1 p_1 u_2 p_2 \ldots$ through P, $\cup_n p_n$ is a $1-1$ function from B onto $\mathcal{A}$. Letting F be the inverse of this function, we have $\mathcal{B}$ such that $\mathcal{A} \cong_F \mathcal{B}$ and $D(\mathcal{B}) = E(\pi)$.

We have determined all of the ingredients of the $(\alpha+1)$-system. We must check the conditions. The first three are obvious. Toward Condition (4), suppose $\sigma u \in P$, where σ ends in p^0, and
$$p^0 \subseteq p^1 \leq_{\beta_1} \ldots \leq_{\beta_{k-1}} p^k ,$$
for $\alpha > \beta_1 > \ldots > \beta_k$. We need p^* such that $\sigma u p^* \in P$ and for $1 \leq i \leq k$, we have $p^i \leq_{\beta_i} p^*$.

There exist $q^i \supseteq p^i$, for $0 \leq i \leq k$, such that for $i > 1$, $q^i \leq_{\beta_i} q^{i-1}$. Let a be the element effectively chosen to be α-free over $ran(p^0)$. First, suppose that for the appropriate e, either u says that $\varphi_e^{\Delta_\alpha^0}(a) \uparrow$, or else u says that $\varphi_e^{\Delta_\alpha^0}(a) = b$, where q^0 does not take b to a. In either of these cases, q^0, possibly extended, is the required p^*. Now, suppose u says that $\varphi_e^{\Delta_\alpha^0}(a) = b$, and q^0 takes b and a further tuple $\bar{b}_1$ to a and $\bar{a}_1$. Since a is α-free over $ran(p^0)$, there is some $q \supseteq p^0$ such that q takes $b, \bar{b}_1$ to $a', \bar{a}_1'$, where $a' \neq a$ and $q^0 \leq_{\beta_1} q$. In this case, q, possibly extended, is the required p^*.

We define the instruction function q as follows. For $\sigma \in P$ of length $2n+1$, ending in p, let a be the element effectively chosen to be α-free over $ran(p)$, and let $q(\sigma)$ be the true statement of form $\varphi_n^{\Delta_\alpha^0}(a) = b$ or $\varphi_n^{\Delta_\alpha^0}(a) \uparrow$. Clearly, q is $\Delta_{\alpha+1}^0$. We are in a position to apply Theorem 14.1. We obtain a run π of (P, q) such that $E(\pi)$ is c.e. By the remarks above, π yields an isomorphism F from $\mathcal{A}$ onto a structure $\mathcal{B}$ such that $D(\mathcal{B}) = E(\pi)$. Then $\mathcal{B}$ is computable. The definitions of P and q guarantee that all of the requirements are satisfied. Therefore, F is not Δ_α^0. This completes the proof of Theorem 17.3.

Remarks: Suppose that for each tuple $\bar{c}$, there is an element a α-free over $\bar{c}$. In the case where $\alpha = 1$, if we assume only that the existential diagram of $\mathcal{A}$

17.2. Δ^0_α STABILITY

is computable, as Goncharov did in Theorem 12.7, then we may not be able to effectively locate the 1-free elements. Thus, we cannot derive Theorem 12.7 from Proposition 17.2 and Theorem 17.3. Below we state a variant of Theorem 17.3 that, in combination with Proposition 17.2, does yield Theorem 12.7.

Theorem 17.4 *Suppose $\mathcal{A}$ is α-friendly, and there is a Δ^0_α procedure which, when applied to a tuple $\bar{c}$, yields an element a that is α-free over $\bar{c}$. Then $\mathcal{A}$ is not Δ^0_α stable.*

Let us see how the hypotheses of Theorem 12.7 yield the hypotheses of Theorem 17.4, in the case where $\alpha = 1$. Any computable structure is 1-friendly. If the existential diagram of $\mathcal{A}$ is computable, then we can effectively determine whether a given element a is defined by a given finitary existential formula $\varphi(\bar{c}, x)$. If, for each $\bar{c}$, there is an element a that is not definable by any finitary existential formula $\varphi(\bar{c}, x)$, then using a Δ^0_2 oracle, we can find such an a, and a is 1-free over $\bar{c}$.

To prove Theorem 17.4, we cannot apply Theorem 14.1, but the modifications are fairly minor. Here we give a brief sketch.

Sketch of proof of Theorem 17.4: In the $(\alpha + 1)$-system from the proof of Theorem 17.3, we change the definition of the tree P, replacing condition (d) by the following:

(d)$'$ $a \notin ran(p_n)$, u_{n+1} has form $\varphi_n^{\Delta^0_\alpha}(a) \uparrow$ or $\varphi_n^{\Delta^0_\alpha}(a) = b$, and if it is $\varphi_n^{\Delta^0_\alpha}(a) = b$, then $p_{n+1}(b) \neq a$. (The difference between (d) and (d)$'$ is that in the latter, the element a need not be α-free over $ran(p_n)$.)

The new tree P is fatter. The terms in U refer to elements that may or may not be α-free. The new instruction function q is defined in the same way as in the earlier proof—using trivial values on the extra nodes. It gives true statements about elements that actually are α-free. The runs of (P, q) are the same as in the proof of Theorem 17.3. With the fatter tree, we no longer have an $(\alpha + 1)$-system—Condition (4) need not hold. Nonetheless, if we think through the proof of Theorem 14.1, we find that we can salvage the conclusion.

From the new tree P and the new instruction function q, we derive a family of trees P^β and corresponding instruction functions q_β, for $1 \leq \beta \leq \alpha + 1$. The new trees are again uniformly c.e., but they are fatter than the trees that we would get starting from the thin tree in the proof of Theorem 17.3. The instruction functions q_β are again Δ^0_β, uniformly in β. For $\beta \geq 2$, q_β involves true statements about elements that are α-free. For $\beta \geq 2$, the runs of (P^β, q_β) are the same as those of old tree and the old instruction function at level β. For $\beta = 1$, we define $q_1(\sigma)$ always to be a picture c for which there is a completion p such that $\sigma c p \in P^1$. We approximate q_2, and we determine σ^+, and a tentative c to serve as $q_1(\sigma)$, as we did in the proof of the metatheorem. However, we do not accept c unless we have also, in enumerating P^1, found an appropriate p to complete the picture. We update our approximations for q_2, and we revise σ^+ and our tentative choice of c until we also have the required p.

17.3 Well orderings

We can apply the general theorems on stability to determine for which ordinals β a given well ordering is Δ^0_α stable. The result below is from [3].

Theorem 17.5 *Suppose α is a computable ordinal, with $\omega^{\delta+n} \leq \alpha < \omega^{\delta+n+1}$, where δ is either 0 or a limit ordinal, and $n < \omega$. Then α is $\Delta^0_{\delta+2n}$ stable, and for $\beta < \delta + 2n$, α is not Δ^0_β stable.*

Proof: First, we show that α is $\Delta^0_{\delta+2n}$ stable. By Proposition 7.2, for each $\gamma < \alpha$, we can find a computable $\Sigma_{\delta+2n}$ formula $\varphi_\gamma(x)$, with no parameters, such that $\varphi_\gamma(x)$ defines γ (in α). Thus, we have a formally $\Sigma^0_{\delta+2n}$ defining family. It follows that α is relatively $\Delta^0_{\delta+2n}$ stable.

Now, using Theorem 17.3, we show that for $\beta < \delta + 2n$, α is not Δ^0_β stable. By Theorem 4.19, there is a computable well ordering of type α in which we can effectively determine the Cantor normal form of intervals. As we observed in Proposition 15.11, this ordering is β-friendly. We must say how to find an element that is β-free over a given tuple $\bar{c}$. We can find an interval I of type $\omega^{\delta+n}$ containing no point of $\bar{c}$.

Case 1: Suppose $n = 0$, so $\beta < \delta$.

We can find elements a and a' of I such that the interval to the left of a in I has type ω^β and the interval to the left of a' in I has type $\omega^\beta \cdot 2$. The intervals to the right of a and a' in I both have type ω^δ. By Lemma 15.9,

$$\omega^\beta \geq_\beta \omega^\beta \cdot 2 \ .$$

It follows that in α, $\bar{c}, a \geq_\beta \bar{c}, a'$. Therefore, a is β-free over $\bar{c}$.

Case 2: Suppose $n = m + 1$, so $\beta \leq \delta + 2m + 1$.

The interval I has type $\omega^{\delta+m+1}$. We can find elements a and a' of I such that the interval to the left of a in I has type $\omega^{\delta+m}$, and the interval to the left of a' in I has type $\omega^{\delta+m} \cdot 2$. The intervals to the right of a and a' in I both have type $\omega^{\delta+m+1}$. By Lemma 15.9,

$$\omega^{\delta+m} \geq_{\delta+2m+1} \omega^{\delta+m} \cdot 2 \ ,$$

so we have

$$\omega^{\delta+m} \geq_\beta \omega^{\delta+m} \cdot 2 \ .$$

Then in α, $\bar{c}, a \geq_\beta \bar{c}, a'$, and a is β-free over $\bar{c}$.

We are in a position to apply Theorem 17.3. We get the fact that α is not Δ^0_β stable. This completes the proof of Theorem 17.5.

17.4 Δ_α^0 categoricity

By Theorem 10.14, for any computable structure $\mathcal{A}$, the following are equivalent:
(1) $\mathcal{A}$ is relatively Δ_α^0 categorical,
(2) $\mathcal{A}$ has a formally Σ_α^0 Scott family..

We have seen that if there is a Σ_α^0 Scott family consisting of computable Σ_α formulas, then there is a c.e. Scott family also consisting of these formulas.

If $\mathcal{A}$ is relatively Δ_α^0 categorical, then it is obviously Δ_α^0 categorical. Our goal is to prove the converse, under some effectiveness assumptions. For the case where $\alpha = 1$, we have the result of Goncharov (Theorem 12.3) saying that if the finitary Π_2 diagram of $\mathcal{A}$ is computable, then $\mathcal{A}$ is computably categorical if and only if it has a formally c.e. Scott family.

We define yet another notion of freeness. For tuples $\bar{c}$ and $\bar{a}$ in $\mathcal{A}$, we say that $\bar{a}$ is α-*free* over $\bar{c}$ if for all $\bar{a}_1$ and all $\beta < \alpha$, there exist $\bar{a}'$ and $\bar{a}_1'$ such that $\bar{c}, \bar{a}, \bar{a}_1 \leq_\beta \bar{c}, \bar{a}', \bar{a}_1'$ and $\bar{c}, \bar{a}' \not\leq_\alpha \bar{c}, \bar{a}$. The result below says that if $\mathcal{A}$ has a tuple $\bar{c}$ over which no further tuple $\bar{a}$ is α-free, then, under some effectiveness conditions, $\mathcal{A}$ has a formally Σ_α^0 Scott family.

Proposition 17.6 *Suppose $\mathcal{A}$ is α-friendly, with computable existential diagram. Suppose that there is a tuple $\bar{c}$ in $\mathcal{A}$ over which no tuple $\bar{a}$ is α-free (in the sense above). Suppose, moreover, that for each $\bar{a}$, we can find a tuple $\bar{a}_1$ and an ordinal $\beta < \alpha$ witnessing the fact that $\bar{a}$ is not α-free over $\bar{c}$. Then $\mathcal{A}$ has a formally Σ_α^0 Scott family, with parameters $\bar{c}$.*

Proof: Let $\bar{a}$ be a tuple from $\mathcal{A}$. We can find a tuple $\bar{a}_1$ and an ordinal $\beta < \alpha$ witnessing the fact that $\bar{a}$ is not α-free over $\bar{c}$. Then by Theorem 15.2, we can find a computable Π_β formula $\varphi_{\bar{a},\bar{a}_1}^\beta(\bar{c}, \bar{x}, \bar{u})$ saying that

$$\bar{c}, \bar{a}, \bar{a}_1 \leq_\beta \bar{c}, \bar{x}, \bar{u} \ .$$

Let $\psi(\bar{c}, \bar{x}) = \exists \bar{u}\, \varphi_{\bar{a},\bar{a}_1}^\beta(\bar{c}, \bar{x}, \bar{u})$. By Proposition 6.10, the set of such formulas $\psi(\bar{c}, \bar{x})$ is the desired Scott family.

The next result says that if for each tuple $\bar{c}$ in $\mathcal{A}$, there is a further tuple that is α-free over $\bar{c}$, then under some effectiveness conditions, $\mathcal{A}$ is not Δ_α^0 categorical.

Theorem 17.7 *Let $\mathcal{A}$ be α-friendly. Suppose that for each tuple $\bar{c}$ in $\mathcal{A}$, we can find a tuple $\bar{a}$ that is α-free over $\bar{c}$. Finally, suppose that the relation $\not\leq_\alpha$ is c.e. Then there is a computable $\mathcal{B} \cong \mathcal{A}$ with no Δ_α^0 isomorphism F from $\mathcal{A}$ onto $\mathcal{B}$.*

Proof: Let B be an infinite computable set of constants. Let $\mathcal{F}$ be the set of finite partial $1-1$ functions from B to $\mathcal{A}$. We produce a computable structure $\mathcal{B}$ with universe B and an isomorphism F from $\mathcal{A}$ onto $\mathcal{B}$ so as to satisfy the following requirements:

R_e: $\varphi_e^{\Delta_\alpha^0}$ is not an isomorphism from $\mathcal{A}$ onto $\mathcal{B}$

We shall carry out what, viewed from level α, is a finite-injury priority construction. The strategy for R_e is as follows. Let $p \in \mathcal{F}$ be the finite part of F^{-1} being preserved for earlier requirements, say p maps $\overline{d}$ to $\overline{c}$. We take the effectively determined tuple $\overline{a}$ that is α-free over $\overline{c}$ in the current sense, and we extend p, mapping $\overline{d}$ and some $\overline{b}$ to $\overline{c}, \overline{a}$. We continue extending until we see that $\varphi_e^{\Delta_\alpha^0}$ takes some $\overline{c}^*, \overline{a}^*$ to $\overline{d}, \overline{b}$. Say we then have $q \supseteq p$ taking $\overline{d}, \overline{b}, \overline{b}_1$ to $\overline{c}, \overline{a}, \overline{a}_1$. Then we can find evidence establishing that one of the following holds:

Case 1: $\overline{c}^*, \overline{a}^* \not\leq_\alpha \overline{c}, \overline{a}$,

Case 2: there exists $\overline{a}'$ such that $\overline{c}, \overline{a}' \not\leq_\alpha \overline{c}, \overline{a}$ and $\overline{c}, \overline{a}' \not\leq_\alpha \overline{c}^*, \overline{a}^*$.

In Case 2, we may ask for more—for any $\beta < \alpha$, there exist $\overline{a}'$ and $\overline{a}'_1$ such that $\overline{c}, \overline{a}' \not\leq_\alpha \overline{c}, \overline{a}$, in addition to the conditions above—$\overline{c}, \overline{a}' \not\leq_\alpha \overline{c}^*, \overline{a}^*$, and $\overline{c}, \overline{a}, \overline{a}_1 \leq_\beta \overline{c}, \overline{a}'\overline{a}'_1$.

Note that if $\overline{c}^*, \overline{a}^* \leq_\alpha \overline{c}, \overline{a}$; i.e., if Case 1 fails, then $\overline{c}, \overline{a}' \not\leq_\alpha \overline{c}, \overline{a}$ implies $\overline{c}, \overline{a}' \not\leq_\alpha \overline{c}^*, \overline{a}^*$, so the existence of $\overline{a}'$, and $\overline{a}'_1$ as in Case 2 follows from the fact that $\overline{a}$ is α-free over $\overline{c}$.

In Case 1, we satisfy R_e by extending q. In Case 2, we replace q by some $q' \supseteq p$ taking $\overline{b}, \overline{b}_1$ to $\overline{a}', \overline{a}'_1$, and we satisfy R_e by extending q'.

We shall define an α-system

$$(L, U, \hat{\ell}, P, E, (\leq_\beta)_{\beta < \alpha})$$

and a Δ_α^0 instruction function q such that a run of (P, q) takes care of all of the requirements, using the strategy above. Let U consist of the finite sets of statements $\varphi_e^{\Delta_\alpha^0}(a) = b$, true or false. Let L be the set of finite sequences

$$\ell = p_0 \overline{b}_0 p_1 \ldots p_{r-1} \overline{b}_{r-1} p_r ,$$

such that
(i) $p_0 = \emptyset$,
(ii) $p_n \subseteq p_{n+1}$,
(iii) $dom(p_n)$ includes the first n elements of $\mathcal{B}$ and $ran(p_n)$ includes the first n elements of $\mathcal{A}$,
(iv) $\overline{b}_n$ is in $dom(p_{n+1})$.

We extend the standard enumeration function and the standard back-and-forth relations to L. For $\ell \in L$ ending in p, let $E(\ell) = E(p)$. For ℓ and ℓ' in L, where ℓ ends in p and ℓ' ends in p', let

$$\ell \leq_\beta \ell' \Leftrightarrow p \leq_\beta p' .$$

17.4. Δ^0_α CATEGORICITY

Let $\hat{\ell} = p_0$, where $p_0 = \emptyset$. We are almost ready to define P, but first, we describe some special subsets of L, corresponding to elements of U. For $u \in U$, let $L(u)$ consist of the elements of L of the form

$$p_0 \overline{b}_0 p_1 \ldots \overline{b}_{r-1} p_r$$

such that for all $e < r$, p_{e+1} *satisfies R_e according to* u, where this means the following. Say that p_e maps $\overline{d}$ to $\overline{c}$, and let $\overline{a}$ be the tuple effectively chosen to be α-free over $\overline{c}$. If u does not include statements putting $\overline{d}, \overline{b}_e$ in the range of $\varphi_e^{\Delta^0_\alpha}$, then p_{e+1} is an extension of p_e taking $\overline{b}_e$ to $\overline{a}$. If u contains statements saying that $\varphi_e^{\Delta^0_\alpha}$ takes some $\overline{c}^*, \overline{a}^*$ to $\overline{d}, \overline{b}_e$, then either p_{e+1} takes $\overline{d}, \overline{b}_e$ to $\overline{c}, \overline{a}$ and

$$\overline{c}^*, \overline{a}^* \not\leq_\alpha \overline{c}, \overline{a} \, ,$$

or else p_{e+1} takes $\overline{d}, \overline{b}$ to $\overline{c}, \overline{a}'$, for some $\overline{a}'$ such that

$$\overline{c}, \overline{a}' \not\leq_\alpha \overline{c}^*, \overline{a}^* \, .$$

Now, let P consist of the finite sequences $\hat{\ell} u_1 \ell_1 u_2 \ell_2 \ldots$ such that
(a) $u_{n+1} \supseteq u_n$,
(b) $\ell_n \in L(u_n)$,
(c) for $\ell_n = p_0 \overline{b}_0 \ldots p_r$, either ℓ_{n+1} has the form

$$p_0 \overline{b}_0 \ldots p_r \overline{b}_r p_{r+1} \, ,$$

or for some $e < r$, p_{e+1} does not appear to satisfy R_e according to u_{n+1} (we say more about this below), and ℓ_{n+1} has the form

$$p_0 \ldots p_e \overline{b}_e p'_{e+1} \, .$$

If p_e takes $\overline{d}$ to $\overline{c}$, $\overline{a}$ is the tuple effectively chosen to be α-free over $\overline{c}$, and p_{e+1} satisfies R_e according to u_n but does not appear to do so according to u_{n+1}, then $\varphi_e^{\Delta^0_\alpha}$ takes some $\overline{c}^*, \overline{a}^*$ to $\overline{d}, \overline{b}_e$, with a halting computation bounded by $n+1$ but not bounded by n, p_e takes $\overline{d}, \overline{b}_e$ to $\overline{c}, \overline{a}$, and (from our enumeration so far), we have not seen that $\overline{c}^*, \overline{a}^* \not\leq_\alpha \overline{c}, \overline{a}$. We have defined all of the ingredients of the α-system. We define the instruction function q such that for $\sigma \in P$ of length $2n+1$, $q(\sigma)$ is the set of true statements $\varphi_e^{\Delta^0_\alpha}(a) = b$ for which the halting computation is bounded by n.

The first three conditions for an α-system clearly hold. Toward Condition (4), take $\sigma u \in P$, where σ ends in $\ell^0 = p_0 \ldots p_{r-1} \overline{b}_{r-1} p_r$. Suppose that

$$\ell^0 \leq_{\beta_0} \ldots \leq_{\beta_{k-1}} \ell^k \, ,$$

for $\alpha > \beta_0 > \ldots > \beta_{k-1} > \beta_k$. Let q^i be the last function in ℓ^i. There exists $p_{r+1} \supseteq p_r$ such that for $1 \leq i \leq k$, we have $q^i \leq_{\beta_i} p_{r+1}$. We may suppose that $ran(p_{r+1})$ includes the tuple $\overline{a}$ effectively chosen to be α-free over $ran(p_r)$. Say that p_{r+1} takes $\overline{b}_r$ to $\overline{a}$. Suppose, furthermore, that the domain and range of p_{r+1} include the first $r+1$ elements of $\mathcal{B}, \mathcal{A}$, respectively.

If we find that
$$p_0\bar{b}_0 \ldots \bar{b}_{r-1}p_r\bar{b}_r p_{r+1} \in L(u) ,$$
then this is the required ℓ^* satisfying Condition (4). If not, then there is some first $e \leq r$ such that p_{e+1} does not appear to satisfy R_e according to u. Say that p_e maps $\bar{d}$ to $\bar{c}$. Then as above, p_{e+1} must take $\bar{b}_e$ to the tuple $\bar{a}$ effectively chosen to be α-free over $\bar{c}$, and u must say that $\varphi_e^{\Delta^0_\alpha}$ takes some $\bar{c}^*, \bar{a}^*$ to $\bar{d}, \bar{b}_e$. Then we can find p'_{e+1} such that $p_0\bar{b}_0 \ldots \bar{b}_e p'_{e+1} \in L(u)$ and $p_{e+1} \leq_{\beta_0} p'_{e+1}$.

In enumerating more of $\not\leq_\alpha$, we may find that
$$\bar{c}^*, \bar{a}^* \not\leq_\alpha \bar{c}, \bar{a} ,$$
and then $p'_{e+1} = p_{r+1}$. Otherwise, since $\bar{a}$ is α-free over $\bar{c}$, we will find p'_{e+1} taking $\bar{d}, \bar{b}_e$, and some $\bar{b}'$ to $\bar{c}, \bar{a}'$, and some $\bar{a}_1$ such that
$$\bar{c}, \bar{a}' \not\leq_\alpha \bar{c}^*, \bar{a}^* \ \& \ \bar{c}, \bar{a}, \bar{a}_1 \leq_\beta \bar{c}, \bar{a}' \bar{a}'_1 .$$

In either case, we have the required ℓ^* satisfying Condition (4).

We are in a position to apply Theorem 14.1. We get a Δ^0_α run
$$\pi = \hat{\ell} u_1 \ell_1 u_2 \ell_2 \ldots$$
of (P, q) such that $E(\pi)$ is c.e. There is a $\Delta^0_{\alpha+1}$ sequence of numbers $(k(n))_{n \in \omega}$ such that $\ell_{k(n)}$ has length $2n+1$, and for all $k \geq k(n)$, $\ell_k \supseteq \ell_{k(n)}$. (This reflects the fact that at level α, the construction is finite-injury.) Let p_n be the last function in ℓ_n. Then $\cup_n p_{k(n)}$ is a $1-1$ function from B onto $\mathcal{A}$. Let F be the inverse, and let $\mathcal{B}$ be the structure induced by F on B. The requirements are all satisfied, and $\mathcal{B}$ is computable since $D(\mathcal{B}) = E(\pi)$.

Remark: The results above, on Δ^0_α categoricity, do not extend Theorem 12.4. The hypotheses of Theorem 17.7, in the case where $\alpha = 1$, are stronger.

17.5 Superatomic Boolean algebras

We can apply the general theorems to determine for which ordinals β a given superatomic Boolean algebra is Δ^0_β categorical. Recall that a superatomic Boolean algebra is a join of m α-atoms, for some $m \in \omega$ and some ordinal α (the Boolean algebra and the ordinal α are both countable). We use the name $I(\omega^\alpha \cdot m)$ (thinking of an interval algebra).

Theorem 17.8 *(a) Suppose that δ is either a limit ordinal or 0, and $k, m < \omega$. Then $I(\omega^{\delta+k+1} \cdot m)$ is $\Delta^0_{\delta+2k+2}$ categorical and not $\Delta^0_{\delta+2k+1}$ categorical.*

(b) Suppose that δ is a limit ordinal, and $m < \omega$. Then $I(\omega^\delta \cdot m)$ is Δ^0_δ categorical and not Δ^0_β categorical for any $\beta < \delta$.

17.5. SUPERATOMIC BOOLEAN ALGEBRAS

Proof for (a): Each tuple $\bar{a}$ in $I(\omega^{\delta+k+1} \cdot m)$ generates a finite sub-algebra. We add a tuple $\bar{a}_1$ to $\bar{a}$ so that each atom of the sub-algebra generated by $\bar{a}, \bar{a}_1$ is a β-atom in the full algebra, for some $\beta \leq \delta + k + 1$. By Proposition 7.3, for any $\beta \leq \delta + k + 1$, we can find a computable $\Pi_{2\beta+1}$ formula $\beta - atom(x)$ saying that x is a β-atom. Then we have a computable $\Sigma_{\delta+2k+2}$ formula $\varphi_{\bar{a}}(\bar{x})$ saying that for some $\bar{u}$, the sub-algebra generated by $\bar{x}, \bar{u}$ has atoms of the proper forms. The set of these formulas $\varphi_{\bar{a}}(\bar{x})$ is a formally $\Sigma^0_{\delta+2k+2}$ Scott family. Therefore, $I(\omega^{\delta+k+1} \cdot m)$ is $\Delta^0_{\delta+2k+2}$ categorical.

To show that $I(\omega^{\delta+k+1} \cdot m)$ is not $\Delta^0_{\delta+2k+2}$ categorical, we use Theorem 17.7. By Proposition 15.15, there exists

$$\mathcal{A} \cong I(\omega^{\delta+k+1} \cdot m)$$

where $\mathcal{A}$ is $(\delta + 2k + 1)$-friendly. Moreover, we may suppose that there is an effective procedure for determining the invariants of any element a in $\mathcal{A}$ (where the invariants are the ordinal β and the number n such that a is a join of n β-atoms). Let $\bar{c}$ be a tuple in $\mathcal{A}$. One of the atoms of the sub-algebra generated by $\bar{c}$ bounds a $(\delta + k + 1)$-atom, call it b. Let $a, a' \leq b$, where a is a $(\delta + k)$-atom and a' is a join of two $(\delta + k)$-atoms. By Proposition 15.14, $a \geq_{\delta+2k+1} a'$, but not $a \leq_{\delta+2k+1} a'$. Then

$$(\mathcal{A}, \bar{c}, a) \geq_{\delta+2k+1} (\mathcal{A}, \bar{c}, \bar{a}') ,$$

but not

$$(\mathcal{A}, \bar{c}, a') \leq_{\delta+2k+1} (\mathcal{A}, \bar{c}, a) .$$

This shows that a is $(\delta + 2k + 1)$-free over $\bar{c}$.

We are in a position to apply Theorem 17.7. We get a computable copy $\mathcal{B}$ of $\mathcal{A}$ with no Δ^0_α isomorphism from $\mathcal{A}$ onto $\mathcal{B}$. This completes the proof of Theorem 17.8 (a).

Chapter 18

Pairs of computable structures

In this chapter, we give results on the following problems.

Problem 1 *Let $\mathcal{A}$ and $\mathcal{B}$ be structures for the same language. Determine for which sets S there is a uniformly computable sequence of structures $(\mathcal{C}_n)_{n \in \omega}$ such that*
$$\mathcal{C}_n \cong \begin{cases} \mathcal{A} & \text{if } n \in S \\ \mathcal{B} & \text{otherwise .} \end{cases}$$

Problem 2 *For each n, let $\mathcal{A}_n$ and $\mathcal{B}_n$ be structures for the same language. Determine for which sets S there is a uniformly computable sequence of structures $(\mathcal{C}_n)_{n \in \omega}$ such that*
$$\mathcal{C}_n \cong \begin{cases} \mathcal{A}_n & \text{if } n \in S \\ \mathcal{B}_n & \text{otherwise .} \end{cases}$$

These problems were the subject of [7]. We begin with some simple examples. Then, using Theorem 14.1, we prove some general results. The general results can be applied in the construction of linear orderings and Boolean algebras with special features. We give some results on quotient structures.

18.1 Simple examples

The first example gives a pair of linear orderings such that the Δ_2^0 sets are the ones coded as in Problem 1.

Proposition 18.1 *For all sets S, the following are equivalent:*
  (1) there is a uniformly computable sequence of linear orderings $(\mathcal{C}_n)_{n \in \omega}$ such that
$$\mathcal{C}_n \cong \begin{cases} \omega & \text{if } n \in S \\ \omega^* & \text{otherwise ,} \end{cases}$$

(2) S is Δ_2^0.

Proof: (1) $\Rightarrow$ (2) We have finitary Σ_2 sentences

$$\varphi = \exists x \, \forall y \, x \leq y$$

saying that there is a first element, and

$$\psi = \exists x \, \forall y \, y \leq x$$

saying that there is a last element. Then for the uniformly computable sequence $(\mathcal{C}_n)_{n \in \omega}$, the sets

$$S = \{n : \mathcal{C}_n \models \varphi\}$$

and

$$\neg S = \{n : \mathcal{C}_n \models \psi\}$$

are both Σ_2^0. Therefore, S is Δ_2^0.

(2) $\Rightarrow$ (1) Let $g(n, s)$ be a total computable function, with value 0 or 1 for each pair (n, s), such that for all n,

$$\lim_{n \to \infty} g(n, s) = \chi_S(n) \, .$$

At stage s, having previously determined the ordering on finitely many elements of $\mathcal{C}_n$, we add a new element on the right if $g(n, s) = 1$ and on the left if $g(n, s) = 0$. After some stage, $g(n, s)$ is constant. Then the ordering that we build has the desired isomorphism type.

The second example gives a pair of linear orderings that can be used to code an arbitrary Π_2^0 set.

Proposition 18.2 *For all sets S, the following are equivalent:*
(1) there is a uniformly computable sequence of linear orderings $(\mathcal{C}_n)_{n \in \omega}$ such that

$$\mathcal{C}_n \cong \begin{cases} \omega & \text{if } n \in S \\ \omega + 1 & \text{otherwise} \end{cases},$$

(2) S is Π_2^0.

Proof: (1) $\Rightarrow$ (2) We have a finitary Π_2 sentence

$$\varphi = \forall x \, \exists y \, x < y$$

saying that there is no last element. Then for the uniformly computable sequence $(\mathcal{C}_n)_{n \in \omega}$, the set

$$S = \{n : \mathcal{C}_n \models \varphi\}$$

is Π_2^0.

(2) $\Rightarrow$ (1) There is a computable function $g(n, s)$, with value 0 or 1 for each pair (n, s), such that

$$n \in S \Leftrightarrow g(n, s) = 1 \text{ for infinitely many } s \, .$$

18.1. SIMPLE EXAMPLES

We suppose that $g(n,0) = 1$. At stage s, having previously determined the ordering on finitely many elements of $\mathcal{C}_n$, we add a new element at the end if $g(n,s) = 1$, and just before the last element if $g(n,s) = 0$. Again $g(n,s)$ is eventually constant, and the ordering has the desired type.

The next example gives a pair of linear orderings that can be used to code an arbitrary Δ_3^0 set.

Proposition 18.3 *For all sets S, the following are equivalent:*
(1) there is a uniformly computable sequence of linear orderings $(\mathcal{C}_n)_{n \in \omega}$ such that
$$\mathcal{C}_n \cong \begin{cases} \omega + \omega + \omega^* & \text{if } n \in S \\ \omega + \omega^* + \omega^* & \text{otherwise}, \end{cases}$$

(2) S is Δ_3^0.

Proof: (1) $\Rightarrow$ (2) We have finitary Σ_3 sentences φ saying that there is a left limit point (an element with no immediate predecessor), and ψ saying that there is a right limit point (an element with no immediate successor). Then for the uniformly computable sequence $(\mathcal{C}_n)_{n \in \omega}$, the sets

$$S = \{n : \mathcal{C}_n \models \varphi\}$$

and

$$\neg S = \{n : \mathcal{C}_n \models \psi\}$$

are both Σ_3^0. Therefore, S is Δ_3^0.

(1) $\Rightarrow$ (2) Since S is Δ_3^0, there is a Δ_2^0 function $g(n,k)$ such that

$$\lim_{k \to \infty} g(n,k) = \chi_S(n) .$$

Then there is a computable function $h(n,k,s)$ such that

$$\lim_{s \to \infty} h(n,k,s) = g(n,k) .$$

We may suppose that $h(n,k,s)$ is total, with value 0 or 1 for each triple (n,k,s).

We focus on a single n and describe the construction of $\mathcal{C}_n$. At stage s, we have a finite ordering, divided into three intervals, which we refer to as *left*, *right*, and *middle*. The elements of the middle are tentatively designated $b_0, \ldots, b_r$, for some r, where for each $k < r$, b_{k+1} lies to the right of $b_0, \ldots, b_k$ if $h(n,k,s) = 1$ and to the left if $h(n,k,s) = 0$.

At stage $s+1$, we proceed as follows.

Case 1: Suppose that for all $k < r$, $h(n,k,s+1) = h(n,k,s)$.

In this case, we add a new element, designated b_{r+1}, to the middle, putting it to the right of $b_0, \ldots, b_r$ if $h(n,r,s+1) = 1$, and to the left of $b_0, \ldots, b_r$ if $h(n,r,s+1) = 0$. In addition, we add an element at the end of the left interval, and one at the beginning of the right interval.

Case 2: Suppose that for some first $k < r$, $h(n, k, s+1) \neq h(n, k, s)$.

In this case, we drop from the middle interval the elements currently designated b_i for $k < i \leq r$. Those to the left of $b_0, \ldots, b_k$ are added to the left interval, and those to the right of $b_0, \ldots, b_k$ are added to the right interval. We add to the middle a new element, now designated b_{k+1}, locating it to the right of $b_0, \ldots, b_k$, if $h(n, k, s+1) = 1$, and to the left if $h(n, k, s+1) = 0$. Finally, we add a new element at the end of the left interval, and one at the beginning of the right interval.

It is not difficult to see that the construction works. We have fixed n. For each k, the value of $h(n, k, s)$ is eventually constant, so there is an element eventually designated b_k. In the end, the middle, consisting of these elements eventually designated b_k for various k, has type ω if $n \in S$ and type ω^* otherwise. Elements that drop out of the middle interval are absorbed into the left interval, which has type ω, or into the right interval, which has type ω^*. Therefore, $\mathcal{C}_n$ has the desired type.

The example below gives a pair of structures that can be used to code an arbitrary Π_3^0 set.

Proposition 18.4 *Let $\mathcal{A}$ and $\mathcal{B}$ be equivalence structures, each with exactly one class of each finite size $m > 1$, such that $\mathcal{A}$ has one infinite class, while $\mathcal{B}$ has only the finite classes. Then for all sets S, the following are equivalent:*
(1) there is a uniformly computable sequence of structures $(\mathcal{C}_n)_{n \in \omega}$ such that

$$\mathcal{C}_n \cong \begin{cases} \mathcal{A} & \text{if } n \in S \\ \mathcal{B} & \text{otherwise} \end{cases},$$

(2) S is Π_3^0.

Exercise: Prove Proposition 18.4, either directly, or using a 3-system.

18.2 General results

We begin with Problem 1. The result below is an immediate consequence of the fact that satisfaction of computable Π_α sentences in a computable structure is Π_α^0, uniformly in the structure (or its computable index).

Proposition 18.5 *Let $(\mathcal{C}_n)_{n \in \omega}$ be a uniformly computable sequence of structures, and let φ be a computable Π_α sentence. Then the set*

$$S = \{n : \mathcal{C}_n \models \varphi\}$$

is Π_α^0.

18.2. GENERAL RESULTS

The next result gives conditions on a pair of structures $\mathcal{A}$ and $\mathcal{B}$, guaranteeing that for an arbitrary Π^0_α set S, there is a uniformly computable sequence of structures $(\mathcal{C}_n)_{n\in\omega}$ coding S. We let $\leq_\beta$ be the standard back-and-forth relations, although we could substitute stronger relations if this made it easier to establish α-friendliness.

Theorem 18.6 *Let $\mathcal{A}_0$ and $\mathcal{A}_1$ be structures such that $\mathcal{A}_1 \leq_\alpha \mathcal{A}_0$ and $\{\mathcal{A}_0, \mathcal{A}_1\}$ is α-friendly. Then for any Π^0_α set S, there is a uniformly computable sequence of structures $(\mathcal{C}_n)_{n\in\omega}$ such that*

$$\mathcal{C}_n \cong \begin{cases} \mathcal{A}_0 & \text{if } n \in S \\ \mathcal{A}_1 & \text{otherwise}. \end{cases}$$

Proof: We may suppose that $\mathcal{A}_0$ and $\mathcal{A}_1$ have the same universe A. Since S is Π^0_α, there is a Δ^0_α function g such that

$$n \in S \Rightarrow \forall s \, g(n,s) = 0$$
$$n \notin S \Rightarrow \exists s_0 \, ((\forall s < s_0) \, g(n,s) = 0 \, \& \, (\forall s \geq s_0) \, g(n,s) = 1) \, .$$

We define an α-system

$$(L, U, \hat{\ell}, P, E, (\leq_\beta)_{\beta < \alpha})$$

(the same for all n), and a Δ^0_α instruction function q_n, with index that we can compute effectively from n. We obtain $\mathcal{C}_n$ from a run of (P, q_n). We use the uniformity in Theorem 14.1 to say that the structures $\mathcal{C}_n$ are uniformly computable.

We focus on a single n. Let C be an infinite computable set of constants, for the universe of $\mathcal{C}_n$. Let $\mathcal{F}$ be the set of finite partial $1-1$ functions p from C to A. Let $U = \{0,1\}$. Let L consist of the pairs in $\{0,1\} \times \mathcal{F}$, and one extra element—$\hat{\ell} = (\_, \emptyset)$. If $(i,p) \in \{0,1\} \times \mathcal{F}$, then (i,p) represents $(\mathcal{A}_i, p)$. For $\hat{\ell}$, the structure is unspecified.

We extend the standard enumeration function and the standard back-and-forth relations on pairs $(\mathcal{A}_i, \bar{a})$ to pairs (i, p) in L. If $\ell = (i,p)$, where p maps $\bar{b}$ to $\bar{a}$, then $E(\ell) = E(\mathcal{A}_i, \bar{a})$. We let $E(\hat{\ell}) = \emptyset$. If $\ell = (i,p)$ and $\ell' = (j,q)$, where p maps $\bar{b}$ to $\bar{a}$ and q maps $\bar{b}'$ to $\bar{a}'$, then $\ell \leq_\beta \ell'$ if and only if $\bar{b} \subseteq \bar{b}'$ and $(i, \bar{a}) \leq_\beta (j, \bar{a}')$. For $0 < \beta$, this is equivalent to saying that the Π_β sentences made true by p in $\mathcal{A}_i$ are made true by q in $\mathcal{A}_j$. We write $\ell \subseteq \ell'$ if $i = j$ and $p \subseteq q$. For all $\ell \in L$, we let $\hat{\ell} \leq_b \ell$, for all $\beta < \alpha$, and we let $\hat{\ell} \subseteq \ell$.

We are ready to define the tree. Let P consist of the finite sequences

$$\sigma = \hat{\ell} u_1 \ell_1 u_2 \ell_2 \ldots$$

where $u_k \in U$, $\ell_k \in L$, and the following conditions hold:
  (a) ℓ_k has the form (u_k, p_k),
  (b) $dom(p_k)$ and $ran(p_k)$ include the first k elements of C, A, respectively,

(c) if $u_k \neq u_{k+1}$, then $u_k = 0$ and $u_{k+1} = 1$,
(d) if $u_k = u_{k+1}$, then $\ell_k \subseteq \ell_{k+1}$, and in any case, $\ell_k \leq_0 \ell_{k+1}$.

Remark: Let $p = \hat{\ell} u_1 \ell_1 u_2 \ell_2 \ldots$ be a path through P, where $\ell_k = (u_k, p_k)$. If $u_k = 0$ for all k, then $F^{-1} = \cup_k p_k$ is a $1-1$ function from C onto $\mathcal{A}_0$. If $u_k = 1$ for $k \geq k^*$, then $F^{-1} = \cup_{k \geq k^*} p_k$ is a $1-1$ function from C onto $\mathcal{A}_1$. In either case, F induces a structure $\mathcal{C}$ on C such that $E(\pi) = D(\mathcal{C})$.

We have described the α-system. Clearly, L, U, and P are c.e., the relations $\leq_\beta$ are c.e. uniformly in β, and E is partial computable. The first three conditions for an α-system are easy to check. We turn to Condition (4). Suppose $\sigma u \in P$, where σ ends in ℓ^0, and

$$\ell^0 \leq_{\beta_0} \ldots \leq_{\beta_{k-1}} \ell^k ,$$

for $\alpha > \beta_0 > \ldots > \beta_k$. Say $\ell^0 = (v, p)$. We need ℓ^* such that $\sigma u \ell^* \in P$ and $\ell^i \leq_{\beta_i} \ell^*$, for $0 \leq i \leq k$. As usual, we have $q \supseteq p$ such that $\ell^i \leq_{\beta_i} (v, q)$. If $u = v$, then (v, q) (or a minor variant in which q is extended to include the necessary elements in the domain and range) is the required ℓ^*. Suppose instead that $v = 0$ and $u = 1$. Since $\mathcal{A}_1 \leq_\alpha \mathcal{A}_0$, there exists $q' \in \mathcal{F}$ such that $(0, q) \leq_{\beta_0} (1, q')$. Then $(1, q')$ (or a minor variant in which q' is extended to include the necessary elements in the domain and range) is the required ℓ^*.

The instruction function q_n is derived from g. Let $\sigma \in P$, where σ has length $2s + 1$. Then $q_n(\sigma) = g(n, s)$, except that if $g(n, s) = 0$ and one of the even terms of σ (those in U) is 1, then $q_n(\sigma) = 1$. This ensures that if $q_n(\sigma) = u$, then $\sigma u \in P$. Again, we can compute a Δ^0_α index for q_n from n. We are in a position to apply Theorem 14.1. For each n, we get a run $\pi_n = u_0 \ell_0 u_1 \ell_1 u_2 \ell_2 \ldots$ of (P, q_n) such that $E(\pi_n)$ is c.e., with an index that we can compute from n. By the remark above, π_n yields $\mathcal{C}_n$, a copy of either $\mathcal{A}_0$ or $\mathcal{A}_1$, such that $D(\mathcal{C}_n) = E(\pi_n)$. Then $\mathcal{C}_n$ is computable, uniformly in n. If $n \in S$, then $u_k = g(n, k) = 0$ for all k, and $\mathcal{C}_n \cong \mathcal{A}_0$. If $n \notin S$, then $u_k = g(n, k) = 1$ for all sufficiently large k, and $\mathcal{C}_n \cong \mathcal{A}_1$. This completes the proof of Theorem 18.6.

The next result gives conditions on a pair of structures $\mathcal{A}_0$ and $\mathcal{A}_1$, guaranteeing that for an arbitrary $\Delta^0_{\alpha+1}$ set S, we can code S into a uniformly computable sequence of structures $(\mathcal{C}_n)_{n \in \omega}$.

Theorem 18.7 Let $\mathcal{A}_0$ and $\mathcal{A}_1$ be structures such that $\mathcal{A}_0 \leq_\alpha \mathcal{A}_1$, $\mathcal{A}_1 \leq_\alpha \mathcal{A}_0$, and $\{\mathcal{A}_0, \mathcal{A}_1\}$ is $(\alpha+1)$-friendly. Then for any $\Delta^0_{\alpha+1}$ set S, there is a uniformly computable sequence of structures $(\mathcal{C}_n)_{n \in \omega}$ such that

$$\mathcal{C}_n \cong \begin{cases} \mathcal{A}_0 & \text{if } n \in S \\ \mathcal{A}_1 & \text{otherwise} . \end{cases}$$

Exercise: Prove Theorem 18.7. This proof is similar to the proof of Theorem 18.6 except that you will need an $(\alpha + 1)$-system, with $\leq_\alpha = \subseteq$. The instruction functions q_n are all constant.

18.3. RESULTS OF FEINER AND THURBER

Having given some results on Problem 1, we turn to Problem 2. Let $(\alpha_n)_{n\in\omega}$ be a computable sequence of computable ordinals (notations, really). We say that a set S is $\Pi^0_{(\alpha_n)}$ if there exists e such that for all n,

$$n \in S \Leftrightarrow \varphi_e^{\Delta^0_{\alpha_n}}(n) \uparrow .$$

Similarly, S is $\Sigma^0_{(\alpha_n)}$ if there exists e such that for all n,

$$n \in S \Leftrightarrow \varphi_e^{\Delta^0_{\alpha_n}}(n) \downarrow .$$

The parentheses around (α_n) are intended as a reminder that the notion depends on the *sequence* $(\alpha_n)_{n\in\omega}$, as opposed to a *single term* α_n. Note that

$$S \text{ is } \Sigma^0_{(\alpha_n)} \Leftrightarrow \neg S \text{ is } \Pi^0_{(\alpha_n)} .$$

The next result is the analogue of Proposition 18.5, true for the same reason.

Proposition 18.8 *Let $(\varphi_n)_{n\in\omega}$ be a computable sequence of computable infinitary sentences, where φ_n is Π_{α_n}. Let $(\mathcal{C}_n)_{n\in\omega}$ be a uniformly computable sequence of structures. Then $S = \{n : \mathcal{C}_n \models \varphi_n\}$ is $\Pi^0_{(\alpha_n)}$.*

The next result is the analogue of Theorem 18.6.

Theorem 18.9 *For each n, let $\mathcal{A}_n$ and $\mathcal{B}_n$ be structures such that $\mathcal{B}_n \leq_{\alpha_n} \mathcal{A}_n$ and $\{\mathcal{A}_n, \mathcal{B}_n\}$ is α_n-friendly, all uniformly in n. If S is $\Pi^0_{(\alpha_n)}$, then there is a uniformly computable sequence $(\mathcal{C}_n)_{n\in\omega}$ such that*

$$\mathcal{C}_n \cong \begin{cases} \mathcal{A}_n & \text{if } n \in S \\ \mathcal{B}_n & \text{otherwise} \end{cases} .$$

Proof: For each n, we define an α_n-system and a $\Delta^0_{\alpha_n}$ instruction function q_n, with indices that we can compute effectively from n. The definitions are as in the proof of Theorem 18.6, except that for a given n, we replace $\mathcal{A}_0$ and $\mathcal{A}_1$ by $\mathcal{A}_n$ and $\mathcal{B}_n$. (Here the α_n system depends on n.) We obtain $\mathcal{C}_n$ from a run of the α_n-system with the instruction function q_n.

18.3 Results of Feiner and Thurber

Theorem 18.9 can be used to prove some results of Feiner [50] and Thurber [150] on Boolean algebras. We consider sets coded in a Boolean algebra by the truth of certain computable infinitary sentences. (These have the same meanings as the sentences from Feiner's original proof, although Feiner used a different kind of language.) For each n, let ψ_n be the natural computable infinitary sentence saying that for all x containing infinitely many atomic elements, no two of which are identified in the nth derivative, there is an atomic $y \leq x$ such that y contains infinitely many such elements. For $n \geq 2$, we can take ψ_n to be a computable Π_{2n+4} sentence.

Lemma 18.10 *For a Boolean algebra $\mathcal{A}$, let*

$$S(\mathcal{A}) = \{n \in \omega : \mathcal{A} \models \psi_n\}.$$

(a) If $\mathcal{A}$ is computable, then $S(\mathcal{A})$ is $\Pi^0_{(2n+4)}$.
(b) If $\mathcal{A}$ is Δ^0_2, then $S(\mathcal{A})$ is $\Pi^0_{(2n+5)}$.

The result below says that every set meeting the conditions of Lemma 18.10 occurs as $S(\mathcal{A})$, for an appropriate Boolean algebra $\mathcal{A}$.

Theorem 18.11 (Thurber) *(a) If S is $\Pi^0_{(2n+4)}$, then there is a computable Boolean algebra $\mathcal{A}$ such that $S(\mathcal{A}) = S$.*
(b) If S is $\Pi^0_{(2n+5)}$, then there is a Δ^0_2 Boolean algebra $\mathcal{A}$ such that $S(\mathcal{A}) = S$.

We sketch the proof of (a), noting that (b) is simply the relativization of (a) to Δ^0_2. A number of results about Boolean algebras have been obtained by switching to the setting of linear orderings. We take that approach here. Recall that for a linear ordering $\mathcal{A}$, $I(\mathcal{A})$ denotes the corresponding interval algebra.

Lemma 18.12 *Given a computable infinitary sentence ψ in the language of Boolean algebras, we can find a computable infinitary sentence ψ^* in the language of orderings, of the same complexity as ψ, such that if $\mathcal{A}$ is a linear ordering and $\mathcal{B} = I(\mathcal{A})$, then*

$$\mathcal{B} \models \psi \Leftrightarrow \mathcal{A} \models \psi^*.$$

The proof is quite straightforward, using induction.

We described Feiner's sentences ψ_n, in the language of Boolean algebras. Consider the corresponding sentences ψ_n^*, in the language of orderings. Using Lemma 18.12, we can reduce the proof of Theorem 18.11(a) to showing that if S is $\Pi^0_{(2n+4)}$, then there is a computable linear ordering $\mathcal{A}$ such that

$$S = \{n : \mathcal{A} \models \psi_n^*\}.$$

The outline is as follows. We let $\mathcal{A} = \Sigma_{n \in \omega} \mathcal{C}_n$, where for each n, $\mathcal{C}_n$ is isomorphic to one of a pair of linear orderings $\mathcal{A}_n$ or $\mathcal{B}_n$. The fact that $\mathcal{A}$ is computable will follow from the fact that the sequence $(\mathcal{C}_n)_{n \in \omega}$ is uniformly computable.

The structures $\mathcal{A}_n$ and $\mathcal{B}_n$ will be shuffle sums; in particular,

$$\mathcal{A}_n \cong \sigma(\omega^\omega + \eta) \quad \text{and} \quad \mathcal{B}_n \cong \sigma(\omega^\omega + \eta, \omega^{n+1} + \eta).$$

Reflecting on the meaning of the sentences, we can see that

$$\sigma(\omega^\omega + \eta) \models \psi_n^* \quad \text{and} \quad \sigma(\omega^\omega + \eta, \omega^{n+1} + \eta) \models \neg \psi_n^*.$$

Moreover, if $\mathcal{A} = \Sigma_{n \in \omega} \mathcal{C}_n$, where for each n, $\mathcal{C}_n$ is a copy of either $\sigma(\omega^\omega + \eta)$ or $\sigma(\omega^\omega + \eta, \omega^{n+1} + \eta)$, then

$$\mathcal{A} \models \psi_n^* \Leftrightarrow \mathcal{C}_n \cong \sigma(\omega^\omega + \eta).$$

18.3. RESULTS OF FEINER AND THURBER

Now, our goal is to produce a uniformly computable sequence of linear orderings $\mathcal{C}_n$ such that

$$\mathcal{C}_n \cong \begin{cases} \sigma(\omega^\omega + \eta) & \text{if } n \in S \\ \sigma(\omega^\omega + \eta, \omega^{n+1} + \eta) & \text{otherwise} \end{cases}.$$

We shall apply Theorem 18.9. We must check that the conditions are satisfied.

Claim 1: For each n, $\sigma(\omega^\omega + \eta) \geq_{2n+4} \sigma(\omega^\omega + \eta, \omega^{n+1} + \eta)$.

Proof of Claim 1: We have

$$\omega^{n+1} \geq_{2n+3} \omega^\omega .$$

Therefore,

$$\omega^{n+1} + \eta \geq_{2n+3} \omega^\omega + \eta .$$

From this, it follows that

$$\sigma(\omega^\omega + \eta) \geq_{2n+4} \sigma(\omega^\omega + \eta, \omega^{n+1} + \eta) .$$

Claim 2: There exist $\mathcal{A}_n \cong \sigma(\omega^\omega + \eta)$ and $\mathcal{B}_n \cong \sigma(\omega^\omega + \eta, \omega^{n+1} + \eta)$ (with extra relations) such that $\{\mathcal{A}_n, \mathcal{B}_n\}$ is $(2n+4)$-friendly, uniformly in n.

Proof of Claim 2: Recall the two examples near the end of Chapter 15, indicating how it may be convenient to use back-and-forth relations stronger than the standard ones to prove the existence of α-friendly structures or families of structures. First, using Proposition 15.11, and proceeding as in Example 1, we obtain, for each n, a pair of nice copies of

$$\omega^\omega + \eta \text{ and } \omega^{n+1} + \eta ,$$

with an extra unary predicate for the dense part, such that the pairs are $(2n+4)$-friendly, uniformly in n. Next, proceeding as in Example 2, we obtain, for each n,

$$\mathcal{A}_n \cong \sigma(\omega^\omega + \eta) \text{ and } \mathcal{B}_n \cong \sigma(\omega^\omega + \eta, \omega^{n+1} + \eta) ,$$

with an extra equivalence relation for which the equivalence classes are the copies of $\omega^\omega + \eta$ and $\omega^{n+1} + \eta$ in the shuffle sums, such that the pairs $\{\mathcal{A}_n, \mathcal{B}_n\}$ are $(2n+4)$-friendly, uniformly in n.

Having proved the two claims, we are in a position to apply Theorem 18.9. We obtain the desired uniformly computable sequence of linear orderings $(\mathcal{C}_n)_{n \in \omega}$. By combining these, as planned, we get a computable linear ordering

$$\mathcal{A} = \Sigma_{n \in \omega} \mathcal{C}_n$$

such that
$$S = \{n : \mathcal{A} \models \psi_n^*\} \ .$$
Then the interval algebra $I(\mathcal{A})$ is the required Boolean algebra. This completes the proof of Theorem 18.11.

Feiner's original proof of the existence of a c.e. quotient Boolean algebra with no computable copy involved a single very complicated construction. Thurber's proof, which we are following, breaks the construction into relatively simple modules. Using general results on pairs, we have shown that the sets that can occur as $\{n : \mathcal{A} \models \psi_n\}$, where $\mathcal{A}$ is a computable Boolean algebra, are just the $\Pi^0_{(2n+4)}$ sets, and the sets that can occur as $\{n : \mathcal{A} \models \psi_n\}$, where $\mathcal{A}$ is a c.e. quotient Boolean algebra (or, equivalently, a Δ^0_2 Boolean algebra), are just the $\Pi^0_{(2n+5)}$ sets. The next lemma says that the two classes of sets are not the same.

Lemma 18.13 *There exists a set S that is $\Pi^0_{(2n+5)}$ and not $\Pi^0_{(2n+4)}$.*

Proof: Let
$$n \in S \Leftrightarrow \varphi_n^{\Delta^0_{2n+4}}(n) \downarrow \ .$$

Feiner's Theorem follows immediately.

Corollary 18.14 (Feiner) *(a) There is a c.e. quotient Boolean algebra with no computable copy.*
(b) For each α such that $1 \leq \alpha < \omega_1^{CK}$, there is a Σ^0_α quotient Boolean algebra with no Δ^0_α copy.

Proof: (a) Let S be $\Pi^0_{(2n+5)}$ and not $\Pi^0_{(2n+4)}$, as in Lemma 18.13. Let $\mathcal{A}$ be as in Theorem 18.11 (b), a Δ^0_2 Boolean algebra such that $S(\mathcal{A}) = S$. By Theorem 9.17, $\mathcal{A}$ is isomorphic to a c.e. quotient algebra. There is no computable copy of $\mathcal{A}$, for if there were, then by Lemma 18.10, S would be $\Pi^0_{(2n+4)}$.

The proof of (b) is the same except that everything is relativized to Δ^0_α.

18.4 Limit structures

We have considered constructions in which, at the top level, we copy parts of two structures, eventually settling on one of them. It is sometimes useful to copy parts of an infinite sequence of structures, without ever settling on just one. Hurlburt [69] proved a fairly general theorem on constructions in which a computable structure is obtained as the "limit" of a highly non-computable sequence of structures. In the proof of the next result, we use Hurlburt's idea, although the conditions of her theorem are not all satisfied.

By Watnik's Theorem (Theorem 9.10), if $\mathcal{A}$ is a Δ^0_3 linear ordering, then $(\omega^* + \omega) \cdot \mathcal{A}$ has a computable copy. The same is true with ω replacing $\omega^* + \omega$. This is Theorem 9.11. The result below generalizes Theorem 9.11, with ω^α instead of ω (see [6] and [5]).

18.4. LIMIT STRUCTURES

Theorem 18.15 *Let α be a computable ordinal, and suppose $\mathcal{A}$ is a $\Delta^0_{2\alpha+1}$ linear ordering. Then $\omega^\alpha \cdot \mathcal{A}$ has a computable copy $\mathcal{B}$. Moreover, there is a $\Delta^0_{\alpha+1}$ function taking $a \in \mathcal{A}$ to the first element of the corresponding copy of ω^α in $\mathcal{B}$.*

Proof: There are two cases, depending on whether $\mathcal{A}$ has a first element. In proving Theorem 9.11 in Chapter 13, we reduced the case where $\mathcal{A}$ has a first element to the case where it does not. We can do the same here, so we suppose that $\mathcal{A}$ has no first element. Let $(\mathcal{C}_n)_{n \in \omega}$ be a $(2\alpha + 1)$-friendly family of orderings, all with a fixed computable universe C, where for each n, $\mathcal{C}_n$ is has type $\omega^\alpha \cdot n$. We may suppose that there is an effective procedure for locating the tuple $\bar{c}_n$ of first elements of copies of ω^α in $\mathcal{C}_n$. Note that $\mathcal{C}_n$ is rigid, and for each element of $\mathcal{C}_n$, there is a computable $\Pi_{2\alpha}$ defining formula $\varphi(\bar{c}_n, x)$, saying that the interval from some c in $\bar{c}_n$ to x has order type β, for some $\beta < \omega^\alpha$.

We define a $(2\alpha + 1)$-system

$$(L, U, P, E, (\leq_\beta)_{\beta < 2\alpha+1})$$

and a $\Delta^0_{2\alpha+1}$ instruction function q. We may suppose that the universe of $\mathcal{A}$ is a computable set $A = \{a_0, a_1, a_2, \ldots\}$. Let B be an infinite computable set of constants, for the universe of $\mathcal{B}$. Let $\mathcal{F}$ be the set of finite partial $1-1$ functions from B to C. Let U be the set of orderings on finite subsets of A. Let L consist of the pairs (n, p), where $n \in \omega$ and $p \in \mathcal{F}$. The "n" indicates that p is being considered as a partial function from B to $\mathcal{C}_n$, providing interpretations for the constants in its domain. Let $\hat{\ell} = (1, \emptyset)$.

If $\ell \in L$, where $\ell = (n, p)$, we let $E(\ell)$ be the set of basic sentences that p makes true in $\mathcal{A}_n$ (a finite set). Suppose $\ell = (n, p)$ and $\ell' = (n', p')$. Then we let

$$\ell \leq_\beta \ell' \Leftrightarrow (\omega^\alpha + \mathcal{A}_n, p) \leq_\beta (\omega^\alpha + \mathcal{A}_{n'}, p').$$

The extra first copy of ω^α makes elements of $ran(p)$ behave more like elements of $\omega^\alpha \cdot \mathcal{A}$, in that none of them lie on a first copy of ω^α. It is not difficult to establish the following.

Lemma 18.16 *Let $\ell = (n, p)$, $\ell' = (n', p')$, where $n \leq n'$, p takes $\bar{b}$ to $\bar{c}_n$, and p' takes $\bar{b}$ to a subset of $\bar{c}_{n'}$, such that $p' \circ p^{-1}$ preserves order. Then $\ell \geq_{2\alpha+1} \ell'$. Moreover, for any $q \supseteq p$ (still in $\mathcal{F}$), there is a unique $q' \supseteq p'$ (also in $\mathcal{F}$), such that q' has the same domain as q, and $(n, q) \leq_{2\alpha} (n', q')$.*

We are ready to define the tree. Let P consist of the finite sequences $\hat{\ell} u_1 \ell_1 u_2 \ell_2 \ldots$, where $\ell_n \in L$, $u_n \in U$, u_n is an ordering on a set including a_k for $k < n$, u_{n+1} is an extension of u_n, with a new first element, and possibly one more new element, and $\ell_n = (m_n, p_n)$ satisfies the following conditions:

(1) m_n is the cardinality of u_n,
(2) $dom(p_n)$ includes the first n constants from B,
(3) $ran(p_n)$ includes $\bar{c}_{m_n}$,
(4) $p_{n+1} \circ p_n^{-1}$ maps $\bar{c}_{m_n}$ into $\bar{c}_{m_{n+1}}$ in the "same" way that the identity function maps $\bar{u}_n$ into $\bar{u}_{n+1}$ (there is a unique isomorphism between $\bar{u}_n$ and

$\bar{c}_{m_n}$, and also between $\bar{u}_{n+1}$ and $\bar{c}_{m_{n+1}}$, since these are pairs of finite orderings of the same size),

(5) for each $k < n+1$, if $\bar{c}_{m_{k'}}$ is the image of $\bar{c}_{m_k}$ under $p_{n+1} \circ p_k^{-1}$, and $\bar{b}$ consists of consists of the elements a_i for $i \leq n$, while $\bar{d}$ is the unque tuple in $\mathcal{A}_{m_{n+1}}$ such that
$$(\mathcal{A}_{m_k}, \bar{c}_{m_k}, \bar{b}) \leq_{2\alpha} (\mathcal{A}_{m_{n+1}}, \bar{c}_{m_{k'}}, \bar{d}),$$
then $ran(p_{n+1})$ includes $\bar{d}$,

(6) $\ell_n \leq_{2\alpha} \ell_{n+1}$.

We define the instruction function q as follows. First, $q(\hat{\ell})$ is the substructure of $\mathcal{A}$ with universe $\{a_0\}$. Suppose
$$\sigma = \hat{\ell} u_1 \ell_1 \ldots u_n \ell_n$$
is in P. If the ordering on u_n agrees with that on $\mathcal{A}$ (i.e., u_n is a substructure of $\mathcal{A}$), then $q(\sigma)$ is the substructure of $\mathcal{A}$ containing the first constant not in u_n, and the first constant lying to the left of all elements of u_n (these may be the same or different). If the ordering on u_n does not agree with that on $\mathcal{A}$, then $q(\sigma)$ is defined trivially to be some u such that $\sigma u \in P$.

The following should be clear from the definitions of P and q.

Lemma 18.17 *For any path $\pi = \hat{\ell} u_1 \ell_1 u_2 \ell_2 \ldots$ through P, if $\mathcal{A}_\pi = \cup_n u_n$, then $E(\pi)$ is the atomic diagram of an ordering*
$$\mathcal{B} \cong \omega^\alpha \cdot \mathcal{A}_\pi .$$
Hence, if π is a run of (P, q), then
$$\mathcal{B} \cong \omega^\alpha \cdot \mathcal{A} .$$

We must show that the conditions for a $2\alpha+1$-system are satisfied. As usual, the first three conditions are clear. Toward Condition (4), suppose $\sigma u \in P$, where σ has length $2n+1$, ending in ℓ^0. Suppose further that
$$\ell^0 \leq_{\beta_0} \ell^1 \leq_{\beta_1} \ldots \leq_{\beta_{k-1}} \ell^k ,$$
for $2\alpha + 1 > \beta_0 > \ldots > \beta_k$. Say $\ell^i = (n_i, p_i)$. There exists $q \supseteq p_0$ such that for all $i = 1, \ldots, k$, $(\omega^\alpha(1 + n_i), p_i) \leq_{\beta_i} (\omega^\alpha(1 + n_0), q)$. Say v is the substructure of $\mathcal{A}$ corresponding to ℓ^0. Then u is an extension with one new element at the beginning, and possibly one more new element. Let n^* be the cardinality of u. The identity embedding of v into u yields a natural embedding of $\mathcal{A}_{n_0}$ into $\mathcal{A}_{n^*}$, so that the first copy of ω^α and the one corresponding to the other new element of u (if any) are the only elements left out of the range. Then $(\omega^\alpha + \mathcal{A}_{n_0}, q) \leq_{2\alpha} (\omega^\alpha + \mathcal{A}_{n^*}, q)$.

We have $\ell' = (n^*, q)$ such that $\ell^i \leq_{\beta_i} \ell'$, for all i such that $0 \leq i \leq k$. However, we cannot claim that $\sigma \ell' \in P$. We extend q to q^* such that for all $k \leq n$, if $\bar{c}_{k'}$ is the image of $\bar{c}_k$ under $q_0 \circ p_k^{-1}$, $\bar{b}$ consists of $\bar{a}_i$ for $i \leq n$, and $\bar{d}$

is the tuple such that $(\mathcal{A}_k, \bar{c}_k, \bar{b}) \leq_{2\alpha} (\mathcal{A}_{n^*}, \bar{c}_{k'}, \bar{d})$, then $\bar{d}$ is in $ran(q^*)$. Now, (n^*, q^*) is the desired ℓ^* satisfying Condition (4).

We are in a position to apply Theorem 14.1. We get a $\Delta^0_{2\alpha+1}$ run π of (P, q) such that $E(\pi)$ is c.e. By Lemma 18.17, $E(\pi)$ is the atomic diagram of a linear ordering $\mathcal{B} \cong \omega^\alpha \cdot \mathcal{A}$. Since $E(\pi)$ is c.e., $\mathcal{B}$ is computable. This completes the proof of Theorem 18.15.

18.5 Quotient orderings

Recall that for linear orderings (in the language with the operation min), we have congruence relations $\sim_\alpha$, where $x \sim_1 y$ if and only if x and y are finitely far apart, and for $\alpha > 1$, $x \sim_\alpha y$ if and only if for some $\beta < \alpha$, the interval between x and y has only finitely many $\sim_\beta$-classes. It follows from Theorem 9.7 that a linear ordering has a Δ^0_2 copy if and only if it is isomorphic to a c.e. quotient structure—the congruence relation is $\sim_1$. It follows from Watnik's Theorem (Theorem 9.10) that a linear ordering has a Δ^0_3 copy if and only if it is isomorphic to a Σ^0_2 quotient structure—again the congruence relation is $\sim_1$.

For each computable ordinal α, $\sim_\alpha$ is definable by a computable $\Sigma_{2\alpha}$ formula. It follows that if $\omega^\alpha \cdot \mathcal{A}$ (or $(\omega^* + \omega)^\alpha \cdot \mathcal{A}$) has a computable copy, then there is a $\Sigma^0_{2\alpha}$ quotient structure isomorphic to $\mathcal{A}$, so that $\mathcal{A}$ has a $\Delta^0_{2\alpha+1}$ copy. Theorem 18.15 gives the converse. The proof yields more. Suppose the linear ordering $\mathcal{A}$ is Δ^0_β for some $\beta < 2\alpha + 1$. Taking the β-system

$$(L, U, P, E, (\leq_\gamma)_{\gamma < \beta})$$

from the proof of Theorem 18.15, together with the instruction function q, which is now Δ^0_β, we obtain a Δ^0_β run $\pi = \hat{\ell}u_1\ell_1u_2\ell_2\ldots$ of (P, q) such that $E(\pi)$ is c.e.

Again $E(\pi)$ is the atomic diagram of a copy $\mathcal{B}$ of $\omega^\alpha \cdot \mathcal{A}$. Now, however, we can show that the congruence relation $\sim_\alpha$ is Δ^0_β. It is enough to show that there is a Δ^0_β procedure for determining, for any β, the first element d of the copy of ω^α that contains b. If b is among the first n constants in B, and $\ell_n = (m_n, p_n)$, then $b \in dom(p_n)$, and there is a tuple $\bar{d}$ in $dom(p_n)$ corresponding to $\bar{c}_{m_n}$. We take the greatest d in $\bar{d}$ such that $d \leq b$. We have the following.

Corollary 18.18 *If $\mathcal{A}$ is a Δ^0_β linear ordering, where $\beta < 2\alpha + 1$, then $\omega^\alpha \cdot \mathcal{A}$ has a computable copy in which the relation $\sim_{\alpha+1}$ is Δ^0_β.*

The information that we have gathered on representing linear orderings as quotient structures is summarized in the next result.

Theorem 18.19 *Let $\mathcal{A}$ be a linear ordering, and let α be a computable ordinal. Then*

(a) $\mathcal{A}$ has a $\Delta^0_{2\alpha+1}$ copy if and only if it is isomorphic to a $\Sigma^0_{2\alpha}$ quotient structure,

(b) $\mathcal{A}$ has a $\Delta^0_{2\alpha+2}$ copy if and only if it is isomorphic to a $\Sigma^0_{2\alpha+1}$ quotient structure,

(c) if α is a limit ordinal, then $\mathcal{A}$ has a Δ^0_α copy if and only if it is isomorphic to a Δ^0_α quotient structure.

Proof: We get (a) from Theorem 18.15 and the comments above. We get (c) from Corollary 18.18, letting $\beta = \alpha$. For (b), suppose that $\mathcal{A}$ is $\Delta^0_{2\alpha+2}$. By Theorem 9.8 (relativized), there is a $\Delta^0_{2\alpha+1}$ ordering

$$\mathcal{B} \cong \omega \cdot \mathcal{A}$$

in which $\sim_1$ is $\Sigma^0_{2\alpha+1}$. By Theorem 18.15, there is a computable ordering

$$\mathcal{C} \cong \omega^\alpha \cdot \mathcal{B},$$

with a $\Delta^0_{2\alpha+1}$ function f taking $b \in \mathcal{B}$ to the first element of the corresponding copy of ω^α in $\mathcal{C}$. Then $\mathcal{C} \cong \omega^{\alpha+1} \cdot \mathcal{A}$. The relation $\sim_\alpha$ is $\Sigma^0_{2\alpha+1}$ on $\mathcal{C}$, since it has a computable $\Sigma_{2\alpha+1}$ definition. We have $c \sim_{\alpha+1} c'$ if and only if there exist $b, b' \in \mathcal{B}$ such that
  (i) $f(b) \sim_\alpha c$,
  (ii) $f(b') \sim_\alpha c'$, and
  (iii) $b \sim_1 b'$.
From this, we can see that $\sim_{\alpha+1}$ is $\Sigma^0_{2\alpha+1}$.

By Theorem 9.17, a Boolean algebra has a Δ^0_2 copy if and only if it is isomorphic to a c.e. quotient algebra. In fact, there are results for Boolean algebras analogous to all of the results for linear orderings in Theorem 18.19.

Theorem 18.20 *Let $\mathcal{A}$ be a Boolean algebra, and let α be a computable ordinal. Then*

(a) $\mathcal{A}$ has a $\Delta^0_{2\alpha+1}$ copy if and only if it is isomorphic to a $\Sigma^0_{2\alpha}$ quotient structure,

(b) $\mathcal{A}$ has a $\Delta^0_{2\alpha+2}$ copy if and only if it is isomorphic to a $\Sigma^0_{2\alpha+1}$ quotient structure,

(c) if α is a limit ordinal, then $\mathcal{A}$ has a Δ^0_α copy if and only if it is isomorphic to a Δ^0_α quotient structure.

Sketch of proof: We could give a direct proof. Here instead, we use the results for linear orderings. Given a Boolean algebra $\mathcal{A}$, we pass to a linear ordering $\mathcal{A}^*$ of the same degree. Applying results from Theorem 18.19, we obtain a computable linear ordering $\mathcal{B}^*$ with a congruence relation $\sim^*$, of the appropriate complexity, such that $\mathcal{A}^* \cong \mathcal{B}^*/\sim^*$ (the language of the linear ordering has the operation symbol min). Now, we let $\mathcal{B}$ be the interval algebra of $\mathcal{B}^*$. We define a congruence relation $\sim$ on $\mathcal{B}$, such that for b and b' in $\mathcal{B}$, $b \sim b'$ if and only if $b \Delta b'$ is a finite join of half-open intervals $[d, d')$ such that $d \sim^* d'$.

In (a), where $\sim^*$ is $\Sigma^0_{2\alpha}$, we can see that $\sim$ is also $\Sigma^0_{2\alpha}$. Similarly, in (b), where $\sim^*$ is $\Sigma^0_{2\alpha+1}$, so is $\sim$. In (c), where $\sim^*$ is Δ^0_α, $\sim$ is clearly Σ^0_α. We consider the complement of $\sim$. We have $b \not\sim b'$ if and only if $b \Delta b'$ is a finite join of half-open intervals at least one of which has the form $[d, d')$, where $d \not\sim^* d'$. Then $\not\sim$ is also Σ^0_α.

18.5. QUOTIENT ORDERINGS

Odintsov and Selivanov [130] have some transfer theorems for Boolean algebras, involving congruence relations corresponding to various natural ideals. involving the congruence relations corresponding to various natural ideals.

Chapter 19

Models of arithmetic

In this chapter, we describe some results on complexity of models of arithmetic. While the standard model, $\mathcal{N}$, is computable, its theory, TA, is not even arithmetical. McAloon asked whether there is a nonstandard model $\mathcal{A}$ of PA such that $\mathcal{A}$ is arithmetical but $Th(\mathcal{A})$ is not. Harrington [65] gave a positive answer. It was for this result that he developed his method of "workers", for nested priority constructions. Here, we shall prove Harrington's Theorem using Theorem 14.4. In addition, we prove a result of Solovay [145] and Marker [107] characterizing the degrees of nonstandard models of TA. We shall state a result of Solovay characterizing the degrees of models of other completions of PA [146], [84], [86], but we give complete proofs only of weaker results in this direction.

Since Chapter 6, the formulas that we have used were, for the most part, infinitary. Throughout the present chapter, the formulas are finitary. A B_n formula is a Boolean combination of Σ_n formulas. Recall, from Chapter 3, that a complete type is the set of all formulas satisfied by some tuple (in some structure). A complete B_n type is the set of all B_n formulas in some complete type.

19.1 Scott sets

We define the notion of a "Scott set". The only connection between this and the notion of a "Scott family", which appeared in Chapters 6, 10, 12 and 17, is that both are due to Dana Scott. A *Scott set* is a set $\mathcal{S} \subseteq P(\omega)$ satisfying the following three conditions:

(1) if $X \in \mathcal{S}$ and $Y \leq_T X$, then $Y \in \mathcal{S}$,
(2) if $X, Y \in \mathcal{S}$, then $X \oplus Y \in \mathcal{S}$,
(3) if $P \subseteq 2^{<\omega}$ is an infinite tree in $\mathcal{S}$, then P has a path in $\mathcal{S}$.

Note: In the presence of the other conditions, Condition 3 is equivalent to the following:

(3)′ if A is a set of axioms in $\mathcal{S}$, then A has a completion in $\mathcal{S}$.

Scott sets arise naturally in connection with completions of PA. Let T be a theory in the language of arithmetic, let $\psi(x)$ be a formula, and let $X \subseteq \omega$. We say that $\psi(x)$ *represents* X *with respect to* T if

$$T \vdash \psi(S^{(n)}(0)) \quad \text{for } n \in X,$$
$$T \vdash \neg\psi(S^{(n)}(0)) \quad \text{for } n \notin X.$$

Let $Rep(T) = \{X : X \text{ is representable with respect to } T\}$. It turns out that if T is PA, or any computably axiomatizable extension of a certain fragment of PA, then $Rep(T)$ is the family of computable sets. We are more interested in the case where T is complete. For $T = TA$, $Rep(T)$ is the family of all arithmetical sets. Scott [140] characterized the sets $\mathcal{S} \subseteq P(\omega)$ of the form $Rep(T)$, where T is a completion of PA.

Theorem 19.1 (Scott) *For $\mathcal{S} \subseteq P(\omega)$, the following are equivalent:*
(a) there is a completion T of PA such that $Rep(T) = \mathcal{S}$,
(b) $\mathcal{S}$ is a countable Scott set.

Sketch of proof: (a) $\Rightarrow$ (b) (i) If X is represented by the formula $\psi(x)$ and $\chi_Y = \varphi_e^X$, then Y is represented by the formula saying that for oracle machine e, with input y and oracle information given by $\psi(x)$, there is some halting computation with output 1. (There is another formula saying that for *all* halting computations, the output is 1.)

(ii) If X and Y are represented by $\psi(x)$ and $\theta(x)$, respectively, then $X \oplus Y$ is represented by the formula saying that for some z, either $2z = x$ and $\psi(z)$, or else $2z + 1 = x$ and $\theta(x)$. (There is another formula saying that for all z, if $2z = x$, then $\psi(z)$ and if $2z + 1 = x$, then $\theta(z)$.)

(iii) Finally, suppose $P \subseteq 2^{<\omega}$ is an infinite tree represented by the formula $\psi(x)$. Then there is a path represented by the formula saying that x is a "finite" sequence (the length of x may not actually be finite), and for all $u < length(x)$,

$$x(u) = \begin{cases} 0 & \text{if } (x|u) \cup \{(u,0)\} \text{ has extensions of all possible lengths in } P, \\ 1 & \text{otherwise.} \end{cases}$$

The sentence saying that P is "infinite" (i.e., P contains sequences of all lengths) may or may not be in T. If not, then the "path" consists of finite initial segments of the left-most node at the last level of P.

(b) $\Rightarrow$ (a) Let $X_1, X_2, X_3, \ldots$ be a list of the sets in $\mathcal{S}$. There is a Π_2 formula $\varphi(x)$, an iteration of the Gödel-Rosser sentence (see the Appendix), such that for all $X \subseteq \omega$,

$$PA \cup \{\varphi(S^{(x)}(0)) : x \in X\} \cup \{\neg\varphi(S^{(x)}(0)) : x \notin X\}$$

is consistent. In fact, for each n, we can find a Π_{n+2} formula $\varphi_n(x)$ such that for any $X \supseteq \omega$ and any completion T of PA,

$$PA \cup (T \cap B_n) \cup \{\varphi_n(S^{(x)}(0)) : x \in X\} \cup \{\neg\varphi_n(S^{(x)}(0)) : x \notin X\}$$

19.1. SCOTT SETS

is consistent.

By Property (3′) of Scott sets, $\mathcal{S}$ contains a completion of PA in which $\varphi_0(x)$ represents X_0. We let $T \cap B_2$ be the B_2 part of this theory. Note that this is in $\mathcal{S}$. Suppose that we have determined $T \cap B_{2n}$ in $\mathcal{S}$, such that for $k < n$, the Π_{2k+2} formula $\varphi_{2k}(x)$ represents X_k. Staying within $\mathcal{S}$, we take a completion of $PA \cup (T \cap B_{2n})$ in which the Π_{2n+2} formula $\varphi_{2n}(x)$ represents X_n, and we let $T \cap B_{2n+2}$ be the B_{2n+2} part of this theory.

We have indicated how Scott sets are associated with completions of PA. They are also associated with models. For a nonstandard model $\mathcal{A}$ of PA, let $SS(\mathcal{A})$ denote the collection of sets of the form

$$d_a = \{n \in \omega : \mathcal{A} \models p_n|a\} \ .$$

We call $SS(\mathcal{A})$ the *Scott set* of $\mathcal{A}$. (Some people call it the *standard system* of $\mathcal{A}$—the name still fits.) The next result says that $SS(\mathcal{A})$ is equal to the set of "standard parts" of sets definable in $\mathcal{A}$, with parameters.

Proposition 19.2 *Let $\mathcal{A}$ be a nonstandard model of PA. Then $X \in SS(\mathcal{A})$ if and only if*

$$X = \{n \in \omega : \mathcal{A} \models \varphi(\bar{c}, S^{(n)}(0))\} \ ,$$

for some formula $\varphi(\bar{u}, x)$, and some tuple $\bar{c}$ in $\mathcal{A}$.

Proof: First, suppose $X \in SS(\mathcal{A})$, say $X = d_a$. Let $\varphi(a, x)$ be the natural formula saying that $p_x|a$. Then

$$n \in d_a \Leftrightarrow \mathcal{A} \models \varphi(a, S^{(n)}(0)) \ ,$$

so d_a is the standard part of $\varphi^{\mathcal{A}}(a, x)$.

Now, take a formula $\varphi(\bar{u}, x)$ and parameters $\bar{c}$ in $\mathcal{A}$. Let $\psi(\bar{c}, u, y)$ be the natural formula saying

$$(\forall x < u)(\varphi(\bar{c}, x) \leftrightarrow p_x|y) \ .$$

For all finite numbers u, we have $\mathcal{A} \models \exists y\, \psi(\bar{c}, u, y)$. Therefore, by Overspill (see the Appendix), there is some infinite number u and some a such that $\mathcal{A} \models \psi(\bar{c}, u, a)$. Then $d_a = \{n \in \omega : \mathcal{A} \models \varphi(\bar{c}, S^{(n)}(0))\}$.

The result below, from [140], [48], says that the Scott set of a model is in fact a Scott set.

Theorem 19.3 (Scott) *If $\mathcal{A}$ is a nonstandard model of PA, then $SS(\mathcal{A})$ is a Scott set. Moreover, if $T = Th(\mathcal{A})$, then $Rep(T) \subseteq SS(\mathcal{A})$.*

Exercise: Prove Theorem 19.3. The proof that $SS(\mathcal{A})$ is a Scott set is similar to the proof of Theorem 19.1. The fact that $Rep(T) \subseteq SS(\mathcal{A})$ follows from Proposition 19.2.

Although we are only interested in countable models, Theorem 19.3 holds for models of arbitrary cardinality. For any countable Scott set S, there is a nonstandard model $\mathcal{A}$ of PA such that $SS(\mathcal{A}) = S$. (For example, take the prime model of a theory T such that $Rep(T) = S$.) Nadel [126] showed that for any Scott set S of cardinality at most $\aleph_1$, there is a model $\mathcal{A}$ of PA such that $SS(\mathcal{A}) = S$. Thus, if we assume the Continuum Hypothesis, then all Scott sets are represented by nonstandard models of PA. Without the Continuum Hypothesis, however, it is unknown whether all Scott sets are represented in this way.

19.2 Enumerations

For any completion T of PA, there is a natural enumeration R of $Rep(T)$ such that R is computable in T. We form a computable list $(\varphi_n(x))_{n\in\omega}$ of the formulas with just x free, and let n be an index for the set represented by the formula $\varphi_n(x)$. If $\mathcal{A}$ is a nonstandard model of PA, then there is a natural enumeration of $SS(\mathcal{A})$; namely,

$$R = \{(a,n) : \mathcal{A} \models p_n | a\} .$$

We refer to this as the *canonical enumeration*.

Proposition 19.4 *If $\mathcal{A}$ is a nonstandard model of PA, then the canonical enumeration of $SS(\mathcal{A})$ is computable in $\mathcal{A}$.*

Proof: The proof of Tennenbaum's Theorem (Theorem 3.21) shows this. To decide whether $\mathcal{A} \models p_n | a$, we remember the division algorithm, and we search $D(\mathcal{A})$ for an open sentence saying $a = p_n q + S^{(r)}(0)$, for some $r < p_n$. The search will halt, and

$$\mathcal{A} \models p_n | a \Leftrightarrow r = 0 .$$

Suppose S is a Scott set. An *effective enumeration* of S is an enumeration R of S, equipped with functions f, g, and h witnessing the fact that S is a Scott set. An effective enumeration R, f, g, h of S is said to be *computable in X* if R, f, g, h are all computable relative to X.

Remarks: (1) If T is a completion of PA, then the proof of Theorem 19.1 yields computable functions making the natural enumeration of $Rep(T)$ into an effective enumeration computable in T.

(2) If $\mathcal{A}$ is a nonstandard model of PA, then we can easily equip the canonical enumeration of $SS(\mathcal{A})$ with functions f, g, h to obtain an effective enumeration that is computable in the complete diagram $D^c(\mathcal{A})$. Then the functions making the enumeration effective may be more complicated than the enumeration itself. The following result says that we can replace an arbitrary enumeration of a Scott set by an effective one without increasing the complexity (see [107]).

19.2. ENUMERATIONS

Theorem 19.5 (Marker) *For any $X \subseteq \omega$ and any Scott set S, if S has an enumeration computable in X, then it has an effective enumeration computable in X.*

Sketch of proof: First, let T be a completion of PA, where $T \neq TA$ and $T \in S$. We construct a model $\mathcal{B}$ of T such that $D^c(\mathcal{B}) \leq_T X$ and $SS(\mathcal{B}) = S$. Using the fact that any partial type in S extends to a complete type in S, we obtain the following.

Lemma 19.6 *If S is a Scott set, and T is a completion of PA in S, then there is a homogeneous model $\mathcal{A}$ of T such that for any complete type Γ consistent with T, Γ is realized in $\mathcal{A}$ if and only if $\Gamma \in S$.*

It is east to see that if $\mathcal{A}$ is the model in Lemma 19.6, then $SS(\mathcal{A}) = S$. (We note that $\mathcal{A}$ also has a property called "recursive saturation", although this fact is of no importance for the proof of Theorem 19.5.) We want $\mathcal{B} \cong \mathcal{A}$ such that $D^c(\mathcal{B}) \leq_T X$. We shall apply Theorem 9.27, relativized to X. Let R be an enumeration of S, computable in X. The following lemma says that we have satisfied the hypotheses of Theorem 9.27 (the result of Goncharov and Peretyat'kin on decidable homogeneous structures).

Lemma 19.7 *If T is a complete theory in S, then there exist a relation E computable in X, and a computable function f, such that*
(a) E is an enumeration of the complete types $\Gamma(\overline{u})$ such that $\Gamma(\overline{u}) \in S$ and $\Gamma(\overline{u})$ is consistent with T, and
(b) for all i and $\varphi(\overline{u}, x)$, if $\exists x\, \varphi(\overline{u}, x) \in E_i(\overline{u})$, then $E_{f(i,\varphi(\overline{u},x))}$ is a complete type that extends $E_i(\overline{u}) \cup \{\varphi(\overline{u}, x)\}$.

Proof sketch: We form a computable list of the pairs of the following two kinds:
(i) $(i, \overline{u})$, where $i \in \omega$ and $\overline{u}$ is a tuple of variables,
(ii) $(m, \varphi(\overline{u}, x))$, where $m \in \omega$ and $\varphi(\overline{u}, x)$ is a formula.
We may suppose that if the nth item on the list has the form $(m, \varphi(\overline{u}, x))$, then $m < n$. We shall determine E such that for each n, we can effectively determine the variables of the type E_n and the index for E_n relative to X. We proceed by induction on n.

Suppose first that the nth item on the list has the form $(i, \overline{u})$. Then E_n will be a type in variables $\overline{u}$, and if R_i is an appropriate type, then $E_n = R_i$. We describe the procedure for computing E_n from R. At stage s, we test $R_i \cap s$ to see if this is an initial segment of a complete type, in variables $\overline{u}$, consistent with T. If so, then we let $E_n \cap s = R_i \cap s$. If at some stage s, $R_i \cap s$ no longer passes the test, then we continue making E_n into an appropriate type. When we come to a formula $\psi(\overline{u})$, we add $\psi(\overline{u})$ to E_n if it is consistent to do so; otherwise, we add $\neg\psi(\overline{u})$.

Now, suppose that the nth item on the list has the form $(m, \varphi(\overline{u}, x))$, where $m < n$. We let $f(m, \varphi(\overline{u}, x)) = n$. Say that we have determined the index for E_m as a type in variables $\overline{u}$. Then we can check whether $\exists x\, \varphi(\overline{u}, x) \in E_m$. If so, then E_n will be a completion of $E_m(\overline{u}) \cup \{\varphi(\overline{u}, x)\}$, computable in E_m. The procedure for computing E_n from R in this case is similar to the one above. When we come to a formula $\psi(\overline{u}, x)$, we add $\psi(\overline{u}, x)$ to E_n if it is consistent to do so; otherwise, we add $\neg \psi(\overline{u}, x)$.

Now, we are in a position to apply Theorem 9.27, relativized to X. We obtain $\mathcal{B} \cong \mathcal{A}$ such that $D^c(\mathcal{B}) \leq_T X$. By Remark (2) above, there is an effective enumeration of $SS(\mathcal{B})$ computable in $D^c(\mathcal{B})$. This completes our sketch of the proof of Theorem 19.5.

The proof above is remarkably indirect—we determine a model $\mathcal{A}$ of PA such that $SS(\mathcal{A}) = S$ and $D^c(\mathcal{A}) \leq_T X$, etc.

Open problem: Find a direct proof of Theorem 19.5.

Using Theorem 19.5, we obtain the following improvement of Theorem 19.1.

Corollary 19.8 (Knight-Marker) *Let R be an enumeration of a Scott set S. Then there is a completion T of PA, such that $Rep(T) = S$ and T is computable in R.*

Proof sketch: By Theorem 19.5, S has an effective enumeration computable in R. We proceed as in Theorem 19.1, using the effective enumeration to determine indices for the fragments of T.

19.3 Structures representing a Scott set

It turns out to be easier to construct a model, controlling the degree, if we specify more than just the theory. Solovay characterized the degrees of (nonstandard) models of a particular completion T of PA by characterizing the degrees of models of T with a given Scott set $S \supseteq Rep(T)$. The basic result for TA says that for a Scott set S containing the arithmetical sets, the degrees of nonstandard models with Scott set S are the same as the degrees of enumerations of S. It follows that the degrees of nonstandard models of TA are the same as the degrees of enumerations of Scott sets containing the arithmetical sets.

We shall obtain results on models of PA from some general results on structures "representing" a given Scott set. These general results apply to models of an arbitrary theory, possibly unrelated to PA. The definition that we give below is from [102]. There is a different definition in [84]. For a structure $\mathcal{A}$ and a Scott set S, we say that $\mathcal{A}$ *represents* S if for all tuples $\overline{c}$ in $\mathcal{A}$ and all complete B_n types $\Gamma(\overline{c}, x)$ such that $\Gamma(\overline{c}, x)$ is consistent with $D^c(\mathcal{A})$ (or with the type of $\overline{c}$),

$$\Gamma(\overline{c}, x) \text{ is realized in } \mathcal{A} \Leftrightarrow \Gamma(\overline{c}, x) \in SS(\mathcal{A}) .$$

19.3. STRUCTURES REPRESENTING A SCOTT SET

Remark: Some structures do not represent any Scott set, and some represent all Scott sets. There are specific examples given in the exercises below.

Exercise: Prove that $\mathcal{N}$ does not represent any Scott set.

Exercise: Prove that $(Q, <)$ represents all Scott sets.

Theorem 19.9 *Suppose $\mathcal{A}$ is a nonstandard model of PA. Then $\mathcal{A}$ represents a unique Scott set, namely $SS(\mathcal{A})$.*

Proof: We first show that $\mathcal{A}$ cannot represent two distinct Scott sets. Suppose that $\mathcal{A}$ represents S_1 and S_2, where these are distinct Scott sets, with X in S_1 and not in S_2. There is a complete B_1 type $\Gamma(x) \in S_1$ such that $\Gamma(x)$ contains the natural formulas saying

$$p_n | x, \quad \text{for } n \in X$$
$$p_n \not| x, \quad \text{for } n \notin X.$$

Since $\mathcal{A}$ represents S_1, it must realize $\Gamma(x)$. Since $\mathcal{A}$ represents S_2, $\Gamma(x) \in S_2$. Now, $X \leq_T \Gamma(x)$, so we have $X \in S_2$, a contradiction.

Next, we show that $\mathcal{A}$ represents the Scott set $SS(\mathcal{A})$. For any tuple $\overline{u}$ of variables, and any finite n, there is a natural formula $Sat_{B_n}(x, \overline{u})$ saying that x is a B_n-formula satisfied by $\overline{u}$. For all models of PA, in particular, for $\mathcal{A}$, for any tuple $\overline{a}$ appropriate to substitute for $\overline{u}$, if k is the Gödel number of a B_n formula $\psi(\overline{u})$, then

$$\mathcal{A} \models Sat_{B_n}(S^{(k)}(0), \overline{a}) \Leftrightarrow \mathcal{A} \models \psi(\overline{a}).$$

Then by Proposition 19.2, the B_n-type of $\overline{a}$ is in $SS(\mathcal{A})$.

Now suppose that $\Gamma(\overline{u}, y)$ is a complete B_n-type, and $\Gamma(\overline{c}, y)$ is consistent with $D^c(\mathcal{A})$, where $\Gamma(\overline{u}, y) \in SS(\mathcal{A})$. Let $\Gamma(\overline{u}, y) = d_a$. Let $\varphi(\overline{c}, a, u)$ say

$$\exists y \, (\forall x < u) \, (\, Sat_{B_n}(x, \overline{c}, y) \leftrightarrow p_x | a \,).$$

For any finite u, $\varphi(\overline{c}, a, u)$ is satisfied in $\mathcal{A}$, since every finite subset of $\Gamma(\overline{c}, y)$ is satisfied. By Overspill, there is some infinite u such that $\varphi(\overline{c}, a, u)$ is satisfied. Then we have b such that for all finite k,

$$\mathcal{A} \models Sat_{B_n}(S^{(k)}(0), \overline{c}, b) \leftrightarrow p_k | a.$$

This means that b realizes $\Gamma(\overline{c}, y)$. Therefore, $\mathcal{A}$ represents $SS(\mathcal{A})$. This completes the proof of Theorem 19.9.

Let T be a complete theory, and let S be a Scott set. For T to have a model representing S, it is necessary (and also sufficient) that $T_n = T \cap \Sigma_n \in S$, for all $n \in \omega$. When this condition holds, we say that S is *appropriate for T*. If T is a completion of PA, then a Scott set S is appropriate for T if and only if $S \supseteq Rep(T)$.

The result below is related to Theorem 9.27 (of Goncharov and Peretyat'kin,) and also to a result of Lerman and Schmerl [94]. It gives some indication of how the notion of a structure representing a Scott set may be used.

Theorem 19.10 *Let S be a countable Scott set with an enumeration $R \leq_T X$. Let $\mathcal{A}$ be a structure representing S, and suppose that the relation*

$$Q = \{(i, \bar{a}) : R_i \text{ is the } B_1 \text{ type of } \bar{a}\}$$

is $\Sigma_2^0(X)$. Then there exists $\mathcal{B} \cong \mathcal{A}$ such that $\mathcal{B} \leq_T X$.

Proof: By Theorem 19.5, S has an effective enumeration computable in X. Moreover, using $\Delta_2^0(X)$, we can pass effectively from an index in one enumeration to an index in the other. Thus, we may suppose that R (the enumeration associated with Q) is an effective enumeration. We shall use the three facts below.

Fact 1: There is a function $g_1 \leq_T X$ such that for all tuples $\bar{a}$ from $\mathcal{A}$, $lim_{s\to\infty} g_1(\bar{a}, s)$ exists and is equal to some i such that $(i, \bar{a}) \in Q$.

Fact 2: There is a function $g_2 \leq_T X$ such that if i is an index for the B_1-type $\Gamma(\bar{u})$ realized by $\bar{c}$, and $\gamma(\bar{u}, x)$ is an existential formula with

$$\exists x\, \gamma(\bar{u}, x) \in \Gamma(\bar{u}),$$

then $g_2(i, \gamma)$ is an index for a B_1 type $\Gamma^*(\bar{u}, x) \supseteq \Gamma(\bar{u}) \cup \{\gamma(\bar{u}, x)\}$, where $\Gamma^*(\bar{c}, x)$ is realized by some $a \in \mathcal{A}$.

Fact 3: There is a function $g_3 \leq_T X$ such that if i is an index for a B_1 type $\Gamma(\bar{u}, x)$ realized by $\bar{c}$ and some a, then $lim_{s\to\infty} g_3(i, \bar{c}, s)$ is such an a.

Facts 1 and 3 are not difficult to verify. We sketch the proof of Fact 2. We take a computable list of all existential formulas $\varphi(\bar{u}, x)$ in variables $\bar{u}, x$, starting with $\gamma(\bar{u}, x)$. We put $\gamma(\bar{u}, x)$ into Γ^*. Then, proceeding down the list, we add $\varphi(\bar{u}, x)$ if it is consistent to do so. We determine the consistency of $\varphi(\bar{u}, x)$ by taking the conjunction $\psi(\bar{u}, x)$ of the existential formulas already in Γ^*, and checking whether

$$\exists x\, (\psi(\bar{u}, x)\ \&\ \varphi(\bar{u}, x)) \in \Gamma(\bar{u}).$$

Finally, we let $\Gamma^*(\bar{u}, x)$ be the B_1 type generated by the existential formulas that we have included and the negations of the ones that we have omitted. Clearly, $\Gamma^*(\bar{u}, x) \leq_T \Gamma(\bar{u})$. Since our enumeration of S is effective, using X, we can pass effectively from an index for Γ to an index for Γ^*. The B_1 type Γ^* has the feature that it is consistent with all completions of Γ. Since $\mathcal{A}$ represents S, it must realize $\Gamma^*(\bar{c}, x)$.

Now, we describe the construction of $\mathcal{B}$, using the three facts. Let B be an infinite computable set, for the universe of $\mathcal{B}$. Let $(a_n)_{n\in\omega}$, $(b_n)_{n\in\omega}$ be computable lists of the elements of $\mathcal{A}$, $\mathcal{B}$, respectively. Fix a computable list of the atomic sentences with constants from B. We determine a function F from

19.3. STRUCTURES REPRESENTING A SCOTT SET

B to $\mathcal{A}$ and we enumerate $D(\mathcal{B})$ so that F will be an isomorphism from $\mathcal{B}$ onto $\mathcal{A}$. There are obvious requirements.

R_{2r}: $a_r \in ran(F)$

R_{2r+1}: $b_r \in dom(F)$

At stage s, we determine a finite subset δ_s of $D(\mathcal{B})$. We also tentatively determine a chain $(p_0, \ldots, p_n)$ of finite partial $1-1$ functions from B to $\mathcal{A}$. We suppose that for $k \leq n$, p_k maps $\bar{b}_k$ to $\bar{a}_k$, where $\bar{b}_0 = \bar{a}_0 = \emptyset$; if $k = 2r+1$, then $\bar{a}_k$ includes a_r; and if $k = 2r+2$, then $\bar{b}_k$ includes b_r. Finally, we tentatively determine a sequence of indices for B_1 types Γ_k, for $0 \leq k \leq n$, and if n is odd, also for $k = n+1$, such that the following conditions hold:

(1) for $k \leq n$, and for $k = n+1$ if k is odd, δ_s is consistent with $\Gamma_k(\bar{b}_k)$, in the sense that $\Gamma_k(\bar{b}_k)$ contains the existential sentence obtained by taking the conjunction of δ_s and quantifying out the constants not in $\bar{b}_k$,

(2) Γ_0 is the B_1 type of $\emptyset$ (the B_1 theory),

(3) for odd $k \leq n$, Γ_k appears to be the B_1 type realized by $\bar{a}_k$, according to g_1 with s; i.e., $g_1(\bar{a}_k, s)$ is our index for Γ_k,

(4) for even k such that $0 < k \leq n+1$, Γ_k is sure to be realized by $\bar{a}_{k-1}$ and some element c, provided that Γ_{k-1} is realized by $\bar{a}_{k-1}$—the function g_2 is used in computing an index for Γ_k,

(5) for even $k = 2r+2 < n$, p_k maps b_r to an element c such that $\bar{a}_{2r}, a_r, c$ appears to realize Γ_k, according to g_3 with s; i.e., if i is our index for Γ_k, then $g_3(i, \bar{a}_{2r}, a_r, s) = c$.

At stage 0, we let $\delta_0 = \emptyset$, we consider the trivial chain (p_0), where $p_0 = \emptyset$, and we let Γ_0 be as in (2) above. At stage $s+1$, we modify the stage s chain of functions $(p_0, \ldots, p_n)$, and we act in one of the following ways:

(a) extend the sequence, adding some p_{n+1},

(b) for some $k \leq n$, drop p_i for all $k \leq i \leq n$, having determined that at least one of the guesses behind p_k was not stable by stage s,

(c) leave the sequence as is, having determined that at least one of the guesses needed for p_{n+1} is not stable by stage $s+1$. Suppose that α is the first atomic sentence on our list such that neither α nor $\neg\alpha$ is in δ_s. We let δ_{s+1} be the result of adding to δ_s either α or $\neg\alpha$, where the choice is made so that for the greatest possible $n' \leq n$, δ_{s+1} is consistent with $\Gamma_k(\bar{b}_k)$, for all $k \leq n'$, and also for $k = n'+1$, if n' is odd. Let $n'' \leq n'$ be greatest such that for all $k \leq n''$, the stage s guesses behind p_k match the stage $s+1$ guesses.

Case 1: Suppose $n'' < n$.

In this case, we drop p_i for $n'' < i \leq n$, keeping $\Gamma_{n''+1}$ if n'' is odd. This is action (b).

Case 2: Suppose $n'' = n$.

In this case, we attempt to determine p_{n+1}, as follows:

Subcase 2 (i): Suppose $n = 2r$.

Let Γ_{n+1} be the type that $\bar{a}_n, a_r$ appears to realize, according to g_1 with $s + 1$. We look for some d, either one of the constants appearing in δ_{s+1} or the first new one, such that $\Gamma_{n+1}(\bar{b}_n, d) \cup \delta_{s+1}$ is consistent. If we find d, then we let p_{n+1} map $\bar{b}_n, d$ to $\bar{a}_n, a_r$. In addition, using g_2, we determine $\Gamma_{n+2} \supseteq \Gamma_{n+1}$ such that $\Gamma_{n+2}(\bar{b}_n, d, b_r)$ is consistent with δ_{s+1}. This is action (a). If we do not find d, then we leave p_{n+1} undefined. This is action (c).

Subcase 2 (ii): Suppose $n = 2r + 1$.

We have determined Γ_n and Γ_{n+1} such that if Γ_n is realized by $\bar{a}_n$, then Γ_{n+1} is realized by $\bar{a}_n$ and some c, and $\Gamma_{n+1}(\bar{b}_n, b_r) \cup \delta_{s+1}$ is consistent. Using g_3 with $s + 1$, we obtain c which seems to serve, and we let p_{n+1} map $\bar{b}_n, b_r$ to $\bar{a}_n, c$. This is action (a).

Claim: All of the requirements are eventually satisfied.

Suppose we have $(p_0, \ldots, p_n)$ taking care of the first n requirements, where $n = 2r$. When our guess at the type Γ_{n+1} of $\bar{a}_{n+1}$ becomes stable, then we will have p_{n+1} taking care of R_{2r}. At that time, we will also have a type Γ_{n+2} that is sure to be realized by $\bar{a}_{n+1}$ and some c. When our guess at such a c becomes stable, then we will have p_{n+2} taking care of R_{2r+1}. This completes the proof.

The hypotheses of Theorem 19.10 may have seemed obscure. In the next few results, we show how they arise.

Theorem 19.11 *Let T be a complete theory, and let $R \leq_T X$ be an enumeration of a Scott set S appropriate for T. Suppose there is a $\Delta_3^0(X)$ function t such that for all n, $t(n)$ is an R-index for $T_n = T \cap S_n$. Then T has a model $\mathcal{A}$, representing S, such that*

$$Q = \{(i, \bar{b}) : R_i \text{ is the } B_1 \text{ type realized by } \bar{b}\}$$

is $\Sigma_2^0(X)$.

Proof: This is a finite injury construction. Let A be an infinite, computable set of constants, for the universe of $\mathcal{A}$. We fix a computable list of triples $(k, j, \bar{c})$, where $k, j \in \omega$ and $\bar{c}$ is a tuple of constants from A such that if $(k, j, \bar{c})$ is n^{th} on the list, then $k \leq n$ and $\bar{c}$ is included among the first n constants.

There are two requirements for each n.

A_n: Assign a B_{n+1} type to a tuple including the first n constants.

19.3. STRUCTURES REPRESENTING A SCOTT SET

Win: For the nth triple $(k,j,\bar{c})$ on the list, if R_j is a complete B_k type, $R_j(\bar{c},x)$ is witnessed or else is inconsistent with the B_{n+1} type assigned to the first n constants.

Approximation of T: Let g be a $\Delta_2^0(X)$ function such that for $n \geq 2$, $\lim_{s\to\infty} g(n,s) = t(n)$. We suppose that for all s,

(1) $g(2,s) = t(2)$,

(2) there is a complete theory T' such that for $2 \leq n \leq s+2$, $T' \cap \Sigma_n$ has R-index $g(n,s)$.

At stage s, we have a finite sequence $t_2, \ell_0, \ldots, t_{r+2}, \ell_r$, for some $r \leq s$. For each $n \leq r$, ℓ_n represents an attempt to satisfy requirements A_n and Win, using the current approximations of T. That is, $t_{n+2} = g(n+2,s)$, and $\ell_n = (i_n, \bar{a}_n)$, where

(1) R_{i_n} is a complete type in which the Σ_{n+2} sentences (with no added constants) are just those of $R_{t_{n+2}}$,

(2) $\bar{a}_n$ is a finite sequence including the first n constants,

(3) for the nth triple $(k,j,\bar{c})$ is a complete B_k type (in the appropriate language) and $R_{i_n}(\bar{a}_n) \cup R_j(\bar{c},x)$ is consistent, then $R_j(\bar{c},a) \subseteq R_{i_n}(\bar{a}_n)$, for some $a \in \bar{a}_n$,

(4) $R_{i_n}(\bar{a}_n) \cap B_{n+1} \subseteq R_{i_{n+1}}(\bar{a}_{n+1})$.

At stage $s+1$, either we extend the sequence

$$t_2, \ell_0, \ldots, t_{r+2}, \ell_r ,$$

or else, for some $n \leq r$, we change t_{n+2}, i_n, and $\bar{a}_n$, and drop any later terms. In either case, we maintain the B_1 type assigned to $\bar{a}_r$.

Case 1: For all $n \leq r$, $g(n+2, s+1) = g(n+2, s)$.

In this case, we extend the sequence $t_2, \ell_0, \ldots, t_{r+2}, \ell_r$, attempting the next pair of requirements, A_{r+1} and Win_{r+1}. Let $t_{r+3} = g(r+3, s+1)$. Using $\Delta_2^0(X)$, we determine i_{r+1} and $\bar{a}_{r+1}$ such that $\bar{a}_{r+1} \supseteq \bar{a}_r$, the first $r+1$ constants are included in $\bar{a}_{r+1}$, $R_{i_{r+1}}$ is a complete type with exactly the same Σ_{n+3} sentences as $R_{t_{r+3}}$, $R_{i_r}(\bar{a}_r) \cap B_{r+1} \subseteq R_{i_{r+1}}(\bar{a}_{r+1})$, and for the $(r+1)^{st}$ triple $(k,j,\bar{c})$, if R_j is a complete B_k type, then either $R_j(\bar{c},a) \subseteq R_{i_{r+1}}(\bar{a}_{r+1})$, for some $a \in \bar{a}_{r+1}$, or else $R_{i_{r+1}}(\bar{a}_{r+1}) \cup R_j(\bar{c},x)$ is inconsistent.

Case 2: For some first $n \leq r$, $g(n+2, s+1) \neq g(n+2, s)$.

In this case, we replace t_{n+2}, i_n, and $\bar{a}_n$, by t, i, and $\bar{a}$, chosen as follows. We let $t = g(n+2, s+1)$. Note that $R_{i_{n-1}}(\bar{a}_{n-1}) \cap B_n \subseteq R_{i_r}(\bar{a}_r)$. The Σ_{n+1} sentences of $R_{t_{n+1}}$, $R_{t_{n+2}}$ and R_t are the same, and R_{t_n} is a complete type with the same Σ_{n+1} sentences. We take i and $\bar{a}$ such that

(i) $\bar{a}_r \subseteq \bar{a}$,

(ii) R_i is a complete type with the same Σ_{n+2} sentences as R_t,

(iii) $R_i(\bar{a}) \supseteq R_{i_r}(\bar{a}_r) \cap B_n$, and

(iv) for the nth triple $(k, j, \bar{c})$, either $R_j(\bar{c}, a) \subseteq R_i(\bar{a})$ for some $a \in \bar{a}$, or else $R_i(\bar{a}) \cup R_j(\bar{c}, x)$ is inconsistent.

We have said how to determine a finite sequence

$$t_2 \ell_0 t_3 \ell_1 t_4 \ell_2 \ldots,$$

for each stage s. For each n, there is some stage after which the first $2n$ terms in the sequence remain fixed. Therefore, the limit exists. This is an infinite sequence

$$t_2 \ell_0 t_3 \ell_1 t_4 \ell_2 \ldots$$

such that $T = \cup_n R_{t_{n+2}}$, and if $\ell_n = (i_n, \bar{a}_n)$, then

$$\cup_n R_{i_n}(\bar{a}_n) \cap B_{n+1}$$

is the complete diagram of a model $\mathcal{A}$ of T, where $\mathcal{A}$ represents $\mathcal{S}$.

Note that i is the R-index of the complete B_1 type realized by $\bar{c}$ if and only if R_i is equal to the appropriate restriction of the type assigned to some $\bar{a} \supseteq \bar{c}$ at some stage in the construction. From this, it follows that

$$Q = \{(i, \bar{a}) : R_i \text{ is the complete } B_1 \text{ type realized by } \bar{a}\}$$

is $\Sigma_2^0(X)$.

The next result has the same conclusion as Theorem 19.11, but the hypotheses on the approximations of the theory are weaker.

Theorem 19.12 *Let T be a complete theory, and let $R \leq_T X$ be an enumeration of a Scott set $\mathcal{S}$ appropriate for T. Suppose that there is a uniform effective procedure which, for each n, locates an R-index for $T_{n+1} = T \cap \Sigma_{n+1}$, using $\Delta_n^0(X)$. Then T has a model $\mathcal{A}$, representing $\mathcal{S}$, such that*

$$Q = \{(i, \bar{b}) : R_i \text{ is the } B_1 \text{ type realized by } \bar{b}\}$$

is $\Sigma_2^0(X)$.

Proof: We use Theorem 14.4, relativized to X'. We define a special $(n + 1)$ system "relativized" to X'. Recall that a special $(n + 1)$-system resembles an ω-system except that Condition (4) is weaker. Our special system will have the form

$$(L, U, \hat{\ell}, P, E, (\leq_n)_{1 \leq n < \omega}).$$

Relativizing to X' means that L, U, and P are $\Sigma_2^0(X)$, E is partial $\Delta_2^0(X)$, and the relations $\leq_n$ are $\Sigma_2^0(X)$ uniformly in n. For readability, we let $\leq_{n+1}$ play the roll of $\leq_n$— $\leq_1$ will be preserved on the bottom level. Thus, Condition (4) says that if $\sigma u \in P$, where σ has length $2n + 1$, ending in ℓ^0,

$$\ell_0 \leq_{m_0} \ell_1 \leq_{m_1} \ldots \leq_{m_{k-1}} \ell_k \ldots,$$

19.3. STRUCTURES REPRESENTING A SCOTT SET

and $n + 2 > m_0 > \ldots > m_k \geq 1$, then there exists ℓ^* such that $\sigma u \ell^* \in P$ and $\ell_i \leq_{m_i} \ell^*$ for $i \leq k$.

We define an instruction function q that is $(n + 1)$ "relative to" X', where this means that there is a uniform effective procedure that, for all n, and for all $\sigma \in P$ of length $2n + 1$, computes $q(\sigma)$ using $\Delta^0_{n+1}(X')$, or $\Delta^0_{n+2}(X)$. The theorem lets us conclude that there is a run π of (P, q) such that $E(\pi)$ is c.e. relative to X', or $\Sigma^0_2(X)$.

Overview: The requirements are as in the previous proof. If σ has length $2n+1$, ending in ℓ^0, and $\ell^0 \leq_{n+1} \ell^1$, then the B_{n+1} type assigned in ℓ^0 must be preserved in ℓ^1. If σ follows q, then ℓ^0 is based on the true Σ_{n+2} theory T_{n+2}, and $q(\sigma)$ is an index for the true Σ_{n+3} theory T_{n+3}.

We let A be an infinite computable set of constants, and we form a computable list of triples representing witnessing tasks, as in the previous proof. Let L consist of the pairs $\ell = (i, \bar{a})$, where R_i is a complete type, consistent with PA, in variables $\bar{x}$ corresponding to $\bar{a}$. If $\ell = (i, \bar{a})$, where $\bar{a}$ has length n, then $E(\ell)$ consists of the pairs $(j, \bar{c})$ such that $j < n$, $\bar{c} \subseteq \bar{a}$, and R_j is the restriction of the complete type R_i to a complete B_1 type in the variables corresponding to $\bar{c}$. Let $\hat{\ell} = (u_0, \emptyset)$, where u_0 is an R-index for T_1. For $\ell, \ell' \in L$, if $\ell = (i, \bar{a})$ and $\ell' = (i', \bar{a}')$, let $\ell \leq_n \ell'$ if $\bar{a} \subseteq \bar{a}'$ and $R_i(\bar{a}) \cap B_n \subseteq R_{i'}(\bar{a}')$. We write $\ell \subseteq \ell'$ if $R_i(\bar{a}) \subseteq R_{i'}(\bar{a}')$.

Remark: If $\ell \leq_n \ell'$ and $k < n$, then there exists ℓ'' such that $\ell \subseteq \ell''$ and $\ell' \leq_k \ell''$.

Let $U = \omega$. We define P so that for each path

$$\pi = \hat{\ell} u_1 \ell_1 u_2 \ell_2 \ldots ,$$

the term u_n is an index for the Σ_{n+1} fragment of a complete theory T', and ℓ_n carries out the nth step in the construction of a model of T' representing $\mathcal{S}$, just as in the previous proof. Let P consist of the finite sequences

$$\hat{\ell} u_1 \ell_1 u_2 \ell_2 \ldots ,$$

where $\hat{\ell}$ and u_0 are as above, u_n is an R-index for a complete Σ_{n+1} theory with $R_{u_n} = R_{u_{n+1}} \cap \Sigma_{n+1}$, and if $\ell_n = (i_n, \bar{a}_n)$, then $\bar{a}_n$ includes the first n constants from A, the Σ_{n+1} sentences of R_{i_n} are the ones in R_{u_n}, and for the nth triple $(k, j, \bar{c})$ on our list, if R_j is a complete B_k type, then either $R_j(\bar{c}, a) \subseteq R_{i_n}(\bar{a}_n)$ for some $a \in \bar{a}_n$, or else $R_{i_n}(\bar{a}_n) \cup R_j(\bar{c}, x)$ is inconsistent, and $\ell_n \leq_n \ell_{n+1}$.

Lemma 19.13 *If $\pi = \hat{\ell} u_1 \ell_1 u_2 \ell_2 \ldots$ is a path through P, where $\ell_n = (i_n, \bar{a}_n)$, then*

$$T(\pi) = \cup_n R_{u_n}$$

is a complete theory extending T_1, and

$$\cup_n (R_{i_n}(\bar{a}_n) \cap B_n)$$

is the complete diagram of a model $\mathcal{A}(\pi)$ of $T(\pi)$ such that $\mathcal{A}$ represents $\mathcal{S}$. Moreover, $E(\pi)$ is the set of pairs $(j, \bar{c})$ such that R_j is the complete B_1 type realized by $\bar{c}$ in $\mathcal{A}(\pi)$.

Proof of Lemma: For all n, R_{u_n} is a complete Σ_{n+1} theory, $T_1 = R_{u_0}$, and for all n, $R_{u_n} \subseteq R_{u_{n+1}}$. Therefore, $T(\pi)$ is a complete theory extending T_1. For $n \geq 1$, ℓ_n is chosen to assign a type to the first n constants, and to carry out the nth witnessing task. Therefore, $\cup_n (R_{i_n}(\bar{a}_n) \cap B_n)$ is the complete diagram of a model $\mathcal{A}_\pi$ of $T(\pi)$ such that $\mathcal{A}_\pi$ represents $\mathcal{S}$. It is clear that $E(\pi)$ is the set of pairs $(i, \bar{a})$ such that R_i is the complete B_1 type realized by $\bar{a}$.

We have defined the ingredients of the relativized special $(n + 1)$-system. Before checking that the conditions are satisfied, we define the special instruction function q. For σ of length $2n+1$ in P, if σ follows q, as it has been defined so far, then $q(\sigma)$ is the R-index for $T_{n+2} = T \cap \Sigma_{n+2}$, determined using $\Delta^0_{n+2}(X)$. In any case, $q(\sigma)$ is some u such that $\sigma u \in P$.

We must make sure that

$$(L, U, \hat{\ell}, P, E, (\leq_n)_{1 \leq n < \omega})$$

satisfies the four conditions for a special $(n + 1)$-system, where the role of $\leq_0$ is played by $\leq_1$, etc. The first three conditions are clear. The following lemma yields Condition (4).

Lemma 19.14 Let $\sigma u \in P$, where σ has length $2n + 1$ and ends in ℓ^0,

$$\ell^0 \leq_{m_0} \cdots \leq_{m_{k-1}} \ell^k ,$$

for $n + 1 > m_0 > \ldots > m_{k-1} > m_k \geq 1$. Then there exists ℓ^* such that $\sigma u \ell^* \in P$ and $\ell^i \leq_{m_i} \ell^*$ for $i \leq k$.

Proof of Lemma: Taking extensions of ℓ^i, and working our way back from $i = k$ to $i = 0$, we arrive at $\ell^{**} \supseteq \ell^0$ such that for $i \leq k$, $\ell^i \leq_{m_i} \ell^{**}$. Say $\ell^0 = (i, \bar{b})$ and $\ell^{**} = (i', \bar{b}')$. By the definition of P, R_u is a complete Σ_{n+2} theory. Since R_u and $R_i(\bar{b})$ agree on Σ_{n+1} sentences in the language of T (without added constants), there is a completion $R_j(\bar{b}')$ of $R_{i'}(\bar{b}') \cap B_n$ with the same Σ_{n+2} sentences as R_u, in the original language.

We are now in a position to apply Theorem 14.4, relativized to X'. We obtain a run π of (P, q) such that $E(\pi)$ is $\Sigma^0_2(X)$. Then $T(\pi) = T$, and by Lemma 19.13, we have a model $\mathcal{A}(\pi)$, representing $\mathcal{S}$, such that

$$E(\pi) = \{(i, \bar{c}) : R_i \text{ is the } B_1 \text{ type realized by } \bar{c}\} .$$

This completes the proof of Theorem 19.12.

The next result, which we state without proof, has the same conclusion as Theorems 19.11 and 19.12. The hypotheses on the approximations of T are still weaker than those in Theorem 19.12.

Theorem 19.15 *Let T be a complete theory, and let S be a Scott set appropriate for T, with an enumeration $R \leq_T X$. Suppose there are functions t_n, $\Delta_n^0(X)$ uniformly in n, such that for each n, $\lim_{s\to\infty} t_n(\sigma)$ is an R-index for $T_n = T \cap \Sigma_n$, and for all s, $t_n(s)$ is an index for a subset of T_n. Then T has a model $\mathcal{A}$, representing S, such that*

$$Q = \{(i, \bar{a}) : R_i \text{ is the } B_1 \text{ type of } \bar{a}\}$$

is $\Sigma_2^0(X)$.

The proof of Theorem 19.15 is more delicate than that for Theorem 19.12. The metatheorems from Chapter 14 do not suffice. Still, the proof has the same flavor. For details, see [86].

Combining Theorems 19.10 and 19.11, we obtain the following.

Corollary 19.16 *Let T be a complete theory, and let $\mathcal{M}$ be a Scott set appropriate for T, with an enumeration $R \leq_T X$. Suppose there is a function t, $\Delta_3^0(X)$ such that for all finite n, $t(n)$ is an R-index for $T_n = T \cap \Sigma_n$. Then T has a model $\mathcal{A}$, representing S, such that $\mathcal{A} \leq_T X$.*

Combining Theorems 19.10 and 19.12, we obtain the following result, stronger than Corollary 19.16.

Corollary 19.17 *Let T be a complete theory, and let S be a Scott set appropriate for T, with an enumeration $R \leq_T X$. Suppose that there is a uniform effective procedure for determining an R-index for $T_{n+1} = T \cap \Sigma_{n+1}$ using $\Delta_n^0(X)$. Then T has a model $\mathcal{A}$, representing S, such that $\mathcal{A} \leq_T X$.*

Combining Theorems 19.10 and 19.15 (left unproved), we obtain the following, strill stronger than Corollary 19.16.

Corollary 19.18 *Let T be a complete theory, and let S be a Scott set appropriate for T, with an enumeration $R \leq_T X$. Suppose that there are functions t_n, $\Delta_n^0(X)$, uniformly in n, such that for each n, $\lim_{s\to\infty} t_n(s)$ is an R-index for $T_n = T \cap \Sigma_n$, and for all s, $t_n(s)$ is an index for a subset of T_n. Then T has a model $\mathcal{A}$, representing S, such that $\mathcal{A} \leq_T X$.*

Remark: In Theorems 19.11, 19.12, and 19.15, and in Corollaries 19.16, 19.17, and 19.18, we may use indices relative to X instead of R-indices, for fragments of the theory .

19.4 Harrington's Theorem

Here is the result of Harrington, answering the question of McAloon that we stated at the beginning of the chapter.

Theorem 19.19 (Harrington) *There is a nonstandard model $\mathcal{A}$ of PA such that $\mathcal{A}$ is Δ_2^0 and $TA \leq_T Th(\mathcal{A})$—the model is arithmetical, while its theory is not.*

In proving Theorem 19.19, we determine the theory first, and then the model. The lemma below gives the existence of a special set $X \equiv_T TA$.

Lemma 19.20 *There is a set $X \equiv_T TA$ such that for all n, we can determine whether $n \in X$, using an oracle for Δ_n^0.*

Proof: Let X be $\emptyset^{(\omega)} = \{\langle m, x\rangle : x \in \emptyset^{(m)}\}$.

The next lemma gives the theory that we want for Harrington's Theorem.

Lemma 19.21 *Suppose that S is a Scott set with a low enumeration R. Then there is a completion T of PA such that S is appropriate for T, $TA \leq_T T$, and there is a uniform effective procedure that, for each n, yields an R-index for the fragment $T_{n+1} = T \cap \Sigma_{n+1}$, using an oracle for Δ_n^0.*

Proof sketch: Let $(\varphi_n)_{n \in \omega}$ be a computable sequence of sentences such that φ_n is Π_{n+1}, and for any completion T of PA, φ_n is consistent with $PA \cup (T \cap B_n)$ (see the Appendix). We use the sentences φ_n to code into our theory the set X from Lemma 19.20. We may assume that R is an effective enumeration of S. (That is, we may replace the given enumeration R be an effective one $R^* \leq_T R$, and using Δ_2^0, we can pass between R-indices and R^*-indices for the same set.)

We let A_1 consist of the axioms of PA, together with φ_0 if $0 \in X$, and $\neg\varphi_0$ if $0 \notin X$. Let $T \cap B_1$ be the result of taking a completion of A_1 in S and restricting to B_1. Having determined $T \cap B_n$, we let A_{n+1} consist of $T \cap B_n$ and the axioms of PA, plus φ_n if $n \in X$ and $\neg\varphi_n$ if $n \notin X$. Let $T \cap B_{n+1}$ be the result of taking a completion of A_{n+1} in S and restricting to B_{n+1}. Given an R-index for $T \cap B_n$, and knowing whether $n \in X$, we can compute (using the functions that witness the effectiveness of the enumeration) first an R-index for A_{n+1}, then an R-index for a completion, and finally an R-index for the restriction of the complete theory to B_{n+1}.

Remark: If T is as in Lemma 19.21, then $T \neq TA$.

The Scott set S does not contain $TA \cap \Sigma_1$. Since S is appropriate for T, it must contain $T \cap \Sigma_1$.

We are ready to complete the proof of Theorem 19.19. Let S be a Scott set with a low enumeration R. Take the completion T of PA from Lemma 19.21, where S is appropriate for T, $TA \leq_T T$, and there is an effective procedure, uniform in n, for determining an R-index for $T_{n+1} = T \cap \Sigma_{n+1}$ using Δ_n^0. Using Theorem 19.12, we obtain a Δ_2^0 model, as required for Theorem 19.19.

We can strengthen Harrington's theorem. Using Theorem 19.17, we obtain a model that is computable in R. Thus, we have the following result, from [84].

Theorem 19.22 *There exists a nonstandard model $\mathcal{A}$ of PA such that $\mathcal{A}$ is low and $TA \leq_T Th(\mathcal{A})$.*

Here is a further generalization of Harrington's Theorem.

Theorem 19.23 *If R is an enumeration of a Scott set S, then there is a nonstandard model $\mathcal{A}$ of PA such that $\mathcal{A} \leq_T R$ and $Th(\mathcal{A})$ is complete $\Delta^0_\omega(R)$.*

Exercise: Prove Theorem 19.23.

19.5 Solovay's Theorems

Here is the characterization of the degrees of nonstandard models of TA.

Theorem 19.24 (Solovay, Marker) *The degrees of nonstandard models of TA are the degrees of enumerations of Scott sets containing the arithmetical sets—these are the Scott sets appropriate for TA.*

Remark: Solovay [145] showed that the degrees of nonstandard models of TA are the degrees of effective enumerations of Scott sets containing the arithmetical sets. Marker's contribution was Theorem 19.5, saying that we can replace an arbitrary enumeration by an effective one.

Proof sketch: By Theorem 3.21, the set of degrees of nonstandard models of TA is closed upwards. The set of degrees of enumerations of Scott sets containing the arithmetical sets is also closed upwards. Therefore, it is enough to show that for any set X, the following are equivalent:

(1) there is a nonstandard model $\mathcal{A}$ of TA such that $\mathcal{A} \leq_T X$,
(2) there is an enumeration R of a Scott set containing the arithmetical sets such that $R \leq_T X$.

(1) $\Rightarrow$ (2) Given a nonstandard model $\mathcal{A}$ of TA, let R be the canonical enumeration of $SS(\mathcal{A})$. Then $R \leq_T \mathcal{A}$, and $SS(\mathcal{A})$ is a Scott set containing the arithmetical sets.

(2) $\Rightarrow$ (1) Let $R \leq_T X$ be an enumeration of a Scott set S containing the arithmetical sets. By Theorem 19.5, we may assume that R, with appropriate functions, is an effective enumeration of S. Using $\Delta^0_3(X)$, we can find indices for $\emptyset, \emptyset', \emptyset''$, etc. Given an R-index for $\emptyset^{(n)}$, we can find an R-index for $TA \cap \Sigma_n$, since $TA \cap \Sigma_n$ is computable in $\emptyset^{(n)}$, uniformly in n. Thus, using $\Delta^0_3(X)$, we can find indices for $TA \cap \Sigma_n$, for all n. Then by Corollary 19.16, TA has a model $\mathcal{A}$, representing S, such that $\mathcal{A} \leq_T X$. The fact that $\mathcal{A}$ represents S guarantees that it is nonstandard—S contains a complete B_1 type $\Gamma(x)$, consistent with $TA \cap \Sigma_2$, such that $\Gamma(x) \supseteq \{x \neq S^{(n)}(0) : n \in \omega\}$.

Here is Solovay's characterization of the degrees of models of an arbitrary completion of PA. (We shall reduce this result to Theorem 19.15, which we did not prove.)

Corollary 19.25 (Solovay) *If T is an arbitrary completion of PA, then for all sets X, the following are equivalent:*

(1) there is a (nonstandard) model $\mathcal{A}$ of T such that $\mathcal{A} \leq_T X$,

(2) there exist an enumeration $R \leq_T X$ of a Scott set appropriate for T, and functions t_n, $\Delta_n^0(X)$ uniformly in n, such that $\lim_{s\to\infty} t_n(s)$ is an R-index for $T_n = T \cap \Sigma_n$ and for all s, $t_n(s)$ is an R-index for a subset of T_n.

Proof from Theorem 19.15: (2) $\Rightarrow$ (1) We use Corollary 19.18, a consequence of Theorem 19.15. Let $R \leq_T X$ be an enumeration of a Scott set $\mathcal{S}$, appropriate for T, and suppose that we have functions t_n, $\Delta_n^0(X)$, uniformly in n, such that for each n, $\lim_{s\to\infty} t_n(s)$ is an R-index for T_n, and for all s, $t_n(s)$ is an R-index for a subset of T_n. We may suppose that R is an effective enumeration of $\mathcal{S}$. (There is an effective enumeration R^* computable in R, and using $\Delta_2^0(X)$, we can pass effectively between R indices and R^* indices.) Then Corollary 19.18 yields a model $\mathcal{A} \leq_T X$.

(1) $\Rightarrow$ (2) Suppose that $\mathcal{A}$ is a (nonstandard) model of T such that $\mathcal{A} \leq_T X$. Let R be the canonical enumeration of $\mathcal{SS}(\mathcal{A})$, where $R \leq_T \mathcal{A}$. Let $Sat_n(x)$ be the usual formula defining truth for Σ_n sentences, and let

$$\varphi_n(u) = \forall x < u\, (p_x | u \to Sat_n(x))\,.$$

It is not difficult to see that the formula $\varphi_n(u)$ is logically equivalent to a formula with n blocks of like quantifiers, in which the first is $\exists$, followed by a formula $\delta(\overline{y}, u)$ with only bounded quantifiers. As in Theorem 3.18, we can use Matijasevic's Theorem to replace $\varphi_n(u)$ by a formula that is genuinely Σ_n, even in the case where $n = 1$.

We assume that $\varphi_n(u)$ is Σ_n. The universe of $\mathcal{A}$ is computable, so it has an inherited ordering of type ω, in addition to the ordering defined in $\mathcal{A}$ by the formula

$$\exists u\, (x + u = y\ \&\ u \neq 0)\,.$$

We are ready to define the functions t_n. For $n = 1$, for all s, we let $t_1(s)$ be the first R-index for T_1. Suppose $n \geq 2$. Let $t_n(0)$ be the first R-index for $\emptyset$. Given $t_n(s) = a$, and using $\Delta_n^0(\mathcal{A})$, we carry out $s + 1$ steps in an enumeration of $D^c(\mathcal{A}) \cap \Sigma_n$, and we look for b among the first $s + 1$ constants (in the inherited ordering of type ω) such that

(i) there is some standard x such that $\mathcal{A} \models p_x | b\ \&\ p_x \nmid a$,

(ii) for all standard x, $\mathcal{A} \models p_x | a \to p_x | b$,

(iii) $\varphi_n(b)$ has appeared by stage s in our enumeration of $D^c(\mathcal{A}) \cap \Sigma_n$.

If we find such a b, then $t_n(s + 1)$ is the first such (in the inherited ordering); otherwise, $t_n(s + 1) = a$. It is not difficult to see that t_n has the required features.

19.6 Sets computable in all models

By Corollary 10.18, a set X is computable in all copies of a given structure $\mathcal{A}$ if and only if $X \oplus (\neg X)$, or χ_X, is enumeration reducible to some complete Σ_1

type realized in $\mathcal{A}$. McAllister [101], [103] proved the following related result.

Theorem 19.26 (McAllister) *For a complete theory T and a Scott set $\mathcal{S}$ appropriate for T, the following are equivalent:*
(1) for all models $\mathcal{A}$ of T representing $\mathcal{S}$, $X \leq_T \mathcal{A}$,
(2) $X \oplus \neg X$ is enumeration reducible to some complete Σ_1 type Γ such that Γ is in $\mathcal{S}$ and Γ is consistent with T.

The proof of Corollary 10.18 involved taking a generic copy of the given structure. Here the proof involves building a generic model representing the given Scott set. The forcing conditions assign B_n types (for various finite n) to tuples of constants.

19.7 Open problems

Lerman [97] had asked whether there is a nonstandard model $\mathcal{A}$ of TA such that there is no enumeration of just the arithmetical sets computable in $\mathcal{A}$. Lachlan and Soare [94] gave an affirmative answer, using Solovay's characterization of the degrees of non-standard models of TA. McAllister [102] extended the result of [94], showing that for a large class of completions T of PA, there is a model $\mathcal{A}$ of T with no enumeration of $Rep(T)$ computable in $\mathcal{A}$. McAllister did not use Solovay's results.

McAllister's methods do not handle all completions of PA, and this leads to the following question, which is still open.

Question [Detlefsen]: Is there a completion T of PA such that for all models $\mathcal{A}$ of T, there is an enumeration of $Rep(T)$ computable in $\mathcal{A}$?

In 1981-1982, Abramson and Knight asked whether there is a nonstandard model $\mathcal{A}$ of TA such that $deg(\mathcal{A})$ is a minimal upper bound for the arithmetical degrees [85]. This question, and several variants, are answered negatively in [86]. The basic result there says that for any completion T of PA, there is another completion T^*, of strictly lower degree, such that $Rep(T^*) = Rep(T)$. Moreover, we can choose T^* such that $(T^*)' \equiv_T T'$. One corollary says that if $\mathcal{A}$ is a nonstandard model of PA, and $T = Th(\mathcal{A})$, then $\mathcal{A}$ is not a minimal upper bound for $\{deg(X) : X \in Rep(T)\}$. The best corollary says that for any nonstandard model $\mathcal{A}$ of PA, there is an isomorphic copy $\mathcal{B}$ of strictly lower degree.

We end with a minimality question that is still open. It is not related to arithmetic.

Question [Marker]: For a family of sets $\mathcal{S}$ with no element of maximal degree, can $\{deg(X) : X \in \mathcal{S}\}$ have a minimal upper bound?

Appendix A

Special classes of structures

Here we summarize some definitions and important facts about certain classes of structures: vector spaces, fields, linear orderings, Boolean algebras, equivalence structures, Abelian p-groups, and models of arithmetic. The axioms and other formulas that we mention here are finitary.

A.1 Vector spaces

We consider vector spaces over a fixed countable field F to be structures of the form $(V, +, 0, (\cdot_s)_{s \in F})$, where V is the set of vectors, $+$ is vector addition, 0 is the zero vector, and for each s, $\cdot_s$ is multiplication by s. Each vector space over F is characterized, up to isomorphism, by its dimension. If F is finite, then the finite-dimensional vector spaces over F are finite. For each finite $n \geq 1$, there is a finitary sentence distinguishing spaces of dimension at most n from those of higher dimension. The infinite spaces, those of infinite dimension, all have the same theory. If F is infinite, then all vector spaces over F are infinite, and all have the same theory. So, whether F is finite or infinite, all infinite spaces over F have the same theory. One way to prove this is to note that any two vector spaces over F have elementary extensions of the same cardinality greater than that of F. The extensions have the same dimension, so they are isomorphic.

The result below says that for vector spaces over a given field, we have "elimination of quantifiers". To achieve some uniformity, we consider only infinite vector spaces, ignoring the finite-dimensional vector spaces over finite fields.

Theorem A.1 *Let F be a field. Then for each formula $\varphi(\overline{x})$, there is a quantifier-free formula $\psi(\overline{x})$, with the same free variables, such that in all infinite vector spaces V over F, $\varphi(\overline{x})$ and $\varphi(\overline{x})$ are equivalent (i.e., they are satisfied by the same tuples). Moreover, if F is computable, then we have an effective procedure for passing from $\varphi(\overline{x})$ to $\psi(\overline{x})$.*

Proof: Proceeding by induction on $\varphi(\overline{x})$, we come to just one case requiring thought—a formula of the form $\exists u\, \varphi(\overline{x}, u)$, where $\varphi(\overline{x}, u)$ is a conjunction of linear equations and inequations in $\bar{x}, u$. We may suppose that u occurs in all conjuncts. In fact, after some simple manipulations, we may suppose that all conjuncts have the form $u = \tau(\bar{x})$ or $u \neq \tau(\bar{x})$. If one of the conjuncts is positive, then we can solve the equation for u, say $u = \tau(\bar{x})$. We obtain an equivalent formula by dropping this conjunct, replacing all occurrences of u in other conjuncts by $\tau(\bar{x})$, and dropping the quantifier $\exists u$. If none of the conjuncts is positive, then we may replace the formula by $\top$.

As a consequence of Theorem A.1, we have the fact that the theory of an infinite vector space is *strongly minimal*; i.e., in any model of the theory, all of the sets definable by finitary formulas, possibly with parameters, are either finite or co-finite.

Corollary A.2 *For any infinite vector space V, any formula $\varphi(\bar{u}, x)$, and any tuple $\bar{c}$ in V appropriate to substitute for $\bar{u}$, $\varphi^V(\bar{c}, x)$ is either finite or co-finite.*

Proof: We may suppose that the formula is quantifier-free. In fact, we may suppose that it is a finite disjunction of formulas that either
 (1) express x as a linear combination of elements of $\bar{c}$.
 (2) say that x is not in some finite set of linear combinations.
If all of the disjuncts are of type (1), then the set of elements satisfying the formula is finite. If some disjunct is of type (2), then the set of elements satisfying the formula is co-finite.

A.2 Fields

There is a familiar finite set of axioms for fields. Below, we describe the additional axioms characterizing the algebraically closed fields of a given characteristic and those characterizing the real closed fields. For more about these classes of fields, see the general text by Chang and Keisler [26], or the original discussion by Tarski [148].

A field has *characteristic p*, where p is a prime, if the field satisfies the additional axiom
$$\overbrace{1 + \ldots + 1}^{p} = 0$$

A field has characteristic 0 if it does not have characteristic p for any prime p. It requires an infinite set of finitary axioms to say this. To describe fields that are algebraically closed, we have another infinite, computable set of axioms, saying that all non-trivial polynomials have roots.

Algebraically closed fields behave much like vector spaces in that, for a given characteristic, each algebraically closed field is characterized, up to isomorphism, by a dimension, called the "transcendence degree", where this is the maximal size of an algebraically independent set. Again, we have effective elimination of quantifiers, and the finitary theory is strongly minimal.

A.2. FIELDS

Theorem A.3 (Tarski) *For any characteristic, for each formula $\varphi(\overline{x})$, we can find a quantifier-free formula $\psi(\overline{x})$, with the same free variables, such that $\varphi(\overline{x})$ and $\psi(\overline{x})$ are equivalent in all algebraically closed fields of the given characteristic.*

Corollary A.4 *For any algebraically closed field F, any formula $\varphi(\overline{u}, x)$, and any tuple $\overline{c}$ in F, appropriate to substitute for $\overline{u}$, the set*

$$\{a : F \models \varphi(\overline{c}, a)\}$$

is finite or co-finite; i.e., the theory of algebraically closed fields of a given characteristic is strongly minimal.

We omit the proofs of Theorem A.3 and Corollary A.4.

There is a finite set of axioms for the theory of ordered fields. The additional axioms, beyond those for fields, say, of the symbol $<$, that it is a linear ordering, and in addition

$$\forall x \forall y \forall z \, (y < z \to y + x < z + x) \text{ and} \forall x \forall y \forall z \, (y < z \, \& \, 0 < x \to yx < zx) \, .$$

An ordered field F is *Archimedean* if for each element a, there is some $n \in \omega$ such that $F \models a < n$. Every Archimedean ordered field can be embedded in the field of real numbers.

For real closed fields, we add a computable set of axioms saying that all non-trivial polynomials of odd degree have roots, and -1 has no square root. For real closed fields, as for algebraically closed fields, we have elimination of quantifiers.

Theorem A.5 (Tarski) *For each formula $\varphi(\overline{x})$, we can find a quantifier-free formula $\psi(\overline{x})$, with the same free variables, such that $\varphi(\overline{x})$ and $\psi(\overline{x})$ are equivalent in all real closed fields.*

For both algebraically closed fields and real closed fields, a quantifier-free formula is a finite disjunction of finite conjunctions of polynomial equations and inequalities. The difference between algebraically closed fields and real closed fields is that for real closed fields, the inequalities involving a single variable are satisfied by a finite union of intervals (with endpoints in the field), not just by a finite set of points.

Corollary A.6 *For a real closed field F, a formula $\varphi(\overline{u}, x)$, and a tuple $\overline{c}$ appropriate to substitute for $\overline{u}$,*

$$\{a : F \models \varphi(\overline{c}, a)\}$$

is a finite join of intervals, possibly trivial, with endpoints in F.

A structure is called *O-minimal* if the sets definable in the structure by finitary formulas, with parameters, are just the finite joins of intervals. Corollary A.6 says that real closed fields are *O*-minimal.

A.3 Orderings

An excellent reference on orderings is the book by Rosenstein [137]. Here we give mainly definitions.

A *partial ordering* is a structure $(A, \leq)$ such that for all $a, b, c \in A$, the following hold:

1. For all $a \in A$, $a \leq a$ (reflexivity).

2. For all $a, b, c \in A$, $(a \leq b \ \& \ b \leq c) \Rightarrow a \leq c$ (transitivity).

A *strict partial ordering* is a structure $(A, <)$ in which (2) holds, but (1) is replaced by the following.

1'. For all $a \in A$, $a \not< a$ (irreflexivity).

Note that from a partial ordering $(A, \leq)$, we may obtain a strict partial ordering $(A, <)$, letting

$$a < b \Leftrightarrow (a \leq b \ \& \ \neg b \leq a) \ .$$

A *linear ordering* is a strict partial ordering $(A, <)$ with the following additional property.

3. For all $a, b \in A$, $a < b$ or $b < a$ or $a = b$ (comparability).

A *well ordering* is a linear ordering $(A, <)$ with the further property below.

4. If S is a non-empty subset of A, then S has a least element; i.e., there exists $a \in S$ such that for all $b \in S$, $a < b$ or $a = b$.

For example, $(Q, <)$ is a linear ordering, but it is not a well ordering—we could take S to be the set of positive numbers in Q. The structure $(\omega, <)$ is a well rdering, while $(Z, <)$ is not (take S to be Z itself). We note that the empty structure (for the language) satisfies, vacuously, the conditions for being a well ordering.

A.3.1 Operations on orderings

We describe some operations on linear orderings, or order types. Let $\mathcal{A} = (A, <_A)$ and $\mathcal{B} = (B, <_B)$ be linear orderings. The first operation is addition. The sum of the two orderings, $\mathcal{A} + \mathcal{B}$, is the result of putting a copy of $\mathcal{B}$ at the end of $\mathcal{A}$. For convenience, we suppose that the universes, A and B, are disjoint. Then we may take the sum to be

$$(A \cup B, <) \ ,$$

A.4. BOOLEAN ALGEBRAS

where $x < y$ if $x \in A$ & $y \in B$, or if $x, y \in A$ and $x <_A y$, or if $x, y \in B$ and $x <_B y$. The next operation is multiplication. The product, $\mathcal{A} \cdot \mathcal{B}$ is the result of replacing each element of $\mathcal{B}$ by a copy of $\mathcal{A}$. We may take the product to be

$$(A \times B, <) ,$$

where $(a, b) < (c, d)$ if $b <_B d$ or if $b = d$ and $a <_A c$.

The next operation is exponentiation. Here the order type of $\mathcal{A}^{\mathcal{B}}$ is not unique. Suppose $A \neq \emptyset$, and let $a \in A$. Let F_a be the set of functions $f : B \to A$ such that $f(b) = a$ for all but finitely many $b \in B$. If $f, g \in F_a$ and $f \neq g$, then there is a $<_B$-greatest b such that $f(b) \neq g(b)$. Let $f <_a g$ if and only if for this b, $f(b) <_A g(b)$.

Remark: The order type of $(F_a, <_a)$ may be different for different choices of a.

Example: Suppose $\mathcal{A}$ has order type 2, and $\mathcal{B}$ has order type ω. If a is the first element of $\mathcal{A}$, then $\mathcal{A}^{\mathcal{B}}$ has order type ω, while if a is the second element, then $\mathcal{A}^{\mathcal{B}}$ has order type ω^* (the reverse of ω).

Next, we define an infinite sum of orderings. Suppose $\mathcal{A}$ is a linear ordering, and for each $a \in \mathcal{A}$, let $\mathcal{B}_a$ be a linear ordering. Then the sum

$$\sum_{a \in \mathcal{A}} \mathcal{B}_a$$

is the result of replacing each $a \in \mathcal{A}$ by a copy of $\mathcal{B}_a$. Note that the product $\mathcal{B} \cdot \mathcal{A}$ is a special case of this sum, in which $\mathcal{B}_a \cong \mathcal{B}$ for all $a \in \mathcal{A}$. Another useful special case of infinite sum is the "shuffle sum". Let $\mathcal{F}$ be a countable family of linear orderings. The *shuffle sum*, denoted by $\sigma(\mathcal{F})$, consists of densely many copies of $\mathcal{A}$ for all $\mathcal{A} \in \mathcal{F}$. To obtain a concrete representation of this order type, we may partition the set Q of rationals into sets $D_{\mathcal{A}}$ corresponding to the various $\mathcal{A}$ in $\mathcal{F}$, where each set $D_{\mathcal{A}}$ is dense in Q; i.e., between any two distinct rationals, there is an element of $D_{\mathcal{A}}$. Then

$$\sigma(\mathcal{F}) \cong \sum_{q \in Q} \mathcal{B}_q ,$$

where if $q \in D_{\mathcal{A}}$, then $\mathcal{B}_q \cong \mathcal{A}$.

A.4 Boolean algebras

An encyclopedic source of information on Boolean algebras is the handbook edited by Monk and Bonnet [24], with its extended introductory section written by Koppelberg.

A *Boolean algebra* is a structure $\mathcal{A} = (A, \vee, \wedge, ', 0, 1)$ with the following properties:

1. ∨ and ∧ are both commutative,

2. ∨ and ∧ are both associative,

3. each of the operations ∨ and ∧ distributes over the other,

4. for all a and b, $a \vee (a \wedge b) = a$ and $a \wedge (a \vee b) = a$ (absorption laws),

5. for all a, $a \vee a' = 1$ and $a \wedge a' = 0$ (complementation laws).

The most familiar Boolean algebras are algebras of sets. Let X be a set, and let S be a non-empty family of subsets of X, closed under union, intersection, and complement (this implies that $\emptyset$ and X are both in S). Then the structure $(S, \cup, \cap, ', \emptyset, X)$, where $A' = X - A$ is an *algebra of sets*. Starting with an arbitrary family S of subsets of X and closing under union, intersection, and complement, we obtain the algebra *generated* by S.

The following result, a (non-topological) version of the Stone Representation Theorem, says that we are safe in using algebras of sets as our only examples of Boolean algebras.

Theorem A.7 *Every Boolean algebra is isomorphic to an algebra of sets.*

Boolean algebras are related to linear orderings. If $\mathcal{A}$ is a linear ordering, then we form the corresponding *interval algebra* $I(\mathcal{A})$. Assuming that $\mathcal{A}$ has a first element, this is the algebra of sets generated by the half-open intervals $[a, b)$, where b is either an element or ∞. If $\mathcal{A}$ has no first element, then we add the intervals $(-\infty, b)$.

Let $\mathcal{A} = (A, \vee, \wedge, ', 0, 1)$ be a Boolean algebra. We have a relation $x \leq y$ defined by the atomic formula $x \wedge y = x$. An *atom* is an element b such that for all $x \leq b$, either $x = 0$ or $x = b$. For example, in the interval algebra of an ordering of type ω, the intervals of form $[n, n+1)$ are atoms. A Boolean algebra is *atomless* if it has no atoms. We get an atomless Boolean algebra by forming $I(\eta)$, where η is the order type of the rationals. An element b is *atomic* if for all $x \leq b$ such that $x \neq 0$, there is an atom $y \leq x$. So, for example, in the interval algebra of an ordering of type $\omega + \eta$, the atomic elements are those that do not contain any dense subinterval.

For any Boolean algebra $\mathcal{A}$, we may define a sequence of congruence relations, with corresponding quotient structures. Let $x \sim_0 y$ if $x = y$, let $x \sim_{\alpha+1} y$ if the corresponding elements of $\mathcal{A}/\sim_\alpha$ differ by finitely many atoms, and for a limit ordinal α, let $x \sim_\alpha y$ if $x \sim_\beta y$ for some $\beta < \alpha$. We define the notion of an α-atom by induction, saying that x is a 0-atom if it is an atom, and for $\alpha > 0$, x is an α-atom if x cannot be expressed as a finite join of β-atoms for $\beta < \alpha$, but for all y, either $x \wedge y$ or $x \wedge y'$ can be expressed in this form.

A Boolean algebra $\mathcal{A}$ is *superatomic* if there exists α such that $\mathcal{A}/\sim_\alpha$ is trivial (i.e., it is the one-element Boolean algebra).

Theorem A.8 *A countable superatomic Boolean algebra has isomorphism type $I(\alpha)$, where α is a countable ordinal.*

Remark: If we express α in Cantor normal form, it is the leading term, of form $\omega^\beta \cdot n$ that determines the isomorphism type of $I(\alpha)$, since for $\gamma < \beta$, γ-atoms can be absorbed in the β-atoms.

A.5 Equivalence structures

An *equivalence structure* has the form $\mathcal{A} = (A, \sim)$, where $\sim$ is an equivalence relation on A; i.e., $\sim$ is a binary relation with the following properties:

1. $\forall x\, \forall y\, \forall z\, ((x \sim y\ \&\ y \sim z) \to x \sim z)$ (transitivity),

2. $\forall x\, \forall y\, (x \sim y \to y \sim x)$ (symmetry),

3. $\forall x\, x \sim x$ (reflexivity).

Two equivalence structures, $\mathcal{A}$ and $\mathcal{B}$, are isomorphic if and only if they have the same number of equivalence classes of each size, finite or infinite.

A.6 Abelian p-groups

For an excellent discussion of Abelian groups, see Kaplansky's little book [73]. Let p be a fixed prime. An *Abelian p-group* is an Abelian group in which each non-zero element has order p^n, for some n. Let G be an Abelian p-group. We define a nested sequence of subgroups G_α, where $G_0 = G$, $G_{\alpha+1} = pG_\alpha$, and for limit α, $G_\alpha = \cap_{\beta<\alpha} G_\beta$. If G is countable, then there is a countable ordinal α such that $G_\alpha = G_{\alpha+1}$. The *length* of G, denoted by $\lambda(G)$, is the first such α. An element a has *height* β if $a \in G_\beta - G_{\beta+1}$. The group G is *reduced* if every non-zero element has a height; i.e., if $G_{\lambda(G)} = \{0\}$.

Let $P = \{a \in G : pa = 0\}$ (the set of elements of order p, plus 0). This is a subgroup of G. For each $\beta < \lambda(G)$, $(G_\beta \cap P)/(G_{\beta+1} \cap P)$ is essentially a vector over Z_p. Let $u_\beta(G)$ be the dimension. The sequence $(u_\beta(G))_{\beta<\lambda(G)}$ is called the *Ulm sequence* of G.

Theorem A.9 (Ulm) *If G_1 and G_2 are countable reduced Abelian p-groups with the same Ulm sequence, then $G_1 \cong G_2$.*

An arbitrary countable Abelian p-group G can be expressed as a direct sum $G_{\lambda(G)} \oplus G_R$, where G_R is reduced. The first direct summand, $G_{\lambda(G)}$ is unique as a set. It consists of the elements that are infinitely divisible, and it is characterized up to isomorphism by a single dimension—the maximal number of algebraically independent elements. The other direct summand, G_R, is unique up to isomorphism. By Ulm's Theorem, the isomorphism type of G_R is determined by the Ulm sequence.

A.7 Models of arithmetic

An excellent reference on models of arithmetic is the book by Kaye [74]. First order Peano arithmetic, PA, is a computably axiomatizable theory with finitely many axioms ensuring that the operations work correctly on the standard numbers, plus an infinite collection of induction axioms, the *induction schema*.

Axioms for PA

1. $\forall x \forall y \, (S(x) = S(y) \to x = y)$,

2. $\forall x \, S(x) \neq 0$,

3. $\forall x \, x + 0 = x$,

4. $\forall x \, (x + S(y) = S(x + y))$,

5. $\forall x \, x \cdot 0 = 0$,

6. $\forall x \, (x \cdot S(y) = x \cdot y + x)$,

7_φ. $\forall \overline{u} \, (\, \varphi(\overline{u}, 0) \, \& \, \forall y \, [\, \varphi(\overline{u}, y) \to \varphi(\overline{u}, S(y)) \,] \,) \to \forall x \, \varphi(\overline{u}, x))$.

The *standard model of arithmetic* is $\mathcal{N} = (\omega, +, \cdot, S, 0)$. We also refer to isomorphic copies of $\mathcal{N}$ as *standard*. A *nonstandard* model of PA is one that is not isomorphic to $\mathcal{N}$. Any model of PA has a natural linear ordering, defined by the formula

$$\exists z \, (\, z \neq 0 \, \& \, x + z = y \,) \, .$$

The interpretations of the terms $S^{(n)}(0)$ are called *finite*, or *standard*, and other elements of the model (if any) are called *infinite*, or *nonstandard*. In a nonstandard model $\mathcal{A}$, the finite elements form an initial segment in the ordering.

The following result, often called "Overspill", is an immediate consequence of the "induction schema"—the set of axioms 7_φ.

Proposition A.10 *Suppose that $\mathcal{A}$ is a nonstandard model of PA. Let $\varphi(\overline{a}, x)$ be a formula with parameters $\overline{a}$ in $\mathcal{A}$. If $\varphi(\overline{a}, x)$ holds for all finite numbers, then it also holds for some infinite number.*

Proof: Suppose $\varphi(\overline{a}, x)$ holds just for the finite elements. Then

$$\mathcal{A} \models \varphi(\overline{a}, 0) \, \& \, \forall y \, (\, \varphi(a, y) \to \varphi(\overline{u}, S(y)) \,) \, ,$$

but not $\mathcal{A} \models \forall x \, \varphi(\overline{a}, x)$, contradicting 7_φ.

By Gödel's Incompleteness Theorem, the axioms of PA do not generate the theory of $\mathcal{N}$. This is witnessed by a sentence φ which refers to itself and says "I am unprovable from the axioms of PA". This sentence is true in $\mathcal{N}$, and is therefore, unprovable from the axioms of PA. A related sentence ψ, due to Gödel and Rosser, refers to itself, saying, "for any proof of me from the axioms

A.7. MODELS OF ARITHMETIC

of PA, there is a smaller proof of my negation". There can be no proof of ψ from the axioms of PA, since then ψ would be true in $\mathcal{N}$, and this means that there is also a proof of $\neg\psi$. There can be no proof of $\neg\psi$ from the axioms of PA, since then $\neg\psi$ would be true in $\mathcal{N}$, and this means that there is also a proof of ψ.

Gödel showed that there is no computable completion of PA. In fact, the sets of sentences provable and refutable from the axioms of PA have no computable separator.

Proposition A.11 *Given a c.e. index for a set S of sentences, all unprovable from the axioms of PA, we can find $x \notin S$ such that x is another unprovable sentence.*

Proof: For each x, we have a natural sentence ψ_x saying that $x \in K$ (where K is the "halting set"). We have

$$x \in K \Leftrightarrow PA \vdash \psi_x.$$

The set $\{x : \psi_x \in S\}$ is c.e., say it is equal to W_e. Now,

$$\begin{aligned} PA \vdash \psi_e &\Leftrightarrow e \in K \\ &\Leftrightarrow e \in W_e \\ &\Leftrightarrow \psi_e \in S. \end{aligned}$$

Therefore, $\psi_e \notin S$, since if $\psi_e \in S$, then ψ_e would be provable, and S contains no provable sentence. It follows that ψ_e is unprovable.

A.7.1 Matijasevic's Theorem

Hilbert's 10th problem called for an algorithm for deciding which polynomial equations, with integer coefficients, have solutions in the integers. Matijasevic [116], [35] showed that Hilbert's 10th Problem is unsolvable (see also [34], [113]). On the way to showing that there is no such algorithm, Matijasevic proved the following remarkable result.

Theorem A.12 (Matijasevic) *Let $\beta(\overline{x})$ be a formula with only bounded quantifiers. Then we can find an existential formula $\varphi(\overline{x})$ and a universal formula $\psi(\overline{x})$ such that both are provably equivalent to $\beta(\overline{x})$ over PA. Moreover, the existential formula $\varphi(\overline{x})$ can be taken to have the form $\exists \overline{u}\, \alpha(\overline{u}, \overline{x})$, where $\alpha(\overline{u}, \overline{x})$ is a polynomial equation.*

We shall not prove Theorem A.12, but we point out the magic in the statement. If $\mathcal{A}$ is a nonstandard model of PA, then for a formula

$$\delta(x) = (\forall u < x)\, \alpha(u),$$

where α is open, the problem of determining whether $\mathcal{A} \models \delta(b)$, using an oracle for the atomic diagram of $\mathcal{A}$, seems, for nonstandard b, to require checking infinitely many open sentences. Theorem A.12 says that we can determine whether b satisfies δ effectively in the atomic diagram of $\mathcal{A}$. We find an existential formula equivalent to δ, and another existential formula equivalent to $\neg \delta$, and search for witnesses that one of these is satisfied by b.

Below, we indicate how Theorem A.12 was applied to Hilbert's 10th Problem.

Corollary A.13 (Matijasevic) *There is no algorithm for deciding which polynomial equations have solutions in the integers.*

Proof: First, note that for any polynomial equation with integer coefficients, we can find another polynomial equation, with natural number coefficients, such that the first has a solution in the integers just in case the second has a solution in the natural numbers. Conversely, for any polynomial equation with natural number coefficients, we can find another polynomial equation, also with natural number coefficients, such that the first has a solution in the natural numbers just in case the second has a solution in the integers. Thus, we may phrase Hilbert's 10th Problem in terms of polynomial equations with natural number coefficients and solutions in the natural numbers.

Since the halting set K is c.e., it is definable in $\mathcal{N}$ by a natural formula of the form $\exists \overline{v}\, \beta(\overline{v}, x)$, where $\beta(\overline{v}, x)$ has only bounded quantifiers. Replacing $\beta(\overline{v}, x)$ by an existential formula $\varphi(\overline{v}, x)$ as in Theorem A.12, we obtain a definition having the form $\exists \overline{v}\, \exists \overline{u}\, \alpha(\overline{u}, \overline{v}, x)$, where $\alpha(\overline{u}, \overline{v}, x)$ is a polynomial equation.

If there were an algorithm of the kind that Hilbert asked for, then we could apply it to the equation $\alpha(\overline{u}, \overline{v}, S^{(n)}(0))$, to decide whether $n \in K$.

A.7.2 Definability of satisfaction

By a result of Tarski, closely related to Gödel's Incompleteness Theorem, there is no global definition of satisfaction in $\mathcal{N}$. However, we can define satisfaction for formulas of a given complexity, uniformly in all models of PA.

Theorem A.14 *For each $n \geq 1$ and each tuple $\overline{x}$ of variables, we can find a (finitary) Σ_n formula $Sat_n(y, \overline{x})$ such that for any model $\mathcal{A}$ of PA, any tuple $\overline{a}$ in $\mathcal{A}$ appropriate to substitute for $\overline{x}$, and any Σ_n formula $\varphi(\overline{x})$, if k is the Gödel number of $\varphi(\overline{x})$, then*

$$\mathcal{A} \models Sat_n(k, \overline{a}) \Leftrightarrow \mathcal{A} \models \varphi(\overline{a}) \ .$$

Proof sketch: It is not difficult to write a formula having the desired meaning, and consisting of n blocks of like quantifiers, beginning with $\exists$, followed by a formula with only bounded quantifiers. Thanks to Theorem A.12, we can replace the bounded complexity formula by one with quantifiers matching the last block, to obtain a formula $Sat_n(u, \overline{x})$ that is genuinely Σ_n.[1]

---

[1] Martin Davis has pointed out that we need Matijasevic's Theorem only for $n = 1$.

A.7.3 Independence

Let Γ be a set of sentences, and let φ be a further sentence. We say that φ is *independent of* Γ if both $\Gamma \cup \{\varphi\}$ and $\Gamma \cup \{\neg\varphi\}$ are consistent. For example, the Gödel-Rosser sentence is independent of PA. Below, we describe some variants of this sentence. The B_n sentences are the Boolean combinations of Σ_n sentences.

Proposition A.15 *For each $n \geq 1$, we can find a Π_n sentence φ such that for any completion T of PA, φ is independent of $PA \cup (T \cap B_{n-1})$.*

Proof sketch: The sentence φ says "for any proof of me from the axioms of PA and true B_{n-1} sentences, there is a smaller proof of my negation". (The formula Sat_{n-1} allows us to refer to the true Σ_{n-1} and Π_{n-1} sentences, and these imply the true B_{n-1} sentences. Fix a completion T of PA. There is no proof of φ from $PA \cup (T \cap B_{n-1})$, for if there were, then all models would recognize the proof, and all would have a smaller proof of the negation. Now, both proofs are finite, so T is actually inconsistent, a contradiction. Similarly, there is no proof of $\neg\varphi$ from $PA \cup (T \cap B_{n-1})$, for if there were, then all models would have a smaller proof of φ. Again both proofs are finite, so T is inconsistent, a contradiction.

Scott [140] proved the ultimate result in this direction, finding formulas $\psi(x)$ that are "independent" (over some set of sentences) in the sense that $\psi(S^{(n)})$ can be made true for any desired set of numbers n.

Proposition A.16 *For each $n \geq 1$, we can find a (finitary) Π_{n+2} formula $\psi_n(x)$ such that for any completion T of PA and any $S \subseteq \omega$, the set*

$$\{\psi_n(k) : k \in S\} \cup \{\neg\varphi_n(k) : k \notin S\}$$

is consistent with $PA \cup (T \cap B_n)$.

Proof sketch: The formula $\psi_n(x)$ has the meaning "for any proof of me from the axioms of PA, true B_n sentences, and true sentences of the form $\psi_n(y)$ or $\neg\psi_n(y)$, for $y < x$, there is a smaller proof of my negation (from these same things)".

Bibliography

[1] Ash, C. J., "Stability of recursive structures in arithmetical degrees", *Annals of Pure and Applied Logic*, vol. 32(1986), pp. 113-135.

[2] Ash, C. J., "Categoricity in hyperarithmetical degrees", *Annals of Pure and Applied Logic*, vol. 34(1987), pp. 1-14.

[3] Ash, C. J., "Recursive labelling systems and stability of recursive structures in hyperarithmetical degrees", *Trans. of the Amer. Math. Soc.*, vol. 298(1986), pp. 497-514. Corrections: *Ibid*, vol. 300(1988), p. 851.

[4] Ash, C. J., "Labelling systems and r.e. structures", *Annals of Pure and Applied Logic*, vol. 47(1990), pp. 99-119.

[5] Ash, C. J., "A construction for recursive linear orderings", *J. Symb. Logic*, vol. 16(1990), pp. 211-234.

[6] Ash, C. J., C. G. Jockusch, and J. F. Knight, "Jumps of orderings", *Trans. of the Amer. Math. Soc.*, vol. 319(1990), pp. 573-599.

[7] Ash, C. J., and J. F. Knight, "Pairs of recursive structures", *Annals of Pure and Applied Logic*, vol. 46(1990), pp. 211-234.

[8] Ash, C. J., and J. F. Knight, "Relatively recursive expansions", *Fund. Math.*, vol. 140(1992), pp. 137-155.

[9] Ash, C. J., and J. F. Knight, "Recursive expansions", *Fund. Math.*, vol 145(1994), pp. 153-169.

[10] Ash, C. J., and J. F. Knight, "A completeness theorem for certain classes of recursive infinitary formulas", *Math. Logic Quarterly*, vol. 40(1994), pp. 173-181.

[11] Ash, C. J., and J. F. Knight, "Ramified systems", *Annals of Pure and Applied Logic*, vol. 70(1994), pp. 205-221.

[12] Ash, C. J., and J. F. Knight, "Mixed systems", *J. Symbolic Logic*, vol. 59(1994), pp. 1383-1399.

[13] Ash, C. J., and J. F. Knight, "Recursive structures and Ershov's hierarchy", *Math. Logic Quarterly*, vol. 42(1996), pp. 461-468.

[14] Ash, C. J., and J. F. Knight, "Possible degrees in recursive copies", *Annals of Pure and Applied Logic*, vol. 75(1995), pp. 215-221.

[15] Ash, C. J., and J. F. Knight, "Possible degrees in recursive copies II", *Annals of Pure and Applied Logic*, vol. 87(1997), pp. 151-165.

[16] Ash, C. J., J. F. Knight, M. Mannasse, and T. Slaman, "Generic copies of countable structures", *Annals of Pure and Applied Logic*, vol. 42(1989), pp. 195-205.

[17] Ash, C. J., J. F. Knight, and J. B. Remmel, "Quasi-simple relations in copies of a given recursive structure", *Annals of Pure and Applied Logic*, vol. 86(1997), pp. 203-218.

[18] Ash, C. J., J. F. Knight, and T. Slaman, "Relatively recursive expansions II", *Fund. Math.*, vol. 142(1993), pp. 147-161.

[19] Ash, C. J., and A. Nerode, "Intrinsically recursive relations", in *Aspects of Effective Algebra*, ed. by J. N. Crossley, Upside Down A Book Co., Steel's Creek, Australia, pp. 26-41.

[20] Barker, E., "Intrinsically Σ_1^0 relations", *Annals of Pure and Applied Logic*, vol. 39(1988), pp. 105-130.

[21] Barker, E., "Back and forth relations for reduced abelian p-groups", *Annals of Pure and Applied Logic*, vol. 75(1995), pp. 223-249.

[22] Barwise, J., "Infinitary logic and admissible sets", *J. Symb. Logic*, vol. 34(1969), pp. 226-252.

[23] Barwise, J., "Back-and-forth through infinitary logic", *Studies in Model Theory*, ed. by M. D. Morley, M.A.A., *Studies in Mathematics*, vol. 8(1973), pp. 5-34.

[24] Benedikt, M., and C. McCoy, "Functions that preserve definability", in preparation.

[25] Boone, W. W., "The word problem", *Annals of Math.* (2), vol. 70(1959), pp. 297-265.

[26] Chang, C. C., and H. J. Keisler, *Model Theory*, North Holland, 1973.

[27] Chisholm, J., "Effective model theory versus recursive model theory", *J. Symb. Logic*, vol. 55(1990), pp. 1168-1191.

[28] Cholak, P., S. S. Goncharov, B. Khoussainov, and R. A. Shore, "Computably categorical structures and expansions by constants, *J. of Symb. Logic*, vol. 64(1999), pp. 13-37.

[29] Cholak, P., R. Downey, and L. Harrington, "Automorphisms of the c.e. sets: Σ_1^1 completeness", in preparation.

[30] Chong, C. T., and Yue Yang, "Σ_2 induction and infinite injury priority argument I. Maximal sets and the jump operator", *J. Symb. Logic*, vol. 63(1998), pp. 797-814.

[31] Chong, C. T., and Yue Yang, "Σ_2 induction and infinite injury priority argument II. Tame Σ_2 coding and the jump operator", *Annals of Pure and Appl. Logic*, vol. 87(1997), pp. 103-116.

[32] Cooper, S. B., "Enumeration reducibility, non-deterministic computations, and relative computability of partial functions", in *Recursion Theory Week (Oberwolfach, 1989)*, ed. by K. Ambos-Spies et al, Springer, 1990, pp. 57-110.

[33] Davey, K. J., "Inseparability in recursive copies", *Annals of Pure and Applied Logic*, vol. 68(1994), pp. 1-52.

[34] Davis, M. D., "Hilbert's tenth problem is unsolvable", *Amer. Math. Monthly*, vol. 80(1973), pp. 233-269.

[35] Davis, M. D., Yu. Matijasevich, and J. Robinson, "Hilbert's tenth problem. Diophantine equations: positive aspects of a negative solution", in *Mathematical Developments Arising from Hilbert's Problems, Proc. Symp. Pure Math.*, ed. by F. E. Browder, Amer. Math. Soc., Providence, R. I., 1976, pp. 323-378.

[36] Downey, R. G., "Computability theory and linear orderings", in *Handbook of Recursive Mathematics*, vol. 2, ed. by Nerode, et al, pp. 823-976, North Holland, 1998.

[37] Downey, R. G., and G. R. Hird, "Automorphisms of supermaximal subspaces", *J. Symb. Logic*, vol. 50(1985), pp. 1-9.

[38] Downey, R. G., and C. G. Jockusch, "Every low Boolean algebra is isomorphic to a recursive one", *Proc. of the Amer. Math. Soc.*, vol. 122(1994), pp. 871-880.

[39] Downey, R., and J. F. Knight, "Orderings with α^{th} jump degree $0^{(\alpha)}$", *Proc. of the Amer. Math. Soc.*, vol, 14(1992), pp. 545-552.

[40] Dzgoev, V. D., "Constructivization of certain structures", Dep. No. 1606-1979, VINITI.

[41] Dzgoev, V. D., and S. S. Goncharov, "Autostable models", *Algebra and Logic*, vol. 19(1980), (English translation) pp. 28-37.

[42] Ershov, Yu. L., *Enumeration Theory*, Novosibirsk, 1974 (Russian).

[43] Ershov, Yu. L., and S. S. Goncharov, *Constructive Models*, Siberian School of Algebra and Logic, 1999 (Russian).

[44] Ershov, Yu. L., "A hierarchy of sets, I", *Algebra and Logic*, vol. 7(1968), pp. 15-47 (English translation).

[45] Ershov, Yu. L., "On a hierarchy of sets, II", *Algebra and Logic*, vol. 7(1968), pp. 47-74 (English translation).

[46] Ershov, Yu. L., "On a hierarchy of sets, III" *Algebra and Logic*, vol. 9(1970), pp. 34-51 (English translation).

[47] Epstein, R. L., R. Haas, and R. Kramer, "Hierarchies of sets and degrees below $0'''$", *Proc. of Logic Year 1979-80, Univ. of Conn.*, ed. by Lerman, Schmerl, and Soare, Springer, pp. 32-48.

[48] Feferman, S., "Arithmetically definable models of formalizable arithmetic", *Notices of Amer. Math. Soc.*, vol. 5(1958), p. 679.

[49] Feiner, L. J., "The Strong Homogeneity Conjecture", *J. Symb. Logic*, vol. 35 (1970), pp. 375-377.

[50] Feiner, L. J., "Hierarchies of Boolean algebras", *J. Symb. Logic*, vol. 35(1974), pp. 365-374.

[51] Friedberg, R. M., "Two recursively enumerable sets of incomparable degrees of unsolvability", *Proc. of National Academy of Sci., U.S.A.* vol. 43(1957), pp. 236-238.

[52] Goncharov, S. S., "Constructive Boolean algebras", in *Third All-Union Conf. in Math. Logic*, Novosibirsk (1974).

[53] Goncharov, S. S., "Autostability and computable families of constructivizations", *Algebra and Logic*, vol.14(1975), (English translation) pp. 392-408.

[54] Goncharov, S. S., "The quantity of non-autoequivalent constructivizations", *Algebra and Logic* vol. 16(1977), (English translation) pp. 169-185.

[55] Goncharov, S. S., "Strong constructivizability of homogeneous models", *Algebra and Logic*, vol. 17(1978), (English translation) pp. 247-263.

[56] Goncharov, S. S., "Autostability of models and Abelian groups", *Algebra and Logic*, vol. 19(1980), (English translation) pp. 13-27.

[57] Goncharov, S. S., "The problem of the number of non-autoequivalent constructivisations", *Algebra and Logic*, vol. 19(1980), (English translation) pp. 401-414.

[58] Goncharov, S. S., "Autostability of models and Abelian groups", *Algebra and Logic*, vol. 19(1980), (English translation) pp. 13-27.

[59] Groszek, M. J., M. E. Mytilinaios, and T. A. Slaman, "The Sacks density theorem and Σ_2 bounding", *J. Symb. Logic*, vol. 61(1996), pp. 450-467.

[60] Groszek, M. J., and M. E. Mytilinaios, "Σ_2 induction and the construction of a high degree", in *Proc. of Recursion Theory Week (Oberwolfach, 1989)*, ed. by K. Ambos-Spies et al, Springer, 1990, pp. 205-221.

[61] Harizanov, V. S., "Some effects of Ash-Nerode and other decidability conditions on degree spectra", *Annals of Pure and Applied Logic*, vol. 55(1991), pp. 51-65.

[62] Harizanov, V. S., "Effectively nowhere simple relations on computable structures", in *Proc. of International Workshop on Computability Theory and Complexity (Kazan, 1997)*, to appear.

[63] Harizanov, V., S., "Pure computable model theory", in *Handbook of Recursive Mathematics, vol. I*, ed. by Ershov, et al, North Holland, 1998, pp. 3-114.

[64] Harizanov, V., S., *Computable Model Theory*, in preparation.

[65] Harrington, L., "Building non-standard models of Peano arithmetic", handwritten notes, 1979.

[66] Hird, G. R., "Recursive properties of relations on models", *Annals of Pure and Appl. Logic*, vol. 63(1993), pp. 241-269.

[67] Hirschfeldt, D., Ph.D. Thesis, Cornell University, 1999.

[68] Hirschfeldt, D., B. Khoussainov, A. Slinko, and R. A. Shore, "Degree spectra and computable categoricity in algebraic structures", in preparation.

[69] Hurlburt, K. R., "Sufficiency conditions for theories with recursive models", *Annals of Pure and Appl. Logic*, vol. 55(1992), pp. 305-320.

[70] Jockusch, C. G., and R. I. Soare, "Π_1^0 classes and degrees of theories", *Trans. Amer. Math. Soc.*, vol. 173(1972), pp. 33-56.

[71] Jockusch, C. G., and R. I. Soare, "Degrees of orderings not isomorphic to recursive linear orderings", *Annals of Pure and Appl. Logic*, vol. 52(1991), pp. 39-64.

[72] Jockusch, C. G., and R. I. Soare, "Boolean algebras, Stone spaces, and the iterated Turing jump", *J. Symb. Logic*, vol. 59(1994), pp. 1121-1138.

[73] Kaplansky, I., *Infinite Abelian Groups*, revised ed., University of Michigan Press, 1969.

[74] Kaye, R., *Models of Peano Arithmetic*, Oxford Science, 1991.

[75] Keisler, H. J., *Model Theory for Infinitary Logic*, North-Holland, 1971.

[76] Khisamiev, N. G., "The arithmetic hierarchy of Abelian groups", *Sibirsk. Mat. Zh.*, vol. 29(1988), pp. 144-159.

[77] Khisamiev, N. G., "Constructive Abelian groups", in *Handbook of Recursive Mathematics, vol. II*, ed. by Ershov et al, Elsevier, 1998, pp. 1177-1231.

[78] Khoussainov, B., and R. Shore, "Computable isomorphisms, degree spectra of relations, and Scott families", *Annals of Pure and Appl. Logic*, to appear.

[79] Kleene, S. C., "On notation for ordinal numbers", *J. Symb. Logic*, vol. 3(1938), pp. 150-155.

[80] Kleene, S. C., *Introduction to Metamathematics*, Van Nostrand, 1952.

[81] Knight, J. F., "Degrees coded in jumps of orderings", *J. Symb. Logic*, vol. 51(1986), pp. 1034-1042.

[82] Knight, J. F., "Effective construction of models", in *Colloquium, '84*, ed. by J. Paris, A. Wilkie, and G. Wilmers, North-Holland, 1986, pp. 105-199.

[83] Knight, J. F., "Degrees of models with prescribed Scott set", in *Classification: Proc. of Joint U.S.-Israel Workshop*, ed. by J. Baldwin, 1987, pp. 182-191.

[84] Knight, J. F., "Requirement systems", *J. Symb. Logic*, vol. 60(1995), pp. 222-245.

[85] Knight, J. F., "Coding a family of sets", *Annals of Pure and Applied Logic*, vol. 97(1998), pp. 127-142.

[86] Knight, J. F., "True approximations and models of arithmetic", in *Models and Computability*, ed. by B. Cooper and J. Truss, Cambridge University Press, 1999, pp. 255-278.

[87] Knight, J. F., "Minimality and completions of Peano arithmetic", submitted to *J. Symb. Logic*.

[88] Knight, J. F., and M. Stob, "Computable Boolean algebras", to appear in *J. Symb. Logic*.

[89] Kontostathis, K., "The combinatorics of the Friedberg-Muchnik Theorem", in *Logical Methods*, ed. by J. Crossley et al, Birkhäuser, 1993, pp. 467-489.

[90] Kreisel, G., "Set theoretic problems suggested by the notion of potential totality", in *Infinitistic Methods*, Oxford, 1961, pp. 103-140.

[91] Kudinov, O. V., "An autostable 1-decidable model without a computable Scott family of ∃ formulas", *Algebra and Logic*, vol. 35(1996), (English translation), pp. 255-260.

[92] Kueker, D., "Definability, automorphisms, and infinitary languages", in *The Syntax and Semantics of Infinitary Languages*, ed. by J. Barwise, Springer-Verlag, 1968, pp. 152-165.

[93] Kunen, K., *Set Theory: An Introduction to Independence Proofs*, North-Holland, 1980.

[94] Lachlan, A. H., and R. I. Soare, "Models of arithmetic and upper bounds of arithmetic sets", *J. Symb. Logic*, vol. 59(1994), pp. 977 - 983.

[95] Lempp, S., and M. Lerman, "Priority arguments using iterated trees of strategies", *Proc. of Recursion Theory Week (Oberwolfach, 1989)*, ed. by K. Ambos-Spies et al, Springer-Verlag, 1990, pp. 277-296.

[96] Lempp, S., and M. Lerman, "A general framework for priority arguments", *Bulletin of Symb. Logic*, vol. 1(1995), pp. 189-201.

[97] Lerman, M., "Upper bounds for the arithmetical degrees", *Annals of Pure and Applied Logic*, vol. 29(1985), pp. 225-254.

[98] Lerman, M., and J. Schmerl, "Theories with recursive models", *J. Symb. Logic*, vol. 44(1979), pp. 59-76.

[99] Lerman, M., "On recursive linear orderings", in *Proc. of Connecticut Logic Year (1979-1980)*, ed. by M. Lerman, J. Schmerl, and R. I. Soare, Springer, 1981, pp. 132-142.

[100] Love, J., "Stability among r.e. quotient algebras", *Annals of Pure and Applied Logic*, vol. 44(1979), pp. 59-76.

[101] McAllister, A. M., *Computability in Structures Representing a Scott Set*, Ph.D. thesis, University of Notre Dame, 1997.

[102] McAllister, A. M., "Completions of PA: models and enumerations of representable sets", *J. Symb. Logic*, vol. 63(1998), pp. 1063-1082.

[103] McAllister, A. M., "Computability in structures representing a Scott set", to appear in *Arch. for Mathematical Logic*.

[104] McCoy, C., "Finite dimension does not relativize", to appear in *Archives for Math. Logic.*

[105] McCoy, C., "Categoricity in linear orderings and Boolean algebras", pre-print.

[106] McCoy, C., Ph.D. thesis, University of Notre Dame, 2000.

[107] Macintyre, A., and D. Marker, "Degrees of recursively saturated models", *Trans. of the Amer. Math. Soc.*, vol. 282(1985), pp. 539-554.

[108] McNicholl, T., "Intrinsic reducibilities", to appear in *Math. Logic Quarterly.*

[109] Makkai, M., "An application of a method of Smullyan to logics on admissible sets", *Bulletin of the Polish Acad. of Sci: Ser. Sci. Math., Astronom., Physics*, vol. 17(1969), pp. 341-346.

[110] Makkai, M., "On the model theory of denumerably long formulas with finite strings of quantifiers", *J. Symb. Logic*, vol. 34(1969), pp. 437-459.

[111] Makkai, M., "Preservation theorems for logic with denumerable conjunctions and disjunctions", *J. Symb. Logic*, vol. 34(1969), pp. 437 - 459.

[112] Makkai, M., "An example concerning Scott heights", *J. Symb. Logic*, vol. 46(1981), pp. 301-318.

[113] Manin, Y. I., *A Course in Mathematical Logic*, Springer-Verlag, 1977.

[114] Manasse, M., *Techniques and Counterexamples in Almost Categorical Recursive Model Theory*, Ph.D. thesis, University of Wisconsin-Madison, 1982.

[115] Marker, D., "Degrees of models of true arithmetic", *Proc. of the Herbrand Symp.*, ed. by J. Stern, North-Holland, 1982, pp. 233-242.

[116] Matijasevic, Yu., "On recursive unsolvability of Hilbert's tenth problem", *Proc. of Fourth Int. Congress on Logic, Methodology, and Philosophy of Science, Bucharest*, 1973, pp. 89-110.

[117] Metakides, G., and A. Nerode, "Effective content of field theory", *Annals of Math. Logic*, vol. 17(1979), pp. 289-320.

[118] Michalski, G., *On Foundations of Recursion Theory*, Ph.D. thesis, University of Notre Dame, 1995.

[119] Millar, T., "Foundations of recursive model theory", *Annals of Math. Logic*, now *Annals of Pure and Applied Logic*, vol. 13(1978), pp. 45-72.

[120] Millar, T., "Decidability and the number of countable models", *Annals of Pure and Applied Logic*, vol. 27(1984), pp. 137-153.

[121] Miller, R., "The Δ_2^0-spectrum of a linear order", preprint.

[122] Morley, M., "Decidable models", *Israel J. Math.*, vol. 25(1976), pp. 233-240.

[123] Moses, M., "Recursive linear orders with recursive successivities", *Annals of Pure and Applied Logic*, vol. 27(1984), pp. 254-264.

[124] Muchnik, A. A., "On the unsolvability of the problem of reducibility in the theory of algorithms", *Dokl. Akad. Nauk SSSR, N. S.*, vol. 108(1956), pp. 194-197.

[125] Mytilinaios, M. E., and T. A. Slaman, "Σ_2 collection and the infinite injury priority method", *J. Symb. Logic*, vol. 53(1988), pp. 212-221.

[126] Nadel, M., "On a problem of Specker and MacDowell", *J. Symb. Logic*, vol. 45(1980), pp. 612-622.

[127] Nadel, M., "$L_{\omega_1\omega}$ and admissible fragments", in *Model-Theoretic Logics*, vol. 45(1980), pp. 612-622.

[128] Novikov, P. S., "On the algorithmic unsolvability of the word problem in group theory", *Trudi Math. Inst. Steklov*, vol. 44(1955) (Russian).

[129] Nurtazin, A. T., "Strong and weak constructivizations and computable families", *Algebra and Logic*, vol. 13(1974), (English translation), pp. 311-323.

[130] Odintsov, S. P., and V. L. Selivanov, "Arithmetic hierarchy and ideals of Boolean algebras", *Sib. Math. J.*, vol. 30(1989), (English translation) pp. 952-960.

[131] Peretyat'kin, M. G., "Criterion for strong constructivizability of a homogeneous model", *Algebra and Logic*, vol. 13(1974), (English translation) pp. 311-323.

[132] Remmel, J. B., "Recursive Boolean algebras with recursive atoms", *J. Symb. Logic*, vol. 46(1981), pp. 595-615.

[133] Ressayre, J. P., "Models with compactness properties relative to an admissible language", *Annals of Math. Logic*, now *Annals of Pure and Applied Logic*, vol. 11(1977), pp. 31-55.

[134] Rogers, H., Jr., *Theory of Recursive Functions and Effective Computability*, McGraw-Hill, New York, 1967.

[135] Rogers, L., "The structure of p-trees: algebraic systems related to Abelian groups", in *Abelian Group Theory: 2nd New Mexico State Univ. Conf.*, Springer-Verlag, 1976, pp. 57-72.

[136] Rogers, L., "Ulm's theorem for partially ordered structures related to simply presented Abelian p-groups", *Trans. of the Amer. Math. Soc.*, vol. 227(1977), pp. 333-343.

[137] Rosenstein, J., *Linear Orderings*, Academic Press, 1982.

[138] Sacks, G. E., *Higher Type Recursion Theory*, Springer-Verlag, 1990.

[139] Scott, D., "Families of sets binumerable in complete extensions of arithmetic", in *Recursive Function Theory*, ed. by J. Dekker, Amer. Math. Soc., 1962, pp. 117-122.

[140] Scott, D., "Logic with denumerably long formulas and finite strings of quantifiers", in *The Theory of Models*, ed. by J. Addison, L. Henkin, and A. Tarski, North-Holland, 1965, pp. 329-341.

[141] Selivanov, V. L., "Numerations of families of general recursive functions", *Algebra and Logic*, vol. 15(1976), (English translation), pp. 128-141.

[142] Shoenfield, J., "Non-bounding constructions", *Annals of Pure and Applied Logic*, vol. 50(1990), pp. 191-205.

[143] Slaman, T., "Relative to any non-recursive set", *Proc. of the Amer. Math. Soc.*, vol. 126(1998), pp. 2117-2122.

[144] Soare, R. I., *Recursively Enumerable Sets and Degrees*, Springer-Verlag, 1987.

[145] Solovay, R., "Degrees of models of true arithmetic", manuscript circulated in 1982.

[146] Solovay, R., personal correspondence, 1991.

[147] Spector, M., "Recursive well orderings", *J. Symb. Logic*, vol. 20(1955), pp. 151-163.

[148] Tarski, A., *A Decision Method for Elementary Algebra and Geometry*, 2nd edition, revised, University of California Press, Berkeley, 1951.

[149] Tennenbaum, S., "Non-Archimedean models for arithmetic", *Notices of the Amer. Math. Soc.*, vol. 6(1959), p. 270.

[150] Thurber, J. J., "Recursive and r.e. quotient Boolean algebras", *Arch. Math. Logic*, vol. 33(1994), pp. 121-129.

[151] Thurber, J. J., "Every low$_2$ Boolean algebra has a recursive copy", *Proc. of the Amer. Math. Soc.*, vol. 123(1995), pp. 3859-3866.

[152] Vaught, R. L., *Topics in the Theory of Arithmetical Classes and Boolean Algebras*, Ph.D. thesis, University of California, Berkeley, 1954.

[153] Watnik, R., "A generalization of Tennenbaum's theorem on effectively finite recursive linear orderings", *J. Symb. Logic*, vol. 49(1984), pp. 563-569.

[154] Wehner, S., "Enumerations, countable structures, and Turing degrees", *Proc. of the Amer. Math. Soc.*, vol. 126(1998), pp. 2131-2139.

Index

1-free, 189, 190, 267
2-free, 189, 190
2-system, 216
3-system, 218, 219
D_n, 8, 19, 80, 81
d-c.e., v, 86, 167, 189, 190
d-c.e. relative to, v
$H(a)$, 71–75, 81, 85
$H(a)(X)$, 84, 85
$I(\mathcal{A})$, 131, 150, 247, 250, 282, 284, 316, 317
$I(\omega_1^{CK}(1+\eta))$, 250
K, 12, 18, 21, 22, 27, 29
$L_{\omega_1\omega}$, vi, 89, 90, 93–96, 98, 101, 102, 119, 121
    admissible fragment of, 123
low_n, 151
n-c.e., 86, 167
n-system, 215, 220–223, 227, 228
O, 57, 61, 62, 64–67, 71, 82
    path through, 66
O-minimal, 313
PA, 53, 137, 291, 292, 296, 297, 303, 306, 307, 318–321
    axioms for, 318
    completion of, 22, 54, 291–297, 306–309, 319, 321
    nonstandard model of, 53, 291, 293, 294, 296, 297, 306, 307, 309, 318, 319
P_{ω_1}, 89, 102, 103
$Rep(T)$, 292–294, 296, 297, 309
    enumeration of, 294, 309
$s - m - n$ Theorem, 12–14, 18
    relativized, 16
TA, 51, 54, 291, 292, 295, 306, 307
    nonstandard model of, 291, 296, 307, 309
W_e^X, 15
X-computable infinitary formula, 118
α-atom, 91, 92, 109, 272, 273, 316, 317
α-c.e., 86, 87, 167, 189, 191, 192
    relative to, 87, 168
α-computable, 87
α-free, 191, 254–262, 264–273
α-friendly, 241–243, 245–247, 250–252, 254, 255, 258–260, 264, 265, 267, 269, 279, 283
α-system, 227, 228, 230, 231, 236, 251, 261–263, 265, 270, 271, 279, 280
Δ_n^0, 22, 23, 25–28, 71, 74, 215, 221–223, 306
    canonical complete oracle, 28
    partial, 28
Δ_α^0, 74, 75, 85, 111, 133–135, 227–230, 253, 255, 257–259, 262, 263, 265–267, 269, 270, 272, 273, 279, 280, 284, 288
    canonical complete oracle, 75, 113
$\Delta_\alpha^0(X)$, 85
    canonical complete oracle, 85
Δ_1^1, 71, 77, 79–83, 121, 123–127, 130, 132–134
Δ_1^1 index, 79
Δ_n^1, 76, 77
$\equiv_T$, 16
Γ-structure, 111, 112, 209
$\leq_T$, 16
$\mathcal{E}$, 55, 62, 135, 136
    automorphisms of, 135

$\mathcal{E}^*$, 55
$\mathcal{N}$, 51, 52, 54, 90, 108, 136, 178, 242, 291, 297, 318–320
$\mathcal{SS}(\mathcal{A})$, 293–297, 307, 308
ω-c.e.
    second meaning, 87
ω-system, 302
ω_1^{CK}, 62, 65, 129, 130, 248, 249
Π_n^0, 22–28, 52, 55, 71, 74, 75
Π_α^0, 74, 109, 111
$\Pi_\alpha^0(X)$, 85
Π_1^1, 71, 79, 121, 123–130, 132–134
Π_1^1 index, 79
Π_n^1, 76, 77
Σ_n^0, 22–28, 52, 55, 71, 74, 75
Σ_α^0, v, 74, 75, 87, 109, 111, 115, 116, 118, 126, 133, 134, 253, 255, 257, 258, 288
    relative to, v
$\Sigma_\alpha^0(X)$, 85
Σ_1^1, 71, 79
Σ_1^1 index, 79
Σ_n^1, 76, 77
$\sigma(\mathcal{F})$, 250, 251, 282, 283, 315
φ_e^X, 15
$\varphi_e^X(n) \downarrow$, 15
$\varphi_e^X(n) \uparrow$, 15
$\varphi_{e,r}^\tau(n) \downarrow$, 15
$\varphi_{e,r}^\tau(n) \uparrow$, 15
$\varphi_{e,r}(n) \downarrow$, 6
$\varphi_{e,r}(n) \uparrow$, 6
φ_e, 5
$\varphi_e(n) \downarrow$, 6
$\varphi_e(n) \uparrow$, 6

Abelian p-group, 92, 109, 132, 137, 139–142, 252, 311, 317
    length of, 132, 139, 141, 317
    reduced, 92, 132, 168, 317
Abelian group
    language of, 36
Abramson, 309
addition of ordinals, 59, 60, 67
admissible set, vi, 121, 123
Alegria, viii
algebra, 37, 210

    linear ordering as, 37
algebra of sets, 316
algebraically closed field, 187, 312
alternating sequence, 214
alternating tree, 214, 215, 221, 227, 231
analytical, 75, 76
analytical hierarchy, 71, 75, 76, 78
Arana, viii
arithmetic, 23
    language of, 36
    model of, vii, 37, 311, 318
    Presburger, 56
    standard model of, 51–53, 90, 108, 136, 137, 242, 291, 318
arithmetical, 21–23, 26, 51, 52, 71, 75–77, 80, 292, 296
    function, 28
    partial function, 28
arithmetical hierarchy, 21–28, 52, 71, 75, 78
arity, 34, 49
Ash, vi, viii, 123, 159, 174, 176, 181, 182, 187, 198, 213, 214, 239, 252
Ash-Nerode Theorem, 183
assignment, 37
atom
    in Boolean algebra, 151, 152, 273, 316
atomic diagram, 42
atomic model, 48, 201
automorphism, 40, 45, 47, 54, 126, 131, 135, 136, 159, 161, 171, 198, 199, 242

back-and-forth property, 41, 47, 96, 98, 126, 152
back-and-forth relation, 236, 239, 244, 245, 247, 250–252, 263, 279
    derived, 251
    standard, 239–242, 252, 254, 263, 266, 270
    stronger, 250, 251
Barker, 159, 213, 252, 253
Barwise, vi, 121, 123

INDEX

337

Benedict, 178
Bonnet, 315
Boolean algebra, vii, 37, 91, 131, 150, 209, 246, 247, 250, 281, 282, 289, 311, 315, 316
  atomic, 281, 316
  atomless, 150, 152, 316
  computably categorical, 208
  language of, 36
  superatomic, vii, 132, 239, 247, 263, 272, 316
Boone, 55
Borel set, 103
Brouwer, 82

c.e., 5, 8–12, 18
  relative to, 17
c.e. relative to, 15, 17
c.e. index, 12, 18
canonical enumeration
  of Scott set, 294, 307, 308
canonical index
  for finite set, 8
Cantor, 41, 60
Cantor normal form, 60, 68, 69, 91, 94, 132, 245–247, 268, 317
cardinal sum
  of structures, 204
categorical
  Δ_α^0, 239, 263, 269, 272, 273
  computably, 197–199, 201, 202, 205, 206, 208, 209, 239, 263, 269
  relatively Δ_α^0, 175, 269
  relatively computably, 177, 197–199, 206
Chang, 312
characteristic function, 3
Chisholm, 160
Cholak, 135, 206
Chong, 214
Church, 4, 62
Church's Thesis, 4–6, 15
  proof by, 5, 6
closure properties

  of class of α-friendly orderings, 252
co-α-c.e., 87
co-c.e., 9, 23, 87
code
  for computation, 8
  for finite sequence, 1
  for finite set, 8
  for formula, 49
  for pair, 7
  for tuple, 8
coherent family of runs, 234, 235
commutative sum, 60, 247
Compactness Theorem, 44, 45, 121
  Barwise-Kreisel, vi, 121, 123, 133, 135
complete
  Δ_n^0 set, 27
  Π_1^1 set
    O as, 83
  Σ_n^0 set, 27
  for a class, 18
complete c.e. set, 18
complete forcing sequence, or c.f.s., 163
complete type, 155
completeness
  expressive, 159
Completeness Theorem, 44
completions, 50
complexity
  of an infinitary formula, 94
  of infinitary formula, 93
computable, 5, 186
  function, 2
    partial, 2, 4
    total, 2, 4
  relation, 4
computable Π_α formula, vi, 105, 112
computable Σ_α formula, vi, 105, 112, 254
computable dimension, 176, 205, 206, 208
  relative, 176
computable infinitary formula, vi, 105, 107–111, 115–118, 121–123,

126–129, 133, 136
  index for, 107
  propositional, 112, 113, 117
computable ordinal, 64, 142
computable relation
  on numbers and functions, 75
computable relative to, 15–18
computable relative to a set, 17
computable relative to a total function, 17
computable relative to a tuple of relations, 17
computable structure, v–vii, 51, 52
computably enumerable, 5
  alternate definitions, 10
congruence relation, 41, 43, 287
conjunction
  empty
    truth value of, 90
connectives
  infinitary, 89
consequence, 128
consistency
  criterion, 42
  of a sentence or set of sentences, 39
consistency criterion, 43
Consistency Criterion, 45, 95
consistency criterion
  for computable infinitary sentence, 122
  for set of finitary sentences, 121
  for set of infinitary sentences, 121
consistency property, 95
constructive ordinal, 64
constructivization, vii
  strong, viii
Continuum Hypothesis, 294
Crossley, viii

Davey, 169, 181, 192, 193, 259
Davis, 320
decidable, 186
decidable structure, 51, 137, 155, 201
decides, 162

Definability of Forcing Lemma, 164
definability of satisfaction, 320
definable
  element, 51
  relation, 51
defining family, 101, 203
  formally Σ_α^0, 177, 179, 264, 265, 268
  formally c.e., 174, 186, 197, 198, 203, 205, 210
denotation
  of term, 37
dense set
  of forcing conditions, 163
Detlefsen, 309
diagram
  atomic, vii, 39, 51
  complete, vii, 39, 51
  open, 39
dimension, 311
disjunction
  empty
    truth value of, 90
division algorithm, 53, 60
Downey, 143, 144, 147, 150, 151
Dzgoev, 207, 208

effective enumeration
  of Scott set
    complexity of, 294
  of Scott set, 294, 308
effectiveness condition, 184
  for Ash-Nerode Theorem, 183
elementary embedding, 47
elementary equivalence, 38
elementary extension, 38
elementary substructure, 38
elimination of quantifiers, 185, 186, 312, 313
  for vector spaces, 311
Embedding Theorem, 151
enumeration, 152, 155, 295
  of a family of sets, 205
  univalent, 205
enumeration function, 215, 225, 227
  standard, 240, 254, 263, 270

enumeration reducible, 18, 19, 166, 170, 178, 308
enumeration relation, 24, 78, 87
   Π_1^1, 79
   Σ_1^1, 79
Enumeration Theorem, 24
equality, 34
equivalence relation, 41, 317
equivalence structure, 137, 278, 311, 317
   invariants, 137
equivalent
   over a set of sentences, 40
Ershov, vii, viii, 86, 87, 167, 189
Ershov's hierarchy, 87, 167, 189
   relativized, 86
existential type, 178
expansion, 38, 46
exponentiation of linear orderings, 59, 315
exponentiation of orderings, 315
exponentiation of ordinals, 59
expressive completeness, vi
extension, 38

Feiner, 143, 150, 281, 282, 284
field, 36, 90, 311–313
   algebraically closed, 47, 137, 185–187, 312, 313
   Archimedean ordered, 90, 108, 313
   Archimedian ordered, 89
   characteristic of, 312
   of algebraic numbers, 47
   of characteristic 0, 312
   of characteristic p, 312
   ordered, 36, 90, 101, 160, 174, 313
   ordered field, 174
   real closed, 187, 313
Fixed-Point Theorem, 67
forcing, vi, 161, 181
forcing conditions, 161
forcing language, 161
forcing relation, $\Vdash$, 162
formally Σ_1^0 relation, 183, 185

formally Σ_α^0, 165
formally Σ_α^0 relation, 253
formally c.e. relation, 182
formula
   P_{ω_1}, 102
   X-computable, 179
   Π_n, 40
   Π_α, 93
   Σ_n, 40
   Σ_α, 93
   atomic, 34
   basic, 35
   finitary open, 93
   finitary, or elementary, 33
   infinitary, 89
      propositional, 102
   of $L_{\omega_1\omega}$, 89
   open, 35
   predicate, 34
   propositional, 33
      atomic, 33
free variables
   of a type, 155
freeness, 184, 189, 191–193
   appropriate for Δ_α^0 categoricity, 269
   appropriate for Δ_α^0 stability, 264
   appropriate for Δ_α separable relations, 259
   appropriate for intrinsically Δ_α relation, 258
   notion appropriate for intrinsically Σ_α relation, 254
Friedberg, 194, 195, 224
Friedberg-Muchnik Theorem, 225
function
   non-computable, 6

games, in connection with ranks, 99
general recursive, 4, 5
generator
   for a principal type, 47, 202
   for complete type, 202
generic copy, 161, 163, 164, 167, 175, 176, 309

generic model representing a Scott set, 309
Goldbach's Conjecture, 2
Goncharov, vii, 150, 155, 156, 174, 176, 197, 199, 202, 203, 205–209, 239, 263, 267, 269, 295, 297
Groszek, 214
group, vii, 37
    computably presented, 55
    finitely presented, 55
    language of, 36
    recursively presented, 55
    with solvable word problem, 55
Gödel, 5, 52
Gödel-Rosser sentence, 292
Gödel, 4, 292, 319
Gödel number, vii, 49, 106
Gödel's Incompleteness Theorem, 52, 318, 320

halting set, 12, 320
Harizanov, vii, 166, 188, 189, 195
Harrington, 135, 291, 305–307
Harrison, 129, 248
height
    in Abelian p-group, 140, 317
    in tree, 140
    of element in well ordering, 58
Henkin, 45, 48
Henkin construction, 45
hierarchy, 24
    difference, 86
    Ershov, 86
    proper, 24, 78
Hilbert, 2, 320
Hilbert's 10th Problem, 2, 3, 6, 52, 319, 320
Hird, 195
Hirschfeldt, 206
homogeneity, 45, 47
homogeneity of computable, or hyperarithmetical structure, 126
homogeneneity, 202
homogeneous, 46, 155
homogeneous structure, 137

characterized by set of types, 47
Hurlburt, viii, 284
hyperarithmetical, 71, 74, 80–83
    relativized, 84
hyperarithmetical formulas, 123
hyperarithmetical hierarchy, vi, 71, 74, 75, 85, 87, 184, 239, 263
    relativized, 86
hyperarithmetical infinitary formula, vi
hyperarithmetically saturated, 127–129, 136

independence
    of a formula, 321
    of a sentence, 321
index
    for Π^1_1 relation, 79
    for Σ^1_1 relation, 79
    for Σ^1_n relation, 77
    for computable infinitary formula, 106
    for partial computable function, 13
    for Turing machine, 5
induction schema, 318
infinitary formula, vi
    computable, 90
infinitary formulas
    hyperarithmetical, 116
instruction function, 214, 215, 228, 231, 265, 266, 270, 279
    (α_n), 236
    Δ^0_2, 216
    Δ^0_3, 218, 219
    Δ^0_α, 262
    as source of high-level information, 215
interval algebra, 131, 150, 247, 250, 272, 282, 316
intrinsically $c.e.$, 182
intrinsically $c.e.$ relation
    relatively, 182
intrinsically d-$c.e.$, 189

INDEX

relatively, 167
intrinsically α-c.e., 189
intrinsically Δ^0_α, 253
intrinsically Σ^0_α, 253, 255
intrinsically c.e., 181, 183
    relatively, 181
intrinsically computable, 181, 185
    relation, 198
    relatively, 181
invariants
    mathematical, 143
isomorphism, 40
Isomorphism Theorem, 151
isomorphism type, 41

Jockusch, 50, 143, 150–152
Jockusch and Soare, 30, 31, 147
jump, 21
    iterated, 21
    of a set, 21
    of a Turing degree, 22

Kaplansky, 317
Kaye, 318
Keisler, 48, 89, 124, 312
    method of expansions, 124
Khisamiev, 139
Khoussainov, 206
Kleene, 4, 13, 24, 57, 61, 62, 67, 74, 78, 82, 84, 131, 227
Kleene symposium, 213
Kleene's Normal Form Theorem, 8
Kleene-Brouwer ordering, 82, 131
Kontostathis, 214
Koppelberg, 315
Kreisel, vi, 121, 123
Kripke, 123
Kripke-Platek set theory, vi, 123
Kudinov, 206
Kueker, 101
König's Lemma, 30

Lachlan, 309
language
    computable, 49
    infinitary

predicate, 89
propositional, 89
of groups, 36
of orderings, 36
predicate, 33, 34
    first order, 33
propositional, 33
lattice of c.e. sets, 55, 135
least, or μ-operator, 8
left limit, 277
Lempp, 152, 155, 214
Lerman, viii, 145, 214, 297, 309
Limit Lemma, 28
linear dependence
    formula defining, 90
linear ordering, vii, 33, 37, 160, 173, 206, 208, 244, 282–284, 287, 311, 314, 316
    as algebra, 287
    dense, without endpoints, 41
    language of, 36
    of type ω or $\omega + 1$, 276
    of type ω or ω^*, 276
    of type $\omega^*+\omega+\omega$ or $\omega^*+\omega^*+\omega$, 277
logical
    connective, 33
    connectives, 34
    constant, 33
    constants, 34
    quantifiers, 34
logically equivalent, 34, 40
Love, 143, 150, 209, 210
low, 22, 30, 53, 151, 306, 307
Low Basis Theorem, 30, 50
low$_n$, 22
Löwenheim-Skolem Theorem, 44
    for infinitary logic, 94

Makkai, 94, 95, 101, 136
Manasse, 160, 185
many-one
    degree, 17
    equivalence, 17
    reducibility, 17
Marker, 54, 291, 295, 296, 307, 309

Matijasevic, 7, 52, 308, 319, 320
McAllister, viii, 309
McAloon, 291, 305
McCoy, viii, 167, 169, 176, 178, 201
McNicholl, ix, 167, 169
Metakides, 177, 187
metatheorem, 213, 235, 237, 252
    for Δ_n^0 construction, 213
    for finite-injury priority constructions, 224
    for priority construction, 227
    for priority constructions, vi, 133
    transfinite, 227
method of workers, 291
Michalski, viii, 224–226
Millar, 155
minimal upper bound
    for set of degrees, 309
Miranda, viii
model
    of a sentence, or set of sentences, 39
Monk, 315
Morley, 155, 213
Muchnik, 194, 195, 224
Mytilinaios, 214

Nadel, 294
Nerode, 159, 177, 181, 182, 187, 198, 239
normal form, 94, 102, 105, 107
    for analytical relation, 76
    in infinitary logic, 93
    prenex, 40
Normal Form Theorem, 94
Novikov, 55
Nurtazin, vii, 201, 206

Oates, viii, 140
object
    of Δ_n^0 construction, 215
object of priority construction
    mathematical formulation, 214
Odintsov, 289
omitting a type, 48
operations

on linear orderings, 59
    on ordinals, 59
oracle, 14
oracle machine, 15–17
order type, 58, 59, 61, 63–66, 68, 69, 314
ordering
    and hyperarithmetical automorphisms, 131
    of type η, 198
    of type $(\omega^* + \omega) \cdot \mathcal{A}$, 218
    of type ω, 181, 197
    of type $\omega + \omega^*$, 182, 192
    of type $\omega \cdot \mathcal{A}$, 219
ordinal, 57
    X-computable, 119
    classification, 58
    computable, 61
    constructive, 62
    countable, 58
    least non-constructive, 62
    limit, 58
    non-computable, existence of, 61
    notation for, 61
    set theoretic definition, 58
    successor, 58
    uncountable, existence of, 58
ordinal notation, vi, 71, 227
Overspill, 53, 293, 297, 318

pairing function, 7
partial ordering, 314
    strict, 314
partition
    of Boolean algebra, 246
path
    through tree, 215
Peano arithmetic
    first order, 318
Peretyat'kin, 155, 156, 202, 295, 297
picture, 230, 235
    completing, 230
    completion of, 223
    in proof of metatheorem, 223
Platek, 123

INDEX   343

prenex normal form, 40
Presburger arithmetic, 56
presentation
    group, 42
prime, 155
prime model, 47, 201, 202
    decidable, 202
primitive recursive, 5
principal type, 201
priority construction
    infinite-injury, 213
priority conjstruction
    object, 213
priority construction, 121, 133, 213
    Δ_n^0, 213, 227
    Δ_α^0, vi
    abstract statement of object, 215
    finite injury, 181, 199
    finite-injury, vi, 184, 213, 216
    general features, 213
product of orderings, 59, 315
product of ordinals, 59, 60
proof, 42, 44
proper hierachy, 75
propositional
    structure, 34
    variables, 33

quantifier elimination, 242
quantifier-free, 312
quantifiers, 34
quasi-simple, 195
quotient
    structure, 42
quotient structure
    Σ_α^0, 111
quotient algebra, 150, 210
    Σ_α^0, 284
    c.e., 284, 288
quotient group, 42, 56
quotient order, 287
quotient ordering, 42, 143, 216, 287
quotient structure, 33, 41, 43, 55,
        56, 143, 209, 275, 287
    Δ_α^0, 111
    Π_n^0, 55

Π_α^0, 111
Σ_2^0, 287
Σ_n^0, 55
Σ_α^0, 111
c.e., 55, 110, 210
complexity of, 55

r.e., 5
rank, 136, 247
R, 98
r, 98
SR, 98
    calculating, 239
    examples, 100
    of $I(\omega_1^{CK}(1+\eta))$, 250
    of $I(\omega^\alpha)$, 250
    of ω_1^{CK}, 248
    of $\omega_1^{CK}(1+\eta)$, 248
    of ω^α, 248
real closed field, 187
Recursion Theorem, 13, 14, 64, 67
    application, 67
recursive
    function
        partial, 4
        total, 4
    relation, 4
recursive saturation, 295
recursively enumerable, 4
reducibility
    many-one, 17
reduct, 38
relation
    computable, 3
    arithmetical
        on numbers and functions, 76
    intrinsically d-c.e., 189
    intrinsically α-c.e., 189
    intrinsically Σ_α^0, 239
    intrinsically c.e., 192, 239
    intrinsically computable, 185–187
    relatively Σ_α^0, 165
    relatively intrinsically d-c.e., 189
    relatively intrinsically α-c.e., 168
    relatively intrinsically Δ_2^0, 192

relatively intrinsically Δ^0_α, 165
relatively intrinsically Σ^0_n, 178
relatively intrinsically Σ^0_α, 165
relatively intrinsically computable, 185
semi-computable, 3
relational structure, 37, 210
relations
    intrinsically computably separable, 192
relatively Δ^0_α categorical, 175
relatively Δ^0_α stable, 177
relatively intrinsically Σ^0_α, 253
relativization
    of a theorem, 15
Remmel, 151
representable set, 292
resplendence, 46
Ressayre, 127
rigid, 100, 198
    almost, 100, 263, 264
Rogers, vi, 5, 71
Rogers, L., 140
Rosenstein, 314
Rosser, 292
run
    of tree with instruction function, 215
    of tree with instruction function, 215, 228

satisfaction
    of predicate formula, 37
    of propositional formula, 34
satisfaction predicate, 52, 110
saturated, 46, 155
saturated model, 47
saturation, 45, 47, 127
Schmerl, 297
Scott, 96, 98, 99, 101, 291, 292, 321
Scott family, formally Σ^0_α, 175
Scott Isomorphism Theorem, 96
Scott family, 96, 97, 99, 174, 199, 269, 291
    examples, 100
    formally Σ^0_1, 199

formally Σ^0_α, 179, 269, 273
formally c.e., 197–202, 205–208, 269
Scott family, formally c.e., 174, 175
Scott Isomorphism Theorem, vi, 101
Scott rank, 98
Scott sentence, 96, 97
Scott set, 291–293
    appropriate for theory, 297, 300, 302, 305, 308
    of a model, 293
    uncountable, 294
Seetapun, 147
Selivanov, 143, 150, 289
semi-characteristic function, 3
semi-computable, 4
sentence, 35
    in infinitary logic, 93
    propositional, 34
separable
    Δ^0_α, 260
separator, 12, 53, 192, 319
    Δ^0_α, 253, 262
    computable, 193
    for a pair of relations, 169
Shoenfield, 214
Shore, viii, 206
shuffle sum
    of orderings, 145, 250, 251, 282, 283, 315
Single-Valuedness Theorem, 84, 124
Skolem, 84
Slaman, 152, 160, 214
Slinko, 206
Smullyan, 94
Soare, 50, 152, 309
Solovay, 54, 291, 296, 307, 308
special (α_n)-instruction function, 303
special (α_n)-system, 236, 302
special α_n-system, 236
Spector, 62, 65, 82, 129
stable
    Δ^0_α, 239, 263, 264
    computably, 197, 198, 203–205, 208, 209, 239, 263
    relatively Δ^0_α, 264

INDEX

relatively computably, 197, 206
standard system, 293
Stone Representation Theorem, 316
strongly minimal, 312, 313
structure
    cardinality of, 36
    computable, vii
    constructivizable, viii
    decidable, vii
    empty, 36
    for predicate language, 36
    interpretation function of, 36
    strongly constructivizable, viii
    universe of, 36
structure representing Scott set, 296, 300
subformula, 34
    of infinitary formula, 92
substructure, 38
sum of orderings
    infinite, 315
sum of orderings, 59, 314, 315
    infinite, 315
sum of ordinals, 59, 60
symbol
    constant, 34
    logical, 33, 34
    non-logical, 33, 34
    operation, or function, 34
    relation, or predicate, 34
symmetric difference, 55

Tarski, 38, 313, 320
Tarski Criterion
    for elementary substructure, 38
Tennenbaum, 53, 294
term, 34
theory, 39
    complete, 39
    completion of, 39
    of a structure, 51
Thurber, viii, 151, 281, 282, 284
transcendence degree, 312
transfer theorem, 143, 144
transfinite
    induction

    proof by, 58
    induction on ordinal notation, 67
    number, 58
    recursion
        definition by, 59
    recursion on ordinal notation, computable, 67
transitive set, 58
tree, 30
    path through, 30
    subtree of, 30
trivial structure, 54, 128, 171
truth
    of a predicate sentence, 37
    of propositional formula, 34
Truth and Forcing Lemma, 164–166
Turing, 4
    degree, 16
    equivalence, 16
    reducibility, 16
Turing degree, 21
Turing machine, 4–6, 8, 12–15
    index for, 6
    self-replicating, 14
    universal, 4, 6
type
    complete, 46–48, 155–157, 291, 295
    complete B_1, 297–299, 301, 302, 304, 307
    complete B_n, 291, 297, 300, 301, 303
    complete Σ_1, 178, 309
    principal, 47, 48
    realized, 46
type 0 object, 23
type 1 object, 23
type 1, 77

Ulm, 317
Ulm sequence, 132, 140, 141, 317
uniformly computable sequence of structures, 279–281
    coding a set, 275

Vandenberg, viii
variable
    bound, 35
    free, 35
        in infinitary formula, 92
    individual, 34
Vaught, 151
vector space, vii, 33, 37, 45, 90, 100, 137, 160, 166, 170, 173, 177, 185–187, 194, 195, 242, 243, 311, 312
    language of, 36
    theory of, 311
Ventsov, 176
Vlach, viii

Watnik, 144, 218, 219, 284, 287
Wehner, 152, 153, 155
well ordering, vii, 57–59, 61, 63, 64, 91, 129, 142, 239, 244, 245, 263, 268, 314
    computable, 61, 63, 65, 129, 268
    hyperarithmetical, 129

Yang, 214

Lightning Source UK Ltd.
Milton Keynes UK
UKOW01n1516210316

270584UK00015B/207/P